I0055744

Casting Process: Preparation of Materials in Liquid State

Casting Process: Preparation of Materials in Liquid State

Edited by **Travis Xavier**

New Jersey

Published by Clanrye International,
55 Van Reypen Street,
Jersey City, NJ 07306, USA
www.clanryeinternational.com

Casting Process: Preparation of Materials in Liquid State
Edited by Travis Xavier

International Standard Book Number: 978-1-63240-093-2 (Hardback)

Printed in the United States of America.

Contents

Preface

Casting is basically described as a manufacturing process through which a liquid material is solidified with the help of a mould which consists of a hollow cavity of the desired shape. This book consists of different science and technology factors that require careful consideration for casting production. It includes contributions by various professionals with extensive experience in their respective areas. This book discusses topics such as simulation of continuous casting process, control of solidification of continuous castings, effect of mold flux in constant casting, segregation in strip casting of steel, and advancements in shell and solid investment mold methods. It also elucidates various issues related to permanent molding of cast iron, pressure control during filling sand molds, wear resistant castings, and progress in the accurate estimation of graphite nodularity in ductile iron castings.

This book is a comprehensive compilation of works of different researchers from varied parts of the world. It includes valuable experiences of the researchers with the sole objective of providing the readers (learners) with a proper knowledge of the concerned field. This book will be beneficial in evoking inspiration and enhancing the knowledge of the interested readers.

In the end, I would like to extend my heartiest thanks to the authors who worked with great determination on their chapters. I also appreciate the publisher's support in the course of the book. I would also like to deeply acknowledge my family who stood by me as a source of inspiration during the project.

Editor

Section 1

Disposable Mold Castings

Chapter 1

New Casting Method of Bionic Non-Smooth Surface on the Complex Casts

Tian Limei, Bu Zhaoguo and Gao Zhihua

Additional information is available at the end of the chapter

1. Introduction

Many studies have shown that the surfaces of most creatures contain non-smooth structures, such as dimple concaves on the Cybister bengalensis and riblets on the shark skin, as shown in Figure 1. Non-smooth structures are formed in special non-smooth surfaces which have specific biological consequences, such as "shark skin effect" see in [1], “lotus leaf effect” see in [2] and “non-smooth surface effect” in references [3-4], all of which are closely related to certain functions. In accordance with the above-mentioned effects, the functions are drag reduction in [1], self cleaning in [2] and anti-adhesion, respectively in [3-4]. For shark skin, Singh, Yoon and Jackson in [5] found the riblets are directed almost parallel to the longitudinal body axis and this effectively reduces drag by 5%–10%.Ren et al [6] also demonstrated that both riblet and dimple concave non-smooth surfaces could be applied in pumps to increase efficiency. Other studies have shown that these non-smooth surfaces (called bionic non-smooth surface) have some certain functions in the fields such as aviation see in [7], pipeline in [8] and antifouling see in [9-10]. As for the riblet on shark skin and dimple concave on cybister bengalensis, when applied in the engineering, they are simplified as the shape of groove and dimple concave, as shown in Figure 2.

The traditional process method of such non-smooth surface includes Electro Discharge Erosion (EDE), Metal Engraving method (ME) and machining, all of which have disadvantages such as high processing cost and low processing efficiency. For example, ME processes rib-like non-smooth surfaces whose cross-section is a triangle on the complex surface casts directly, as shown in Figure 3. There are several disadvantages as follows: first, because of the limitations of tool radius, it is impossible to process sharp angles on the surface, the eventual angles of the cross-section are round on the surface, which is different from the original design ideas; second, the surface of rib-like non-smooth structures is rough and it needs other post-reprocessing like polishing, which increases the processing cost; third, residual stress on the surface are produced, which affects the quality of the casts. Most important of all, when the casts have complex surface such as impeller flow of centrifugal pump, it is difficult to form continuous rib-like non-smooth surfaces from flow channel entrance to the exit due to the complex flow of impeller. However, this is different from the design idea that the bionic non-smooth surfaces should be arranged in the entire flow channel.

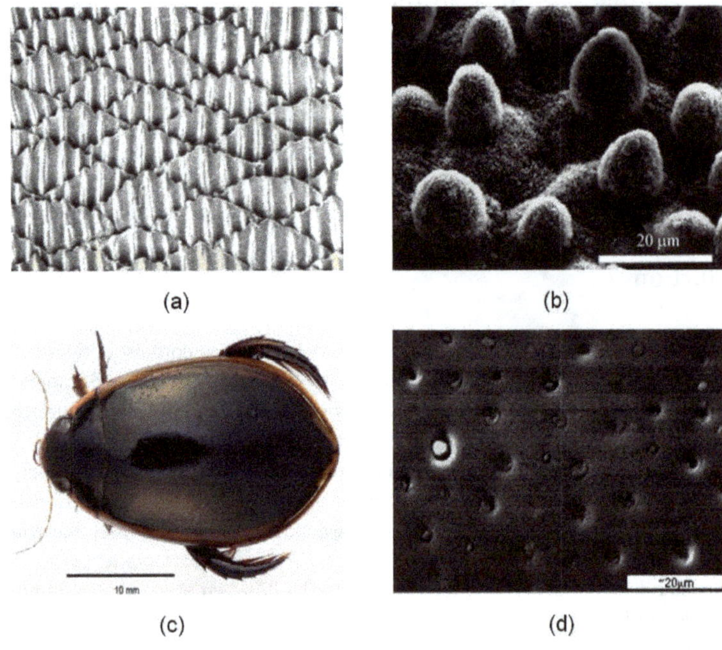

Figure 1. Non-smooth structures of some typical living creatures' skin. (a) riblets and grooves found on the shark skin [1]. (b) mastoid and micro-nano composite structure of lotus-leaf [2]. (c) Cybister bengalensis. (d) Dimple concave non-smooth structures on the back of Cybister bengalensis [3].

Figure 2. Simplified non-smooth structures. (a) Groove mimic from the shark skin. (b) Dimple concave mimic from the Cybister bengalensis skin.

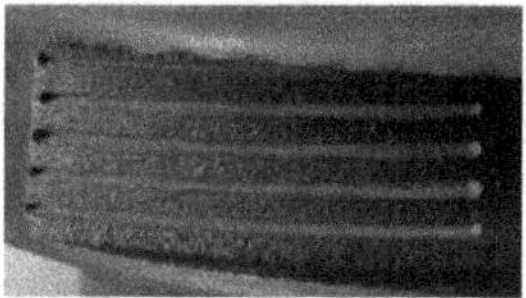

Figure 3. Engraving rib-like non-smooth structures on the complex surface.

In addition to traditional processing method, biology constrain forming technology based on the mechanism of bionic manufacturing see in [11-13] is used to produce a kind of biomimetic skin, but it is suitable for the polymer materials rather than for metal surface process. In light of this, this chapter investigates a new casting method to process non-smooth structures on the complex cast surface.

2. Method and Mechanism

2.1. Disadvantages of traditional casting method of rib-like bionic non-smooth surface

Traditional casting method to form such rib-like non-smooth surface has following disadvantages.

1) It is difficult to form small narrow rib-like structures continuously, as shown in Figure 4(a). Because the cross-section of Structure 6 is equilateral triangle with very small and sharp angels on Casting Mold 5, as shown in Figure 4(c). As Convex Section 4 in Figure 5 is formed by sand sculpturing on Casting Mold 3, it is difficult to depart the molds smoothly in the process of demolding. As a result, the part or the whole of Convex Section 4 collapses and the narrow rib-like non-smooth structures will not be formed or formed incompletely.

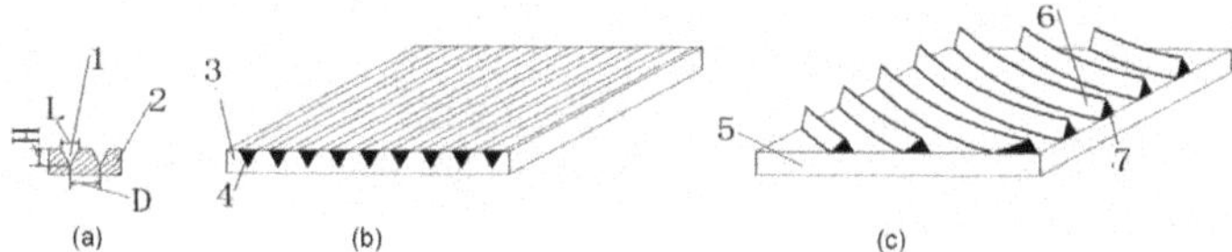

Figure 4. The casting mechanical diagram of rib-like non-smooth structures. (a) the cross-section of rib-like structures. (b) rib-like structures on the rubber plank. (c) paste non-smooth structures to form bionic non-smooth surface on the sand core.

In figure 4(a) L and H are the length and height of rib-like structures respectively, and D is the distance between them. 1 refers to the cross-section of rib-like non-smooth structure which is a triangle; 2 refers to the casting surface; 3 is the rubber plank; 4 is rubber rib-like structures cut from the plank; 5 is the sand core; 6 is the insulation coating materials painted on the rubber rib-like structure; 7 is rib-like non-smooth structures arranged on the sand core to form bionic non-smooth surface.

2) The sharp angle in design will not be formed. Many experiments have shown that, compared with other non-smooth rib-like structures whose cross-section is a triangle, those with a sharp angle have the best drag reduction. However, traditional casting method will not form a sharp angle directly in the demolding process.

2.2. The casting mechanism of one-time casting molding (OTCM) rib-like bionic non-smooth surface

In light of the disadvantages of the above-mentioned traditional casting method, a new casting method is investigated which is called one-time casting molding method (OTCM). During the casting process, two intermediate media are used, namely, hard rubber and high temperature insulation coating material (HTICM). The former is used to form the shape of rib-like non-smooth structures and the latter paints on the surface of rubber non-smooth structures in order to prevent the melted iron from contacting the rubber directly and keeps the rubber from melting within 1 or 2 seconds, so non-smooth structures would be formed as expected. It is especially helpful for the formation of a sharp angle of rib-like structures. After the metal liquids cool with a rapid decline of temperature, the rib-like non-smooth surface is formed. The mechanism of this method is shown in Figure 5. In order to form a triangular cross-section of rib-like non-smooth structure (indicated by 2) on the cast (indicated by 1), the same shape of the convex section (indicated by 4) should be formed on the sand core (indicated by 3), as shown in Figure 5(a) and 5(b).

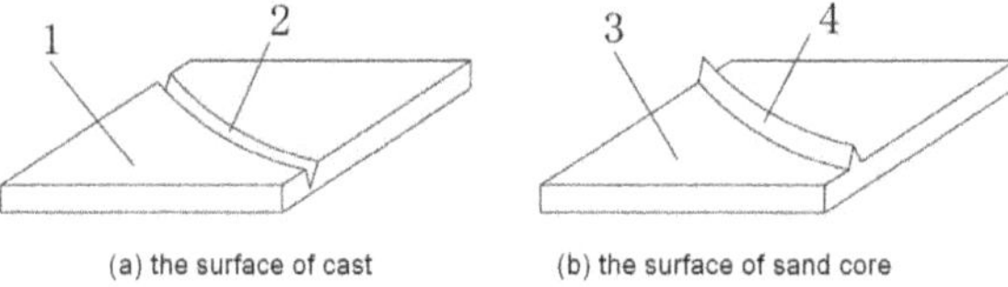

Figure 5. Schematic diagram of OTCM. 1 refers to cast; 2 is the triangle cross-section of the rib-like non-smooth structure; 3 is sand core; 4 is the rubber triangular convex section.

The steps of casting are as follows: ◉cutting many rib-like structures on the rubber plank, whose shape is the same as the final target cast; ◉brushing insulation materials evenly on the non-smooth structures, and only brushing two sides of the rib-like structures; ◉pasting those rib-like structures on the sand core according to the designed direction, location and curvature to form rib-like bionic non-smooth surface; ◉air-drying and trimming the non-smooth structures; ◉installing the sand core with non-smooth surface to the cast mold; ◉casting and demolding. In the casting period, both the hard rubber and HTICM are used as shape media. In the continual high temperature of metal liquids, HTICM become powder and triangular rubber convex melt and disappear, so the narrow triangular rib-like non-smooth surface is formed desirably. By using this method, rib-like non-smooth surface can be achieved even in complex casts. In order to describe the method of OTCM in detail, the impeller of centrifugal pump is selected to show the procedure of casting. The flow of impeller is an irregular complex surface; it is difficult to process such non-smooth surface desirably by using traditional casting or machining method. Non-smooth structure is designed as shown in Figure 4a. The cross-section form of it is triangle, the width of triangle is L, and L=0.5~3.0mm, the height H= (1~1.5) L, the distance between two rib-like structure is D, and D= (1~3) L. The method of OTCM process should be followed in the flow chart, as shown in Figure 6.

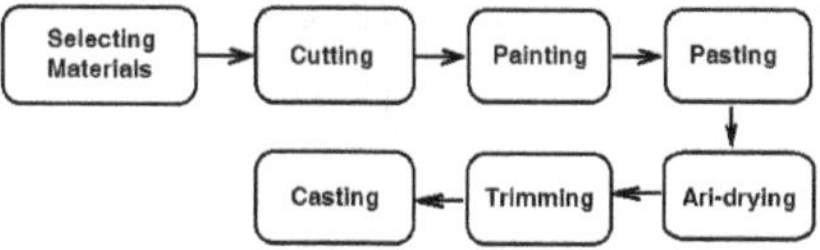

Figure 6. The flow chart of casting process.

2.3. Process

1) Selecting Materials

Two intermediate media, namely, hard rubber and HTICM, should be selected. They are the key factors to obtaining non-smooth surface.

a) Hard rubber

The thickness of hard rubber should more than the height of non-smooth structures. Since the height of this designed non-smooth structure is H=(1~1.5)L, and L=0.5~3.0mm; that means the maximum height is 4.5mm, so the thickness of hard rubber plank should be the 5mm. The hardness of rubber should be moderate; that means, on one hand, it should be convenient to machining, and on the other hand, the deformation should not be large at a high temperature in order to ensure the designed size of triangle. Considering this, the hard rubbers which had hardness values HD is 30 are selected, as shown in Figure 7(a). They are hard black materials at the room temperature; they have good chemical stability, excellent resistance to chemical corrosion and organic solvent resistance, low water absorption, high tensile strength and excellent electrical insulation; most important of all, they should have good machinability performance.

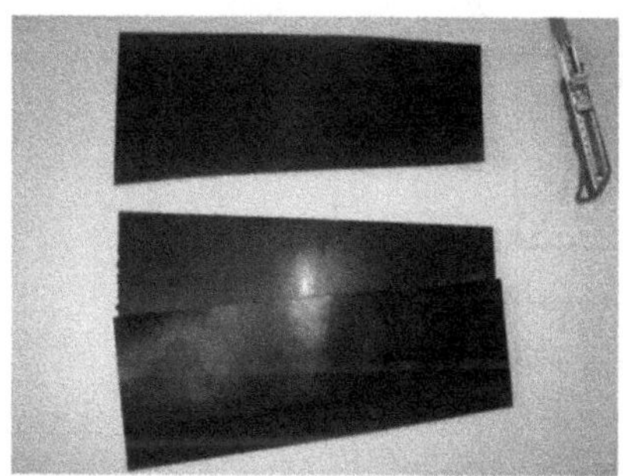

(a) Hard rubber

(b) convex section cutting from hard rubber

Figure 7. Hard rubbers.

b) High temperature insulation coating material (HTICM)

HTICM is a kind of coating, which can protect the surface at a special high temperature for a long time. It has excellent heat resistance compared with ordinary paint. The silicon element or inorganic high temperature insulation paint are used widely. In this chapter, the later insulation paint is selected, and in a certain condition, it still has protective function even at the temperature of 1700°.

2) Cutting

Triangular convex rib-like non-smooth structures which are in agreement with the designed ones should be cut on the selected hard rubber, as shown in Figure 7b. This process is very important, since it is related to the final quality of non-smooth surface on the casts. The non-smooth convex section structures can be machined by the special designed tools, and can also be cut by hand. Considering that some factors will affect the width of rib-like non-smooth structure, such as the thermal expansion and deformation of rib-like triangle in the process of casting and a layer of HTICM painted on the both sides of rib-like triangular convex section, as shown in Figure 8 the size of rib-like convex section should be smaller than that of the designed one, and the difference between them are called reserved size (refer to s, is shown in Figure 8) the size of it depend on the thickness of the coating layer and cross-section of non-smooth structure, and it can be expressed as following formula.$s = l / \cos\alpha$ The relationship between reserved size and thickness of coating layer is shown as Figure 8. Here l is the thickness of the HITCM layer, $\alpha$is the angle between l and s, which is related to the shape of rib-like non-smooth structure. For example, in this chapter, the cross-section of rib-like non-smooth structure is an equilateral triangle, and the reserved size$S = 2l / \sqrt{3}$. So, in order to produce the designed width of rib-like non-smooth structures, the width of rubber convex section L can be expressed asL=D-2S.

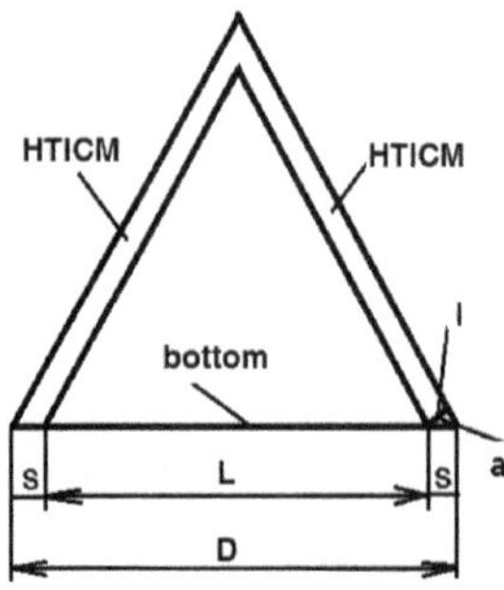

Figure 8. The schematic diagram of HTICM on the rubber convex section. Here, l is the thickness of the layer of HITCM, S refers to reserved size, L is the length of rubber convex section and D is the designed size of rib-like non-smooth structure.

3) Painting

A small brush is used to paint a layer of HITCM evenly on both sides of rib-like rubber convex section, as shown in Figure 8. The thickness of the layer of HITCM is about 0.1mm. This process should be quick to avoid paint drop on the surface of convex section due to the solidification of HITCM.

4) Pasting

A layer of emulsion is painted on the bottom of rib-like non-smooth convex section and then paste them on the two sides of sand core according to the designed direction, location and curvature to form rib-like bionic non-smooth surface, as shown in Figure 9. It is very important to paste the convex section along the flow of impeller, since it is the key steps related to the quality of the final casts.

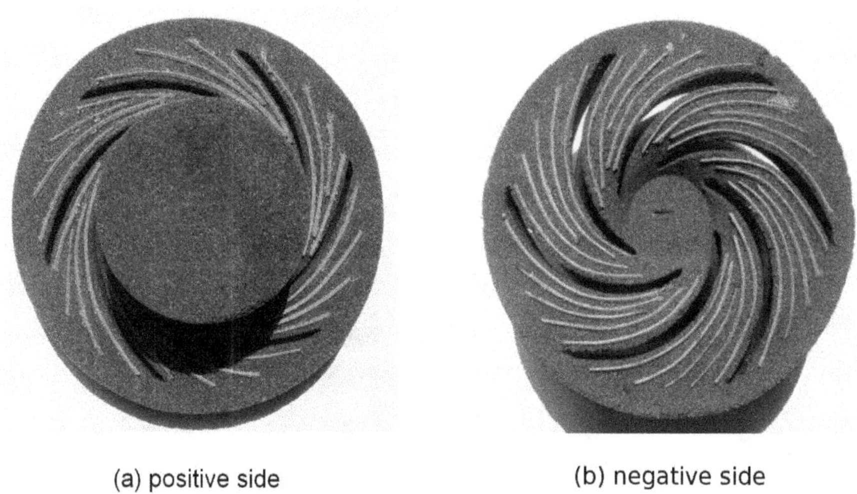

(a) positive side (b) negative side

Figure 9. Pasting non-smooth structure on the two side of sand core.

5) Air-drying

Air-drying the sand core at the room temperature. Since heating and drying method will make the rubber deformed, which will affect the size of non-smooth structures, the quality of non-smooth structure will not be guaranteed.

6) Trimming

Removing extra rubber with a knife. Check whether it has already met the designed requirements. Figure 10 shows the samples of sand core, which is formed after the above series of processes.

Figure 10. Examples of sand core which have rubber non-smooth structures pasted on their surface.

7) Casting

Assembling the above mentioned sand core on the cast molding and pour the metal liquids. HITCM keeps the rubber convex section from being gasified in 1 or 2 seconds, during which the temperature of metal liquids in other places decline rapidly. At the continual high temperature of metal liquids, the sand molding, HITCM, and rubber convex section melt and disappear, and the rib-like non-smooth surface will be formed desirably, the sample of rib-like non smooth surface impeller cast is shown as Figure 11. As the size of non-smooth structure is very small, it does not affect rigidity and life span of the pump impeller. At the temperature of 1000°, the rubber convex is gasified and volatilized completely. HITCMs are mainly silicon, and only a small quantity of them is painted on the surface of rubber convex. After casting, most of them become powder and remain on the surface of casting. And the mechanic properties of casting will not be affected.

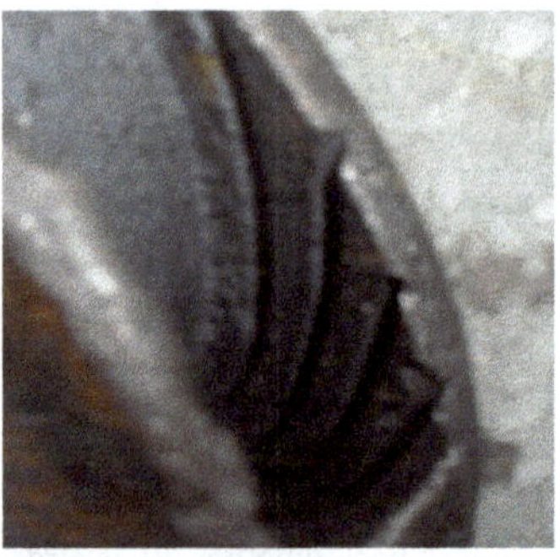

Figure 11. The rib-like non-smooth surface impeller cast.

2.4. processing dimple concave non-smooth surface on the impeller

By using the above mechanism of OTCM, the dimple concave non-smooth structures can also be processed on the complex surface of impeller. However, the process is slightly different from above process of rib-like non-smooth surface. Obtain dimple concave non-smooth surface needs a lot of convex domes pasted on the sand core. The process should be included as follows.

1) selecting the intermediate media

In this process, three intermediate media are needed to form convex domes, namely, Mastic 704, plexiglass, and HITCM. Mastic 704 is a kind of silicone with such characteristics as anti-aging, anti-acid and anti-alkali, high temperature resistance (up to 250°) and with capabilities of no corrosion, insulation and good curing performance, which is a very important feature to form convex domes. Plexiglass, with the chemical name of polymethylmethacrylate, is a kind of polymer polymerized by the methacrylate. Its characteristics are: (a) high transparency. It is currently the best transparent polymer material with light transmission rate reaching 92%, higher than that of glass; (b) high mechanical strength. It is a kind of long chain polymer, and molecular chains are soft, so its mechanical strength is very high with the tensile and impact resistance 7-18 times higher than that of the ordinary glass; (c) easy to process. It can be processed by machining. Above characteristics make plexiglass ideal materials which help form convex domes. In this chapter, the thickness of plexiglass plate is 3mm. The third material needed is HITCM, the same materials used to help cast rib-like non-smooth structures.

2) Drilling dimple concave on the plexiglass plate

Selecting the appropriate drill bit and drilling dimple concaves on the plexiglass plate with an electrical drill. It is very important to select a right drill bit, since it is conductive to controlling the shape and size of dimple concaves.

3) filling Mastic 704 into the dimple concaves on the plexiglass plate

In this process, the whole dimple concaves should be filled completely by Mastic 704 so that no small holes will appear when the mastic is solidified. Only in this way can the desired convex domes be produced.

4) solidifying and forming

Solidifying Mastic 704 fully as soon as possible at the room temperature to form the desired shape and size of convex domes, they are shown in Figure 12.

5) heating and removing

Heating the plexiglass at the temperature of 80° so that it will be softened and deformed. This characteristic of plexiglass helps the convex domes which are formed by Mastic 704 to be removed easily and completely.

Figure 12. Solidifying Mastic 704 on the plexiglass plate.

6) pasting convex domes on the sand core

Pasting these convex domes on the two side of sand core according to the designed direction, location and curvature to form non-smooth surface, as shown in Figure 13.

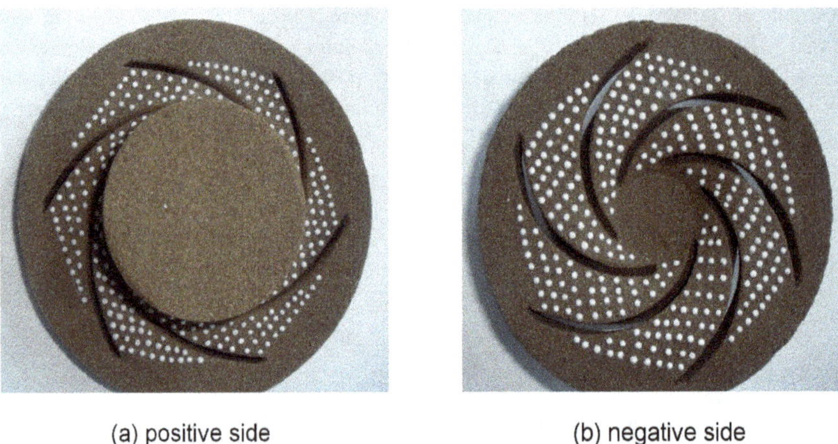

(a) positive side (b) negative side

Figure 13. Pasting non-smooth structures on the two sides of the sand core.

7) painting

Painting HITCM on the convex domes and trying not to paint them on the sand core as much as possible.

8) Assembling and casting

Assembling the sand core with convex domes non-smooth surface to the cast mold, as shown in Figure 14 and then casting.

Figure 14. Assembling and casting.

3. Results and error analysis of OTMC

According to this method, small rib-like bionic and dimple concave non-smooth surface can be formed on the complex cast, as shown in Figure 11 and 14c respectively. As for the rib-like non-smooth structures, its sharp corner can be formed desirably when the cross-section is a triangle. Compared with other processing methods, OTCM is advantageous in forming accurately and continuously, and most important of all, it will not produce residual stress on the casts.

Take the rib-like non-smooth structures as an example to conduct error analysis of OTMC. Here, the non-smooth structures are casted on the complex surface of water centrifugal pump impeller and their cross-section is an equilateral triangle. The designed size on the casts is as follows: the width of rib is 3 mm, the height of rib is 3 mm, and the distance between the two ribs is 6 mm. Measured by the calipers, the size of the final cast is: L=2.96 mm, H=3.03 mm and D=6.03 mm, as shown in Figure 15. This error is in the permissible designed range, and it can be accepted.

Figure 15. The size of the final cast. L is the width of rib, H is the height of the rib, and D is the distance between the two ribs.

4. Conclusion

It can be concluded from the above description and analysis that the process of OTMC should have a small adjustment according to the shape of non-smooth structures, although its mechanism is the same. The main process of OTMC should include: selecting material, forming non-smooth structures, painting HITCM, pasting non-smooth structures on the sand core to form non-smooth surface, drying, trimming and casting. The selection of intermediate media is very important, especially the selection of HITCM, because it is a key step to help form the desired non-smooth surface on the complex casts. In order to obtain desired non-smooth surface on the complex surface, several technical requirements should be met in the process of casting.

1. The surface of rubber convex section and the plexiglass plate should not be polluted by oil or other impurities which may affect the surface quality.
2. The highest resistant temperature of HITCM should not be lower than 1700°.
3. The shape and size of rubber convex section and convex domes should meet the designed requirements.
4. The non-smooth structures should be lined in accordance with the designed direction, location and curvature to form bionic non-smooth surface.
5. The process of casting should agree to the casting technical requirements set by different institutions.

By adopting this method, bionic rib-like non-smooth surface of impeller have been produced in large quantity and with a relatively low productive efficiency. However, production of dimple concave non-smooth surface on the complex casts in quantity still needs improving in the future.

Acknowledgements

The authors are grateful to National Natural Science Foundation of China (Grant No. 51105168), the International (Regional) Cooperation and Exchange of the National Natural Science Foundation of China (Grant No. 50920105504), the Key Program of the National Natural Science Foundation of China (Grant No.50635030), and the research and special founding project of Ministry of Land and public service sectors (Grant No. 20101108206-02).

Author details

Tian Limei[1*], Bu Zhaoguo[1,2] and Gao Zhihua[1]

*Address all correspondence to: lmtian@jlu.edu.cn

1 Key Laboratory of Bionic Engineering of China Ministry of Education, Jilin University, Changchun, China

2 FAW Wuxi Fuel Injection Equipment Research Institute, Wuxi, China

References

[1] Philip, B. (1999). Shark Skin and Other Solutions. *Nature*, 400, 507.

[2] Barthlott, W., & Neinhuis, C. (1997). Purity of the sacred lotus or escape from contamination in biological surfaces. *Planta*, 202, 1-8.

[3] Changhai, Z., Luquan, R., Rui, Z., Shujie, W., & Weifu, Z. (2006). The relationship between the body surface structure of Cybister bengalensis and its unction of reducing resistance. *Journal of Northest Normal University (Nature Science Edition)*, 38, 109-113.

[4] Luquan, R. (2009). Progress in the bionic study on anti-adhesion and resistance reduction of terrain machines. *Sci China Ser E-Tech Sci.*, 52, 273-284.

[5] Singh, R. A., Yoon, E. S., & Jackson, R. L. (2009). Bimimetics. The science of imitating nature. *Tribology & Lubrication Technology*, 41-49.

[6] Ren, L. Q., Peng, Z. Y., Chen, Q. H., Zhao, G. R., & Wang, T. J. (2007). Experimental study on efficiency enhancement of centrifugal pump by bionic non-smooth technique. *Journal of Jilin University*, (Eng. and Tech.Ed.), 37, 575-581, (in Chinese).

[7] Guangji, Li., Xia, Pu., Chaoyuan, Lei., et al. (2008). Brief introduction to the research on biomimetic drag-reduction materials with non-smooth surface. *Materials Research and Application. J.*, 2(4), 455-459.

[8] Vaidyanathan, K., Gell, M., & Jordan, E. (2000). Mechanisms of spallation of electron bearn physical vapor deposited thermal barrier coatings with and without platinum aluminide bond coat ridges. *Surf Coat Techn. J.*, 28, 133-134.

[9] Aimei, Luo., Cunguo, Lin., Wang, Li., et al. (2009). Micromorphology observation of shark skins and evaluation of antifouling ability. *Marine Environmental Science. J.*, 28(6), 715-718.

[10] Yanlei, Peng., Cunguo, Lin., & Wang, Li. (2009). Study on Micromorphology of Shark Skins and Its Antifouling Performance. *Paint & Coatings Industry. J.*, 39(12), 40-43.

[11] Deyuan, Zhang., Jun, Cai., Xiang, Li., et al. (2010). Bioforming Methods of Bionic Manufacturing Mechanical. *Chinese Journal of Mechanical Engineering. J.*, 46(5), 88-92.

[12] Xin, Han., Deyuan, Zhang., Xiang, Li., et al. (2008). Large sharkskin replication preparation of bionic friction reduction surface research. *Chinese Science Bulletin. J.*, 53(7), 838-842.

[13] Xin, Han., & Deyuan, Zhang. (2008). Research of Sharkskin replication process. *Sci China Ser E-Tech Sci. J.*, 38(1), 9-15.

Chapter 2

Sand Mold Press Casting with Metal Pressure Control System

Ryosuke Tasaki, Yoshiyuki Noda,
Kunihiro Hashimoto and Kazuhiko Terashima

Additional information is available at the end of the chapter

1. Introduction

A new casting method, called the press casting process, has been developed by our group in recent years. In this process, the ladle first pours molten metal into the lower (drag) mold. After pouring, the upper (cope) mold is lowered to press the metal into the cavity. This process has enabled us to enhance the production yield rate from 70% to over 95%, because a sprue cup and runner are not required in the casting plan [1]. In the casting process, molten metal must be precisely and quickly poured into the lower mold. Weight controls of the pouring process have been proposed in very interesting recent studies by Noda et al. [2]. However, in the pressing part of the casting process, casting defects can be caused by the pattern of pressing velocity. For example, the brake drum shown in Fig. 1 was produced with the press casting method. Since the molten metal was pressed at high speed, the product had a rough surface. This type of surface defect in which molten metal seeps through sand particles of the greensand mold and then solidifies, is called Metal Penetration. Metal penetration is most likely caused by the high pressure that molten metal generates, and it necessitates an additional step of surface finishing at the least. Thus, the product quality must be stabilized by the suppression of excess pressure in the high-speed press. For short-cycle-time of production, a high-speed pressing control that considers the fluid pressure in the mold is needed. Pressure control techniques have been proposed for different casting methods [3-4]. In the injection molding process, the pressure control problem has been successfully resolved by computer simulation analysis using optimization technique by Hu et al. [5] and Terashima et al. [6]. Furthermore, a model based on PID gain selection has been proposed for pressure control in the filling process. Although the pressure in the mold must be detected in order to control the process adequately using feedback control, it is difficult to measure the fluid pressure, because the high temperature of the molten metal (T ≥ 1200 K) precludes the use of a pressure sensor. Thus, in our previous papers by Tasaki et al. [7], the pressure during pressing at a lower pressing velocity was estimated by using a simply constructed model of molten metal's pressure based on analytical results of CFD: Computational Fluid Dynamics. A new sequential pressing control, namely, a feed forward method using a novel simplified press model, has been reported by the authors of Ref. [7]. It has been shown that this method is very effective for adjusting pressure in the mold. However, in the previous paper, the actual unstationary flow and the temperature drop during pressing was not considered; a detailed analysis that considers the temperature change during pressing is required to reliably predict and control the process behaviors.

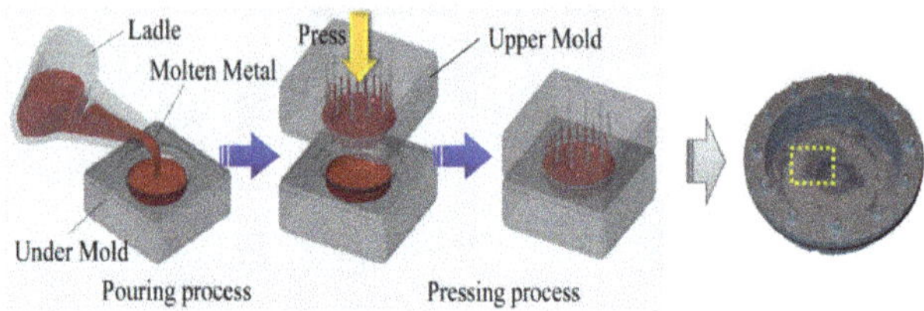

Figure 1. Pouring and pressing processes in press casting.

In this chapter, a novel mathematical model with the pressure loss term of fluid in vertical unstationary flow is derived by assuming that the incompressible viscous flow depends on the temperature drop of the molten metal. The model error for the real fluid's pressure is minimized by the use of parameter identification for the friction coefficient at the wall surface (the sole unknown parameter). Furthermore, the designed velocity of the switching pattern is sequentially calculated by using the maximum values of static, dynamic, and friction pressure, depending on the situation in each flow path during the press. An optimum design and a robust design of pressing velocity using a switching control are proposed for satisfying pressure constraint and shortening the operation time. As a final step in this study, we used CFD to check the control performance using control inputs of the obtained multi-step pressing pattern without a trial-and-error process.

2. Pressing Process in Press Casting

The upper mold consists of a greensand mold and a molding box. The convex part of the upper mold has several passages that are called overflow area, as shown in Fig. 2. Molten metal that exceeds the product volume flows into the overflow areas during pressing. These areas are the only parts of the casting plan that provide the effect of head pressure. As the diagram shows, these are long and narrow channels. When fluid flows into such the area, high pressurization will cause a casting defect. Therefore, it is important to control the pressing velocity in order to suppress the rapid increase in pressure that occurs in high-speed pressing. The upper mold moves up and down by means of a press cylinder and servomotor. The position of the upper mold can be continuously measured due to encoder set in the servo cylinder to control molten metal pressure.

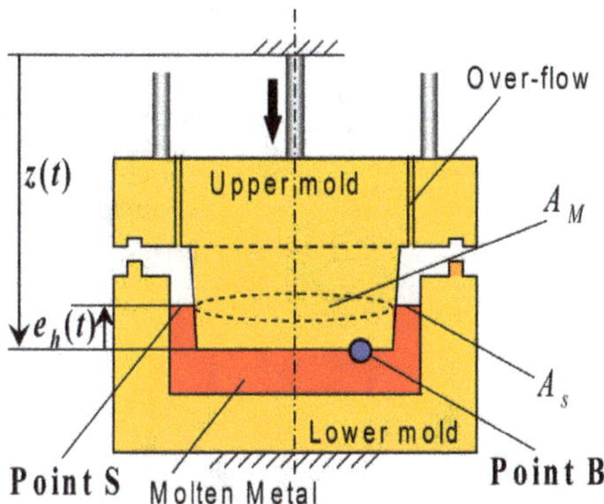

Figure 2. Diagrammatic illustration of pressing.

3. Modeling and Switching Control of Pressure

The online estimation of pressure inside the mold is necessary in the press casting system. The CFD analysis, based on the exact model of a Navier-Stokes equation, is very effective for analyzing fluid behavior offline and is useful for predicting the behavior and optimizing a casting plan. However, it is not sufficient for the design of a pressing velocity control or for the production of various mold shapes, because the exact model calculation would take too much time. Therefore, construction of a novel simple mathematical model for the control design in real time is needed in order to realize real-time pressure control. A simplified mold shape is shown in Fig. 3, where b_i and d_i are the height and the diameter, respectively. P_B is the pressure of the molten metal on a defect generation part where the pressure will cause a defect. The pressure fluctuation during pressing is approximated by a brief pressure model for an ideal fluid; i.e., an incompressible and viscous fluid is assumed. Here, $e_h(t)$ in Fig. 3 is the fluid level from under the surface of the upper mold. The head pressure of P_B is directly derived from $e_h(t)$. The press distance $z(t)$ of the upper mold is the distance that the upper mold must travel until it makes the bottom thickness of the product with the poured fluid in the lower mold. By increasing the pressing velocity or the flowing fluid velocity, the dynamical pressure changes rapidly by the effect of liquidity pressure. The hydrodynamic pressure for the peak fluid height is then involved in determining P_B. Therefore, P_B depends on the head and hydrodynamic pressure determined by using Bernoulli's theorem, and the pressure loss by viscosity flow friction is represented by the following equation:

$$P_b(t) = \rho g e_h(t) + \frac{\rho}{2}\left(1 + \lambda(T)\frac{l(e_h)}{d(e_h)}\right)\dot{e}_h(t)^2 \tag{1}$$

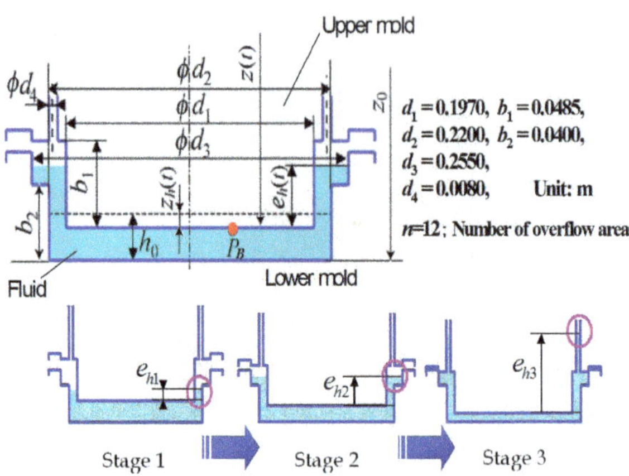

Figure 3. Mold shape and flow pass change.

where ρ[kg/m^3] is the density of fluid and g[m/s^2] is the acceleration of gravity. The adjustable parameter λ is the coefficient of the fluid friction depending on the fluid temperature, and $l(e_h)$ is the mold wall height of the part that causes shear stress in the vertical direction. The surface area of the flow channel decided by the mold shape is represented by $D_i(i=1,2,3)$, and D_i changes as $D_1 = d_2 - d_1$; $D_2 = d_3 - d_1$; $D_1 = d_4/n$ during pressing. The number of overflow areas is represented by n. By the second term on the right-hand side in Eq. 1, the pressure P_B will rise rapidly due to the increasing fluid velocity when fluid flows into the overflow areas. The validity of the proposed pressure model as expressed by Eq. 1 was checked with several CFD simulations in our previous paper[7] under such the condition that temperature of molten metal is constant. The friction coefficient λ was then uniquely identified by a parameter identification fitting with the results derived from the CFD model. We have proposed a switching control for the pressing velocity to suppress the pressure increase. Thus, the pressing velocity necessary to suppress the pressure for defect-free production must be determined and implemented. Here, a multi-switching velocity pattern can be obtained using the following equation, and derived from the pressure model.

$$\dot{z}_k = \sqrt{\frac{2(P_{Blim} - \rho g h_{uk})}{\rho \max\left(A_{Sk}^2 / A_{Mk}^2\right)(1 + \lambda h_{uk} / D_k)}} \tag{2}$$

where the $k^{th}(k = 0,1,..)$-step velocity is decided in order that the maximum velocity satisfies the desired pressure constraint P_{Blim}[Pa]. Because the diameter D_k and square ratio of surface area $(A_{Sk}/A_{Mk})^2$ discontinuously change by each stage during pressing as shown in Fig. 3, a multi-switch velocity control is adopted. The number k of steps of pressing velocity with multi-switching can be determined by the mold shape in the case of Fig. 4, with the maximum value of k being 3. $\dot{z}_0$is the initial pressing velocity up until the point when the bottom surface of the upper mold contacts the top surface of the poured fluid. Derivation of Eq. 2 is straightforwardly calculated, and is omitted due to the paper space limitation.

When the pressing velocity changes from $\dot{z}_k$ to $\dot{z}_{k+1}$[m/s], the pressing distance z[m] is given by information of the mold shape and poured fluid volume. The design of the sequential velocity pattern such as the multi-switch point and each velocity must be adapted to particular mold shape. In the next pressing simulation, a switching velocity input is sequentially designed as shown in Fig. 4, where the press velocity pattern is formed as a trapezoidal shape by the switching position H_{uk} and the pressing acceleration a[m/s^2]. The control performance using the switching velocity of Eq. 2 designed by the proposed simple model was reasonably validated by CFD simulation as shown in Fig. 5. Although the flowing fluid has 3 flow pass stages during pressing for the mold shown in Fig. 3, the designed switching velocity pattern switches only 1 time. This is meant to set a maximum velocity of 50[mm/s] for $\dot{z}_1$at the 1st stage and$\dot{z}_2$at the 2nd stage to suppress extremely turbulent flow. $\dot{z}_3$at the 3rd stage of the narrow flow pass is then set to 6.9[mm/s]. Here the pressing acceleration is set to 1.5[m/s^2], and the total distance of pressing is 15[mm]. The molten metal properties in these simulations are shown in Table 1. The both pressure fluctuations as show in Fig.5 are satisfied under the pressure constraint value assumed as 10[kPa]. This pressure constraint value has been previously decided by using both the actual experimental test and the CFD analysis results of the press with several constant velocity patterns using molten metal.

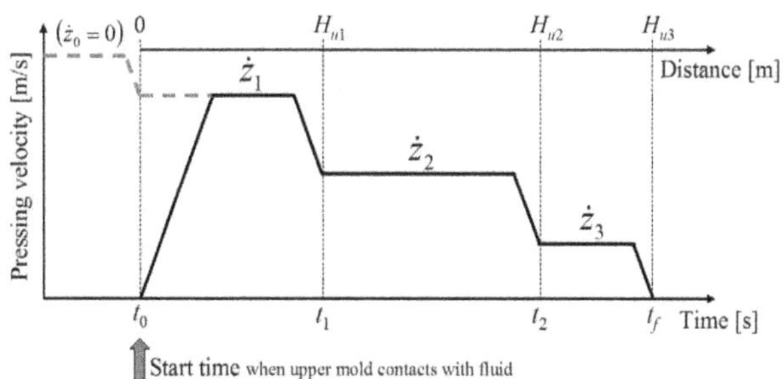

Figure 4. Pressing input shaped by trapezoidal velocities.

In the next chapter, parameter identification of λ [-] will be shown for each simulation condition to consider the pressure increase suppression for viscous fluid with a temperature decrease. Pressure has been rapidly increased while liquid flows into narrow pass $d(e_h)$[m] such that stage-3 in Fig. 3. As seen from Eq. 1, the effect of λ on the variation of pressure $P_b(t)$ becomes larger with the increase of liquid level e_h[m] and flow velocity $\dot{e}_h$ [m/s]. Thus, exact value of $\lambda(T)$ must be given for the region of e_h and $\dot{e}_h$, because our purpose is to suppress the maximum pressure value. Therefore the fitting identification should be considered for only the flow during a short period in stage-3. Then, $T(t)$ is the lowest temperature during pressing because of the end time of pressing.

Density (const.)	7000 [kg/m³]
Viscosity (*T*=1673)	0.02 [Pa•s]
Viscosity (*T*=1423)	0.2 [Pa•s]
Specific heat	771 [J/(kg•K)]
Thermal conductivity	29.93 [W/(m•K)]
Coefficient of heat transfer	1000 [W/(m²•K)]
Liquidus temperature	1473 [K]
Surface tension coefficient	1.8 [-]
Contact angle	90 [deg]

Table 1. Molten metal properties.

4. Parameter Identification

Several parameter identifications of the fluid friction coefficient $\lambda(T_{end})$ at end time of pressing for various upper mold velocities have been carried out by comparing the proposed model with the CFD model analysis. The conditions of molten metal in these identification simulations are shown in Table 1. For the assumpution of temperature-drop cases, initial temperatures are set at 1673, 1623 and 1573[K] respectively. Although the

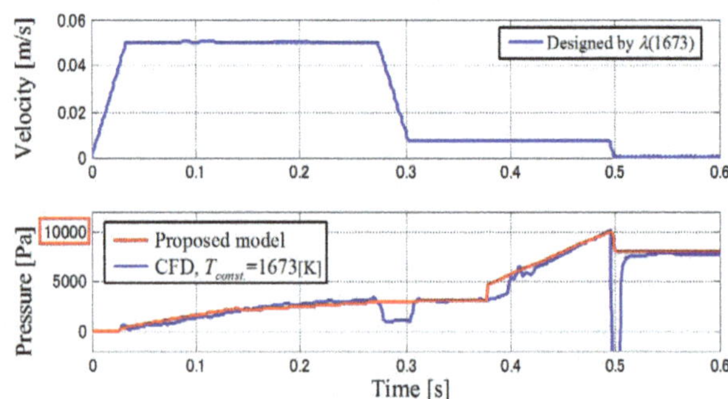

Figure 5. Pressure suppression ($T_{const.}$=1673[K]).

inverse trend of relative change between temperature-drop and viscosity-increase have been clarified, it seems difficult to obtain theoretical equation analytically on the relative change for a wide range of temperature variations and variety of materials. In the temperature drop from 1673 to 1423[K], the viscosity increase is arbitrarily assumed as the linearly dependence changing from 0.02 to 0.20[Pa•s]. Here, the maximum value of the pressure behavior by Eq. 1 of the proposed model is uniquly fitted to the results of the CFD model simuation.

In each case, the time-invariant parameter $\lambda(T_{end})$ have been identified as shown in Fig. 6. Using the designed velocity pattern in Fig. 5 conducted under the condition of constant temperature during pressing, the pressure behavior considering the fluid's heat flow to the molds exceeds the pressure constraint (top in Fig. 6) because of the higher viscosity(bottom in Fig. 6) as shown in Fig. 6. As seen from Fig. 6(a), (b) and (c), the lower temperature at end time induces the larger the value of λ. The temperature drop from start to end of pressing is almost 50[K] in these results. The pressure increase during pressing due to the larger value of $\lambda(T_{end})$ with the decreased temperature is confirmed. The simulation results of a simple model such that $\lambda(T_{end})$ is given as a constant value by fitting almost explains the results of the CFD model. Therefore, it is expected that we can conduct the control design using this simple pressure model under the restricted temperature change.

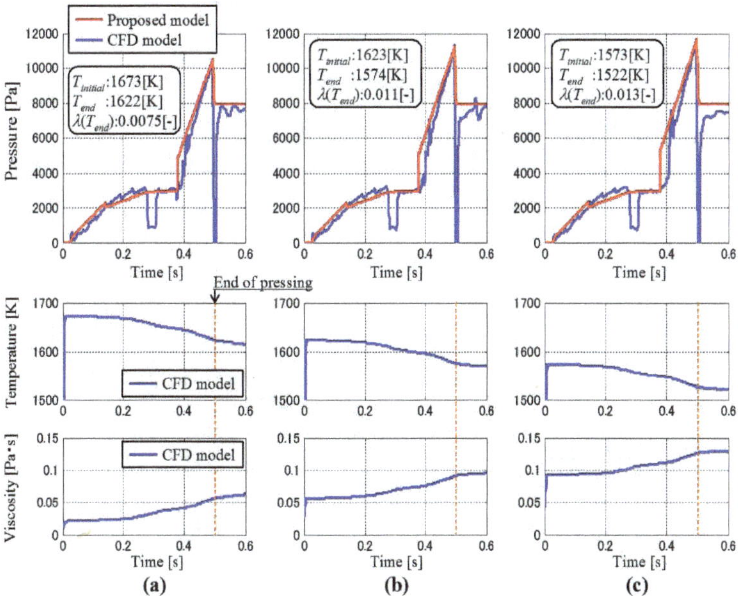

Figure 6. Parameter identification results.

5. Proposed Control Design and Results for Pressure Suppression

In this section, the proposed sequential switch velocity control considering the viscosity increase related to the temperature drop during pressing will be checked by using CFD model simulation with heat flow calculation.

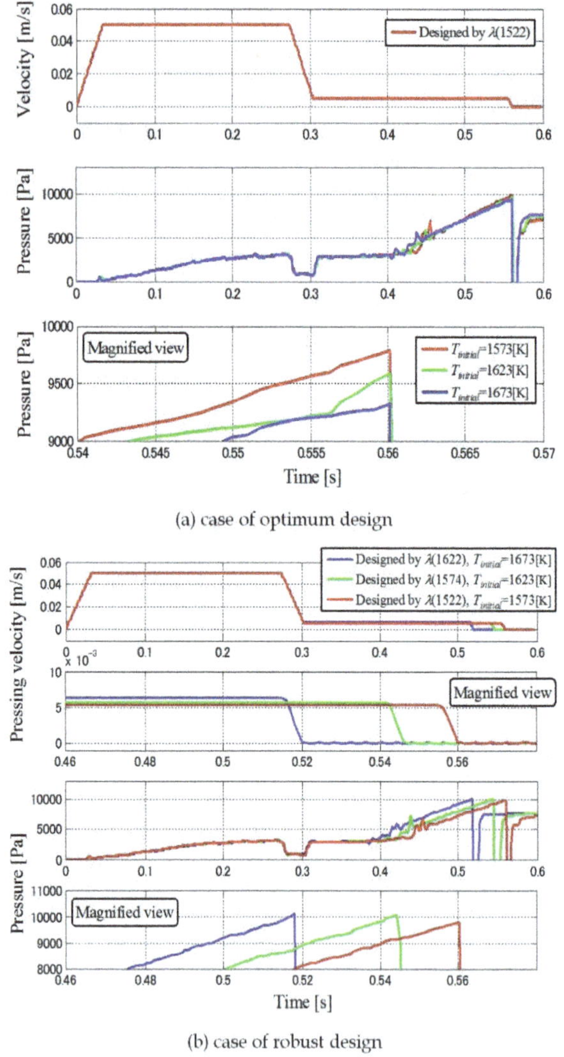

Figure 7. Pressure suppression simulation using CFD simulator with designed velocities.

As example, for the designed pressing velocity patterns using $\lambda(T_{end})$ derived by the previous simulations, where T_{end} =1622, 1574 and 1522[K], the pressure suppression results for the each temperature condition of $T_{initial}$ =1673, 1623 and 1573[K] were checked for a upper pressure constraint: 10[kPa]. Here, the optimum design and robust design are introduced by using the proposed switching control method. Fig. 7(a, upper) shows a comparison of the designed velocity patterns and the magnified view. These lines show the desighed opitimum velosity patterns in the each case of temperature drop. The switched velocities (2nd constant velocity) are slightly different as 6.2, 5.5 and 5.2[mm/s], for the influence of the viscosity increase with the temperature drop. The end time of pressing are then 0.520, 0.546 and 0.560[s] respectively, and the biggest difference of the pressing time is only 0.040[s]. These velocity patterns which differs slightly, guarantees the exact suppression of pressure less than the constraint value as shown in Fig. 7(a, lower). Fig. 7(a, bottom) shows the magnified view of the pressure peak part at the end time of pressing. On the other hand, Fig. 7(b) shows pressure suppression varidation for a robust design of pressing velocity. The designed velocity by $\lambda(T_{end}$=1522) in case of lowest temperature has been checked for $T_{initial}$ = 1673, 1623 and 1573[K]. As seen from Fig. 7(b, bottom), each muximum pressure value is suppressed under the upper constraint of pressure with some allowance. However, the end time is a little bit late compared with the optimum design case. As seen from this result, both methods satisfies the pressure suppression. However, optimum design satisfies both requirements of pressure constraint and shortening the operation time. On the other hand, robust design satisfies only pressure constraint, although this is useful, when temperature drop is not exactly known, but knows the least temperature for all batch operations. These analyses presented that the proposed control to suppress the maximum pressure of viscous flow with temperature drop can design the press switching velosity pattern optimally and robustly, for such the case that temperature drop from start time to end time of press is about 50[K].

6. Summary 1

In this section, we proposed an optimum control method of molten metal's pressure for a high-speed pressing process that limits pressure increase in casting mold. Influence of viscosity increase by temperature drop can be applied to the sequential pressing velocity design. The control design was conducted simply and theoretically, and included a novel mathematical model of molten metal's pressure considering viscous flow. The friction coefficient depending on temperature is meant to generate higher pressure than that in the case modeled without temperature drop during pressing. Using the pressure constraint and information on the mold shape, an optimum velocity design and robust velocity design using multi-switching velocity were derived respectively without trial-and- error adjustment. Finally, the obtained velocity reference's ability to control pressure fluctuation and to realize short cycle time was validated by the CFD simulations. In the near future, the proposed pressure model for optimizing the pressing process will be modified with the theoretical function models on temperature and viscosity-change, and futheremore real experiments will be done.

7. Experimental confirmation of physical metal penetration generation

In this section, we tried several molten metal experiments to clarify the mechanism of physical metal penetration growth and the boundary condition of physical metal penetration generation, and to validate the control performance of the feedforward method using the proposed pressing input design. Several experimental confirmations for the proposed pressure control method with a mathematical model of molten metal pressure were achieved for brake-drum production. The press casting productions with reasonable casting quality for each pressing temperature has been demonstrated through molten metal experiments.

7.1. Physical metal penetration and molten metal's pressure

Liquidus temperature of iron metal is about 1400[K], and the casting mould commonly used is heat-resistant green sand mould, for its advantages of high efficiencies of moulding and recycling. However, some defects are often caused by high pressurized molten metal [10]. Pressurized molten metal soaks into the sand mould surface, and then solidifies and form the physical metal penetration. Physical metal penetration as a typical defect related to higher pressurization inside the mould is offen ocurr on the casting surface. The metal penetration generated on complex shape product such as the products with tight, thin and multilayer walls, is difficult to be removed, while in the case of simple shape product, the defect can be removed by later surface processing. If the defect generation can be prohibited by pressing velocity adjustment, the sound iron castings can be obtained.

7.2. Mechanism of physical metal penetration

Physical factor caused metal penetration is explained by a diagrammatic illustration (Fig. 8) of interfacial surface between the molten metal and the sand mould, and a balance between two sides competing pressure on the boundary [11]. Fig. 8 also shows the relationship between the pressure balance and the metal penetration growth. In Fig. 8, on one side, the molten metal acts as a static pressure, P_{st} (Pa), a dynamic pressure, P_{dyn} (Pa), and a pressure, P_{exp} (Pa), because of expansion during solidification, which can force the liquid into the interstices of the sand grains. On the other side, due to the suppression effect of infiltration, the frictional loss pressure between the liquid metal and the sand grains, P_f (Pa), the pressure resulting from the expansion of the mould gasses, P_{gas} (Pa), and the pressure in the capillary, P_g (Pa), are all acted on the boundary surface. The governing equation that describes the pressure balance at the mould and metal interface can be written as:

$$P_g + P_f + P_{gas} = P_{st} + P_{dyn} + P_{\exp} \tag{3}$$

where the molten metal soaks into sand surface in the case that the right hand side of this equation is larger than the left hand side. As a result, the metal penetration defect is generated. Depending on the contact angle of iron and sand, the capillary pressure can be changed to be negative or positive as shown in Fig. 9. Thus, the pressure has both of beneficial or detrimental effects in preventing penetration at the same time. So capillary pressure can be negligible in Eq. (3). The P_{exp} can be eliminated in the case of the casting with open type mould as shown in Fig. 10. This means that P_{exp} is strongly related to the casting process design. Furthermore, using a slower filling velocity and selecting the moulding material that does not contain the component which can generate gasses, P_{dyn} and P_{gas} are then both negligible. Here, we obtain a simplified relational equation of pressure balance:

$$P_f = P_{st} \tag{4}$$

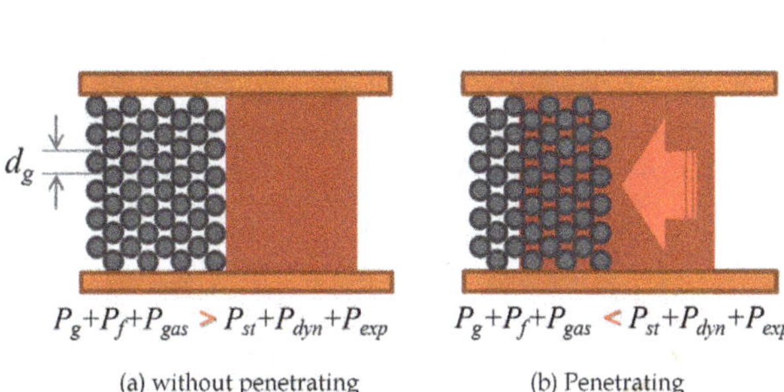

Figure 8. Pressure balance and penetration defect.

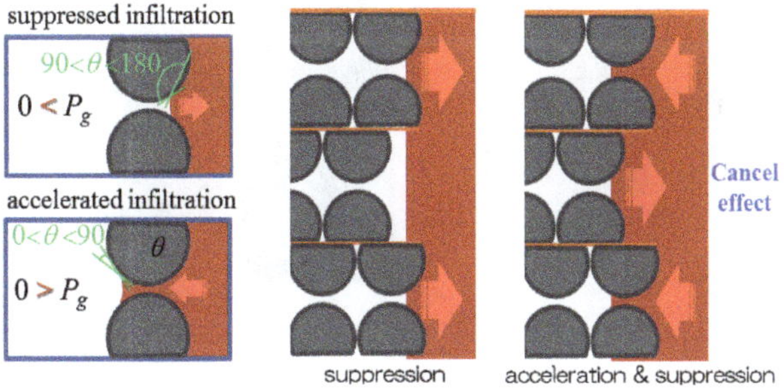

Figure 9. Cancel effect of capillary pressure.

7.3. Penetration phenomena under static pressure

In the conventional gravity casting, molten metal infiltrating into sand particles is generally generated when the high static pressure is added inside the mould. Sound casting products with metal penetration-free are designed such as whose maximum height of liquid head is under the allowable static pressure after filling. But, in the sand press casting case, it is confirmed that the penetration defect on the product surface is generated, even if the mould with a low static pressure is utilized. This indicates that the influences such as the dynamic pressure and the pressure due to the viscous friction depending on temperature drop, must be considered.

To observe the penetration growth under the force of gravity, a test experiment has been achieved with molten metal. A suggested casting mould shape and the casting are shown in Fig. 10. The molten metal was poured into the casting mould quickly at 1,400 ℃, and kept at 1673[K] until the end of filling. The casting mould is 1,000 mm in height and Φ45 mm in diameter. Here, the static pressure at the depth of H_l (m), P_{stb} is simply written as

$$P_{stb} = \rho g H_l \tag{5}$$

Where $\rho = 7{,}000$ (kg m^{-3}) is the density of molten metal, $g = 9.8$ (m s^{-2}) is gravity acceleration, H_l (m) is the vertical depth from the top of the casting product. Equation (5) can give the static pressure value easily, then a pressure constraint value for preventing the penetrated surface can be derived directly according to the maximum depth without metal penetration defect.

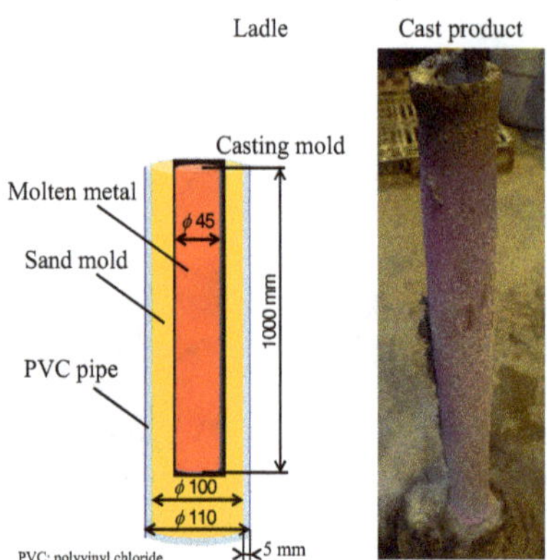

Figure 10. Gravity casting test with opened mold.

The mould release agent covering the casting pattern before moulding was not used in order to prevent the loss of the surface tension; the caking additive of the sand mould was selected for keeping steady the molten metal's properties. A cylindrical casting mold with the diameter of 45(mm) was selected for restricting the temperature distribution of molten metal during pouring. The elimination of physical factors for the penetration generation is considered as follows:

The surface of the product is observed by using the optical microscope. To investigate thoroughly under the casting surface, the cylindrical product is sliced along the direction perpendicularly to its axis, and the cut specimens at each depth, H_l, are pictured respectively. The penetration growths in the early phase and the final phase of solidification are confirmed clearly from Figs. 11(a) and (b). In Fig. 11(a) (H_l =150 (mm)), early phase of penetration in casting surface is identified, and some small sand grains are wrapped in cast metal. In the case of Fig. 11(b) (H_l = 950 (mm)) or bottom part of the product, the infiltration depth of molten metal to sand particles is over 1 (mm). Here, a high pressure loading about 65 (kPa) is estimated by calculating Eq. (5) of the static pressure. In the same way for H_l = 150 (mm), P_{bottom} is obtained as 10 (kPa). Therefore, molten metal's pressure, 10 (kPa), on the sand surface means an upper limit of penetration generation in this casting condition.

Metal penetration growths for each depth, H_l, with a 50 (mm) increment from 50 to 850 (mm) are shown in Fig. 12. Maximum infiltration depth observed in the investigation area of Fig. 12 is increased with vertical depth, H_l. The sand particles inside the metal cannot be removed easily by the next process such as the blast finishing and the grinding. Thus metal penetration defect must be prevented completely, and liquid pressure constraint 10 (kPa) in the case of H_l = 150 (mm) is set for defect-free production.

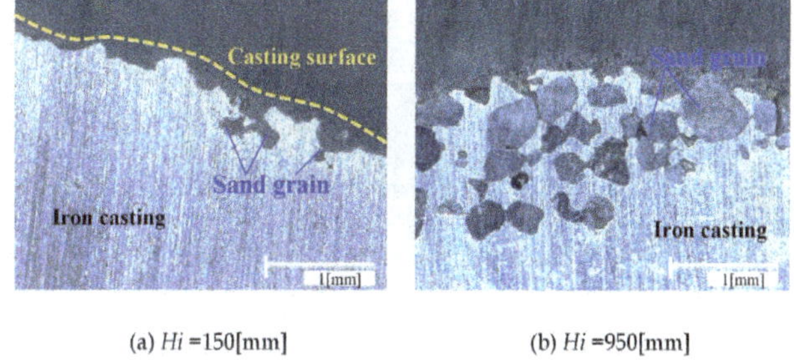

Figure 11. Penetrated surface observation on casting skin.

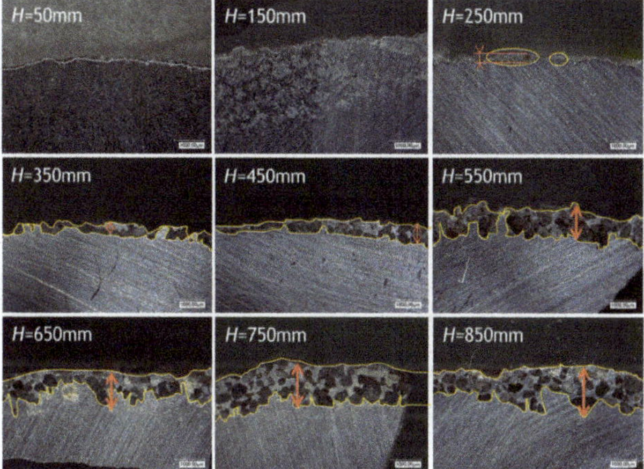

Figure 12. Metal penetration growths for each vertical depth.

7.4. Designed pressing velocity pattern

Substituting the obtained pressure constraint in the previous chapter and mould shape information of target cast product of the drum brake to Eq. (2) in previous chapter, the multi-step velocity pattern is sequentially calculated. Here, the pouring temperature is set to 1,400℃. The initial pressing velocity $\dot{z}_0$ until the upper mould contacts with top surface of poured molten metal, is the maximum pressing velocity, 375 (mm s^{-1}), of press machine. Each velocity for each flow situation are represented in Table 1. The vertical movement is driven accurately by servo cylinder and physical guide bars. The press casting equipment is shown in Fig. 13.

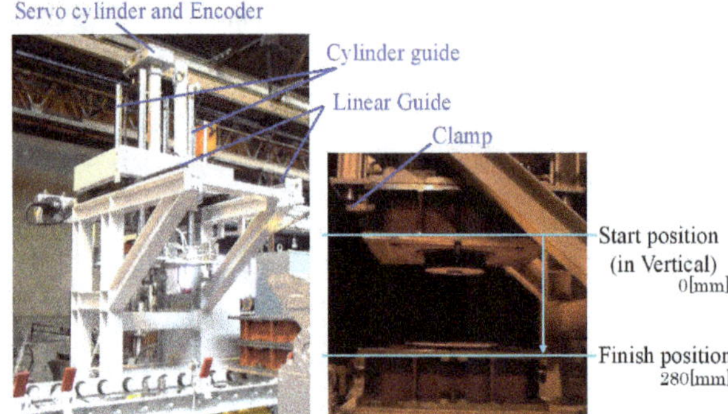

Figure 13. Press casting equipment and mold holding part.

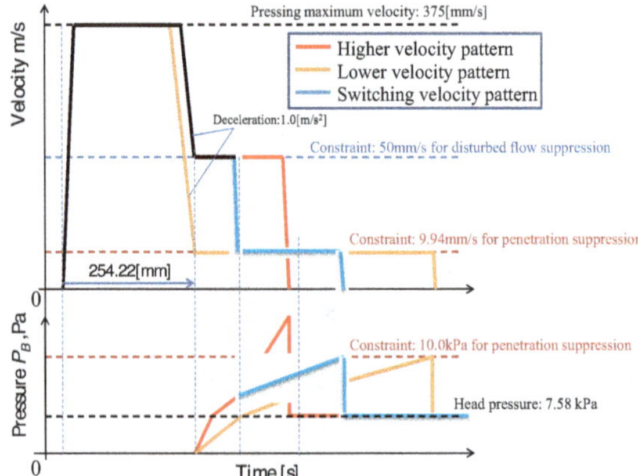

Figure 14. Designed pressing velocity patterns.

The multi-step velocity pattern is shown in Fig. 14. The acceleration of pressing movement is ideally assumed as constant 1 (m s^{-2}). The time constant of this drive system can be set to zero, because the identified exact value is 0.002 s or negligible. Therefore, step type velocity input is shaped as multi-overlapped trapezoid.

For discontinuous flow depended on mould shape, the pressing velocity, 50 (mm s^{-1}), was set in the case of wide liquid surface area. Here the first and second switching velocities, $\dot{z}_1$ and $\dot{z}_2$ calculated by considering the pressure constraint, are higher values in brackets of Table 2 This means that the pressing in the wide flow path must consider an upper limit velocity to prevent the disturbance flow causing overflow to the outside of mould. The velocity constraint was given by experimental trial and error process. Pressure suppression was evaluated by comparing with other conditions shown in Fig. 14.

	$\dot{z}_0$	$\dot{z}_1$	$\dot{z}_2$	$\dot{z}_3$	**Stop**
Pressing Velocity [mm/s]	375.00	50.00 (344.6)	50.00 (542.3)	9.94	0.00
Switching Position [mm]	0.00	254.22	254.22	279.44	280.20

Table 2. Multi-step velocities related to discontinuous change of flow passage.

7.5. Press casting experiments

Effectiveness of the pressure control with multi-step velocity design is confirmed by observing the casting surface. The surface roughness of tested specimens under the given conditions is shown in Fig. 15. In the case of higher velocity pressing (HV: $\dot{z}_1 = \dot{z}_2 = \dot{z}_3$ =50.00[mm/s]), the product surface is the roughest. Dash line circles on the surface show the infiltrated sand particles. This result indicates that the metal penetration defect is clearly generated by pressing with high pressure over 10 kPa. Both in the case of lower velocity (LV: $\dot{z}_1 = \dot{z}_2 = \dot{z}_3$ =9.94[mm/s]) and proposed switching velocity (SV: $\dot{z}_1 = \dot{z}_2$ =50.00[mm/s], $\dot{z}_3$ =9.94[mm/s]), sound products of smoothed surface or defect-free production can be obtained. The pictures of magnified product surface in Figs. 15(a) to (c) are given under the experimental condition of higher pressing temperature 1,400℃ (HT). Here the pressing temperature is adjusted by monitoring with a sensor and naturally cooling the molten metal with the pouring temperature, about 20~30 degrees higher than the pressing temperature. Fig. 15 show that the different surface state does not depend on temperature.

Fig. 16 shows the overview of the casting pressed by the switching velocity pattern (SV). From these photos, better product of SV-HT is clearly verified, because the switch velocity is designed just for the higher temperature 1,400℃. There is a tiny penetration in casting of SV-LT. Higher pressure at the same pressing is generated with higher viscous flow related to lower temperature.

Consequently, the proposed pressing pattern shows defect-free production in the short filling time as almost same as the highest pressing pattern considered with the disturbance flow suppression. The time difference between the cases of HV-LT and SV-LT is only 0.07 (s). This result shows 2 (s) shorter than the case of LV-LT with well production. Furthermore, the comparative validation of the different temperature in Fig. 16 shows that the pressing velocity is designed properly for the monitored poured liquid temperature immediately before pressing. The proposed press casting production considering molten metal's pressure suppression will meet the requirement for practical use with temperature variation range.

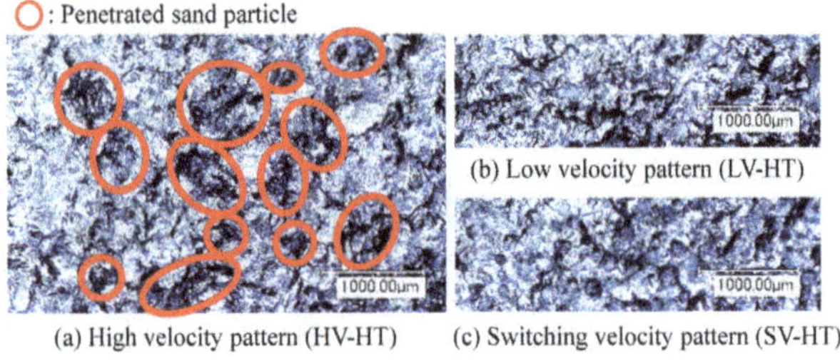

Figure 15. Product surface observations for penetration defect.

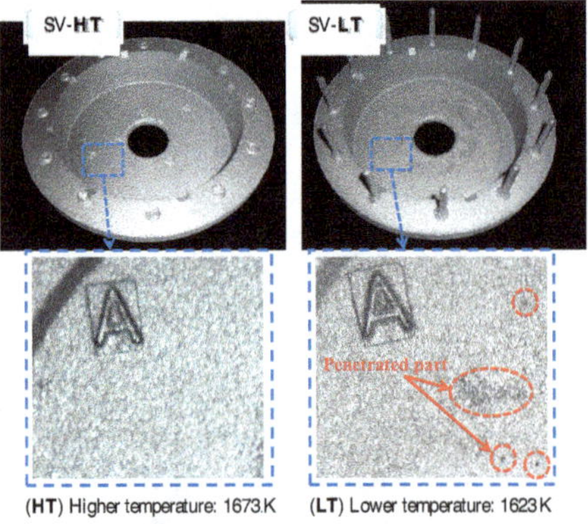

Figure 16. Casting product in case of different temperature conditions.

7.6. Summary 2

The pressing velocity control was proposed in order to suppress increasing pressure with short filling time. A pressure limitation of the penetration generation has been confirmed by a gravity casting experiment for a relation analysis between the static (head) pressure and the infiltrated metal length. Next, by applying the obtained constraint pressure for defect-free to the theoretical control design method with pressing velocity adjustment, the effectiveness of the proposed control method is validated by molten metal experiment. The final results showed that the proposed pressing control realizes sound cast production in almost the same filling time with the high speed pressing, which can cause defect. These confirmation results indicate that the press casting process with our proposed control technique can be adapted properly for environment change such as temperature drop in continual process.

8. Modelling and Control Unstationary Flow

The online estimation of pressure inside the mold is necessary in the press casting system. The CFD analysis, based on the exact model of a Navier-Stokes equation, is very effective for analyzing fluid behavior offline and is useful for predicting the behavior and optimizing of a casting plan [8-9]. However, it is not sufficient for the design of a pressing velocity control or for the production of various mold shapes, because the exact model calculation would take too much time. Therefore, construction of a novel simple mathematical model for the control design in real time is needed in order to realize real-time pressure control.

To analyze flowing liquid motion during pressing, several experiments with colored water and an acrylic mold have been carried out as shown in Fig. 17. The nature of flow will dictate the rectangular Cartesian, cylindrical and spherical coordinates etc. In 3D flow, velocity components exist and change in all three dimensions, and are very complicated to study. In the majority of engineering problems, it may be sufficient to consider 2D flows. Therefore the acrylic mold shaped flat is prepared for flow observation of liquid. The main purpose of our study on the press casting process is to suppress the defect generation of casting product. Air Entrainment during filling is one of the most important problems to solve for flow behavior by adjustment of pressing velocity. If the air is included in molten metal, it will stay and be the porosity defect. By the past experimental result, upper mold velocity less than 50 mm s^{-1} of pressing without air entrainment has been confirmed. From this fact, the pressure model construction is considered for only stationary flow in vertical without air entrainment, or the pressing velocity lower than the upper limit for the defect-free for air porosity.

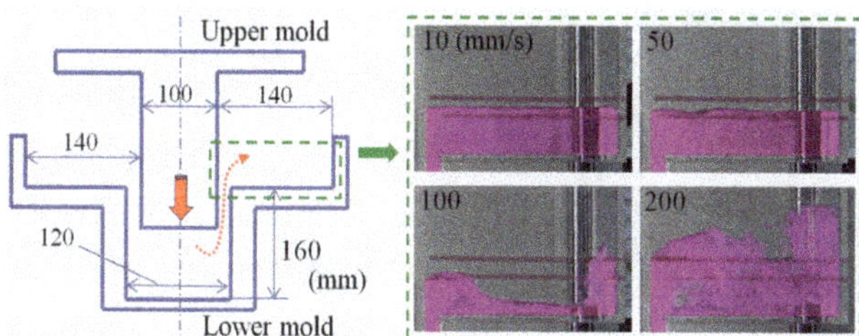

Figure 17. Observational experiment of unstationary flow.

8.1. Pressure Model of Unstationary Flow

Fig. 18 shows the rising flow during pressing and each stream line of molten metal's flow. The unstationary Bernoulli equation for two points: S and B on a given stream line in the flow of an incompressible fluid in the presence of gravity is

$$\int_B^S \frac{\partial U}{\partial t} ds + \frac{1}{2}U_S^2 + \frac{P_S}{\rho} + ge_S = \frac{1}{2}U_B^2 + \frac{P_B}{\rho} + ge_B, \tag{6}$$

where ρ kg m^{-3} is the density of fluid and g m s^{-2} is the acceleration of gravity. The integral is taken along the stream line, and cannot be easily evaluated in general. For the rising flow in press casting, the integral can be quite closely approximated by an integral along the vertical axis. In the case of Fig. 18, the stream line is taken to vertically extend from the bottom surface of upper mold to the free surface of fluid. Placing the origin to the bottom of upper mold surface, substituting $P_S = 0$ (based on gauge pressure) and $e_h = e_S$-e_B, and neglecting $\dot{e}_B^2$ as $\dot{e}_S^2 \dot{e}_B^2$, then the Eq. (6) simplifies to

$$P_B = \rho\left(\ddot{e}_h e_h + \frac{1}{2}\dot{e}_h^2 + ge_h\right) \tag{7}$$

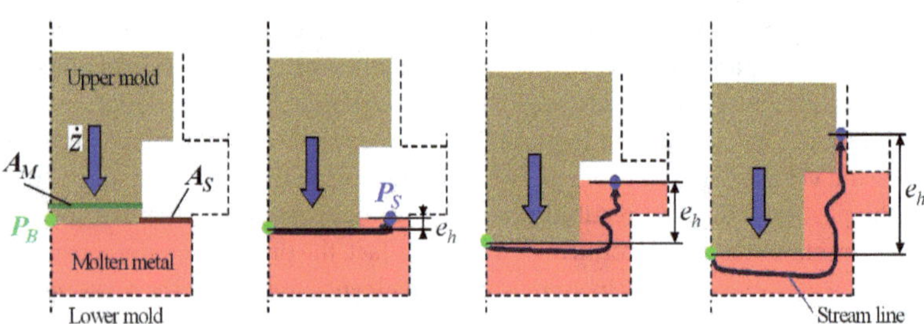

Figure 18. Change of stream line of rising liquid.

The fluid velocity $\dot{e}_h$ m s^{-1} at the free surface A_S m^2 relates the mold surface area A_M m^2 at the same height with the free surface and the pressing velocity $\dot{z}$ m s^{-1} as shown in Fig. 18.

Here, rewriting the extended Bernoulli equation in terms of z m and considering with the initial volume of fluid poured in the lower mold, one obtains

$$P_B = \rho \frac{A_M^2(e_h)}{A_S^2(e_h)}\left(\ddot{z}z + \frac{1}{2}\dot{z}^2\right) + \rho g f(V_p, z) + \Delta p(T, e_h) \tag{8}$$

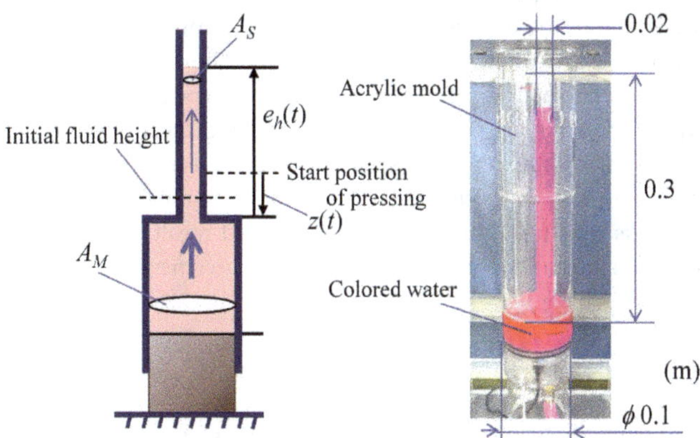

Figure 19. Mold shape for a part of overflow.

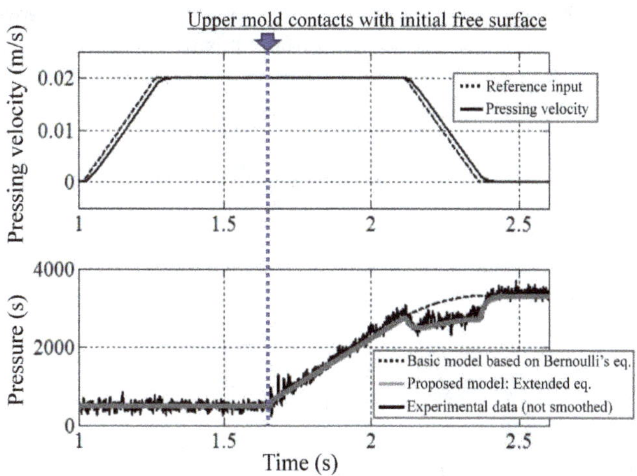

Figure 20. Comparative result between proposed mathematical model and measured pressures.

where $\Delta p(T, e_h)$ means a pressure loss depended on liquid temperature change on flow from upstream to downstream and the vertical flow length e_h contacting with the wall.

To confirm the proposed pressure model for pressed liquid, several experiments using simplified shape mold and water have been carried out. The acrylic mold and its shape are shown in Fig. 19. The vertical movement of the upper mold is derived accurately for reference input of velocity curve by servo-press system. In the experiment as shown in Fig. 20, the actual pressing velocity (solid line) is reshaped for reference input (dashed line). This slight difference is due to the driving motor characteristic approximated by first order lag element with the time constant: 0.020 s. As an example of the confirmation result with proposed model, pressure behavior measured by piezoelectric-type pressure sensor (AP-10S, by KEYENCE Corp.) is shown in Fig. 20 (lower), solid line. Here, the maximum pressing velocity is set to 20 mm s^{-1}, and total moving displacement of press is 22 mm. The dashed line in Fig. 20 (lower) is the pressure calculated result with Bernoulli's equation for steadyfluid flow as described. As seen from this figure, the calculated result of the proposed pressure model considering the unstationary flow, is in excellent agreement with actual pressure behavior during pressing.

8.2. Viscous Influence

In a practical situation, the temperature decrease due to the heat transfer between the molten metal and the mold surface should be considered as an important influence on liquid pressure during pressing. For decreasing temperature, the viscosity increase and higher pressure are then generated, and therefore the penetration defect occurs. Generating the shearing force on the wall surface of the flow path, a point at the upstream is pressurized higher than one at the downstream. Considering the pressure difference between P_B at the bottom of the upper mold and P_S at the free surface, it is written as $\Delta p = P_B - P_S$. Here, the equilibrium relation of force between the shearing force F_w and Δp is derived as following equation by considering the frictional loss pressure.

$$F_w = \frac{\pi}{4}(d)^2 \Delta p = \pi d l \tau \tag{9}$$

Here, using the friction coefficient λ depended on molten metal's temperature T (K), Δp Pa can be represented by the following equation:

$$\Delta p(T, e_h) = \rho \frac{A_M^2(e_h)}{A_S^2(e_h)} \frac{\lambda(T) l(e_h)}{2d(e_h)} \dot{z}^2 \tag{10}$$

After substituting Eq. (10) to Eq. (9), the proposed pressure model conformable to the complex model of CFD is constructed by depending on liquid temperature to express more precisely the molten metal's pressure. Here, $\lambda(T)$ means the coefficient of fluid friction depending on the fluid temperature; it will be sole unknown parameter of the proposed model. $l(e_h)$ is the mold wall length of the part that causes shear stress in the vertical direction. $D_i(i = 1, 2, 3)$ represents the surface area of the flow channel decided by the mold shape as shown in Fig. 3, and D_i will change as $D_1 = d_2 d_1$, $D_2 = d_3 - d_1$, $D_1 = d_4/n$ during pressing. n is the number of overflow areas. By the pressing velocity term in the newly proposed pressure model, it is easily understood that P_B will be rapidly rising due to the increasing fluid velocity when fluid flows into narrow flow path areas.

8.3. Optimized pressure control with continuous velocity input of pressing / Summary 3

In this section, a mathematical modeling and a switching control for pressure suppression of pressurized molten metal were discussed for defect-free production using the press casting. For the complex liquid flow inside vertical path during pressing, the liquid's pressure model for the control design was newly proposed via the unstationary Bernoulli equation, and was represented in excellent agreement with actual pressure behavior measured by a piezoelectric-type pressure sensor. Next, the sequential pressing control design with switching velocity for the high-speed pressing process that limits pressure increase, was applied with considering the influence of viscous change by temperature drop. Using the pressure constraint and information on the mold shape, an optimum velocity design and robust velocity design were derived respectively without trial-and-error adjustment. Consequently, the effectiveness of the pressing control with reasonable pressure suppression has been demonstrated through the CFD. In the near future, the proposed pressure model for optimizing the pressing process will be modified with the theoretical function models on temperature and viscosity-change, and furthermore real experiments with molten metal will be done.

Author details

Ryosuke Tasaki[1*], Yoshiyuki Noda[2], Kunihiro Hashimoto[3] and Kazuhiko Terashima[1]

*Address all correspondence to: tasaki@syscon.pse.tut.ac.jp

1 Department of mechanical engineering, Toyohashi university of Technology, Japan

2 Department of mechanical system engineering, Yamanashi University, Kohu-city, Japan

3 Sintokogio, Ltd., Japan

References

[1] Terashima, K., Noda, Y., Kaneto, K., Ota, K., Hashimoto, K., Iwasaki, J., Hagata, Y., Suzuki, M., & Suzuki, Y. (2009). Novel creation and control of sand mold press casting "post-filled formed casting process. *Foundry Trade Journal International (The Journal of The Institute of Cast Metals Engineers)*, 183(3670), 314-318.

[2] Terashima, K., Noda, Y., Kaneto, K., Ota, K., Hashimoto, K., Iwasaki, J., Hagata, Y., Suzuki, M., & Suzuki, Y. (2009). Novel creation and control of san mold press casting "post-filled formed casting process". *Hommes & Fonderie*, 396(396), 17-27.

[3] Noda, Y., & Terashima, K. (2007). Modeling and feedforward flow rate control of automatic pouring system with real ladle. *Journal of Robotics and Mechatronics*, 19(2), 205-211.

[4] Noda, Y., Yamamoto, K., & Terashima, K. (2008). Pouring control with prediction of filling weight in tilting-ladle-type automatic pouring system. *International Journal of Cast Metals Research, Science and Engineering of Cast Metals, Solidification and Casting Processes, AFC-10 Special*, 21(1-4), 287-292.

[5] Hu, J. V. J. H. (1994). Dynamic modeling and control of packing pressure in injection molding. *Journal of Engineering Materials and Technology*, 116(2), 244-249.

[6] Tasaki, R., Noda, Y., & Terashima, K. (2008). Sequence control of pressing velocity for pressure in press casting process using greensand mould. *International Journal of Cast Metals Research, Science and Engineering of Cast Metals, Solidification and Casting Processes, AFC-10 Special*, 21(1-4), 269-274.

[7] Tasaki, R., Noda, Y., Terashima, K., & Hashimoto, K. (2009). Pressing velocity control considering liquid temperature change in press casting process. *Proc. of IFAC Workshop on Automation in Mining, Mineral and Metal Processing*, 65.

Chapter 3

Evaluation and Modification of the Block Mould Casting Process Enabling the Flexible Production of Small Batches of Complex Castings

Sebastian F. Fischer and Andreas Bührig-Polaczek

Additional information is available at the end of the chapter

1. Introduction

In the current literature about casting processes, the block mould casting is hardly addressed although this process has numerous global applications. Almost all metallic dental implants are manufactured using this process [1-3]. This method is also regularly used in the jewellery industry [4]. The block mould casting process is particularly important for manufacturing metallic foams since it is one of the few process routes for producing cellular structures enabling uniform, open pored foams to be reproduced [5]. As the largest global producer of metallic open pored foams, the company ERG also uses the block mould casting process but rarely communicates details of the casting process. Due to the high degree of freedom the block mould casting process is very suitable for the production of bio-inspired technical devices [6].

The broad objective of this chapter consists of markedly shifting the focus of designers and manufacturers of castings to the block mould casting process. In conjunction with Rapid-Prototyping patterns, this casting method enables extremely complex castings to be variably and flexibly manufactured to their final near net shape [1, 7, 8]. A definite structuring of the pattern's surface is transferred to the casting's surface as a consequence of this method's very accurate reproduction whereby flows around the casting can be optimised in a functionally integrated way [4, 9].

Moreover, this chapter should provide the user with the possibility of optimising the block mould casting process with the aid of the depicted test results. As a consequence of the firing process, cracks can be initiated in the mould by means of which casting defects occur to the point of mould leakage. By means of optimising the mould material's water content and temperature, the mixing duration, the firing temperature and also by adding supplements, the tendency for mould cracking is minimised by using consummate mould material manufacturing. Tests to elevate the cooling rates of the block mould casting process provide improvements in the mechanical properties of the cast metallic components.

2. State of the art

2.1. Classification of the Block Mould Casting Processes

The block mould casting process ranks among the precision casting methods in which the better known investment casting is also included. This is probably the reason why it is frequently misleadingly referred to in the literature as investment casting process. These two methods both employ the lost pattern technique. The patterns are either melted or burnt out of the mould after the moulding material has cured and are thus subsequently no longer available for mould manufacturing [1, 7, 10]. This contrasts with, for example, the widely used sand casting process, in which multiple use patterns are employed. These patterns are parted in order that they can again be used after forming the positive impression in the sand mould. Since the patterns of precision casting methods are removed from the mould by means of melting or vaporising, they do not have to exhibit mould parting. Owing to this, very complex, final near net shaped casting geometries can also be produced which possess undercuts. Apart from these advantages, precision castings exhibit a very low surface roughness compared to sand cast components which can also considerably reduced the castings' machining [4]. The difference between investment and block mould casting processes arises in the moulding material used and therefore on the mould. The patterns for the investment casting process are dipped into a ceramic slurry and subsequently sanded using ceramic granules. After the slurry has dried, this procedure is repeated until up to approx. 8 to 13 layers exist on the pattern. To manufacture a block mould, the pattern is directly embedded in a ceramic slurry where gypsum- or phosphor-bonded investments are mainly employed as the mould material [1, 11]. Thus in comparison to the investment casting process, a great deal of process time and expenditure can be saved during the block mould casting process. In addition to this, the block mould's castings are easier to demould.

2.2. Process Steps

Figure 1a) schematically depicts the sequences of the block mould casting process. The elements of the pattern are produced by injecting wax into a matrix, usually made of aluminium or steel, and are soldered with the help of bee-glue. Depending on the component to be cast, a wax base or a plate is soldered at a wax chute. If a large number of small castings are to be produced, it is sensible to fasten the patterns of the castings onto a wax base. For the production of larger castings or metal foams, a wax plate is suggested since boxes of perforated steel cuvettes are used. This cuvette stabilises the ceramic moulding material, especially during the firing process. Without this reinforcement, the mould would be damaged because of phase transitions in the moulding material which occur during firing (figure 1b).

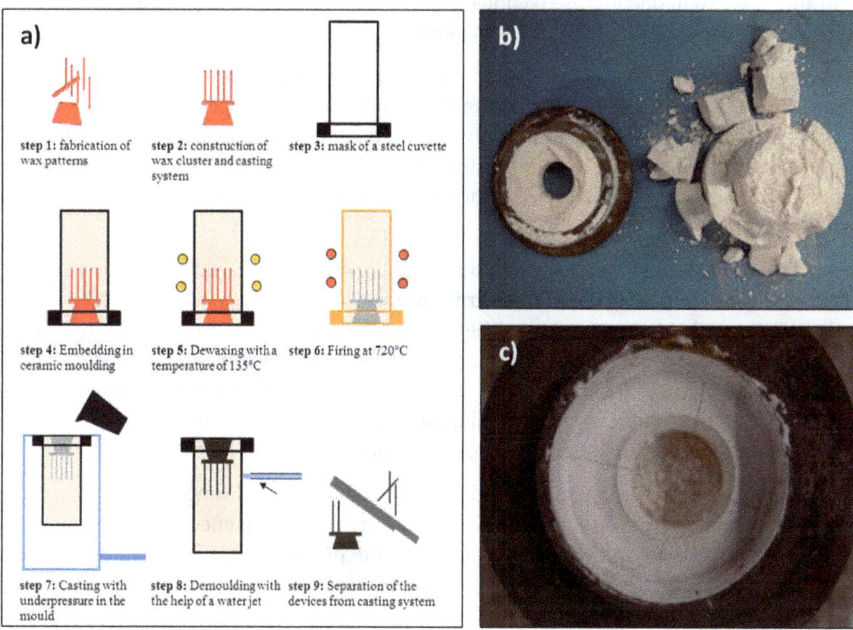

Figure 1. a) Basic steps of the block mould casting process. b) Fired block mould without stabilising cuvette. c) Block mould with cracks after the firing process.

The perforated steel cuvette is masked by plastic foil; the base is closed by a rubber plug. The inner area of the plug is filled with liquid wax via the steel cuvette and the soldered wax cluster is fixed in it. After the solidification of the fixing wax, the steel cuvette is filled with liquid ceramic slurry. The mixing of the moulding material and the filling of the steel cuvette should be carried out inside a vacuum chamber to lower the gas content in the slurry. The gas would otherwise precipitate during the moulding material's setting and lower the strength of the block casting mould. In addition to this, the surface quality of the casting will be reduced [12, 13]. The gas can also be removed from the gypsum by applying a vacuum to the moulds for some minutes after they have been filled with the slurry. In this case, the vacuum treatment should be completed prior to the start of the moulding material's setting process. After a drying period, which depends on the moulding materialused, the block casting moulds are usually dewaxed in an oven at 110 °C to 150 °C. When the moulds are completely dewaxed, the firing process is started in which heating and holding steps are adapted to the moulding material used to prevent cracking due to a too rapid heating rate. After completion of the firing process, the casting temperature is adjusted and held for at least three to four hours. Depending on the fineness of the component to be cast, a vacuum can be generated in the mould to assist the mould filling capacity of the melt (step 7, figure

1). In this way, component cross-sections smaller than 500 μm can be filled such as those which exist in, for example, open-pored metallic foams [11].

Another possibility for supporting the mould filling is to centrifugally load the block mould during the casting process. By maintaining the mould's rotational speed during the solidification phase, this also expedites the supply and the deposition of gases and impurities at the interface between the mould and the casting [1, 2]. After the mould has cooled down, the casting can be recovered by water jet or mould solving agents and removal from the casting system by a saw.

The quality of the mould, and therefore that of the casting, is decisively determined by the moulding material's mixing and by the firing process (steps 4 and 6, figure 1a). The mould tends to form cracks (figure 1c) during the firing process which can produce casting defects [3, 4, 14]. As a worst case, the liquid metal does not remain within the cavity but runs out of the mould through the cracks. In order to minimise the tendency for the block casting mould to crack, the moulding material's manufacture is generally optimised with respect to the water's temperature and quantity, mixing duration and the addition of supplements.

Since ceramic moulding compounds result in low cooling rates during solidification, it is, moreover, expedient to implement measures to elevate the cooling effect of the block mould in order to improve the mechanical properties of the metallic components since these are closely connected with the casting's cooling conditions.

2.3. Moulding and Casting Materials

A basic dilemma arises when a moulding material used in the lost mould process is chosen and prepared. The moulding material should exhibit a certain strain to withstand the load during the mould's handling, thermal stresses and the load due to the casting process. After the mould has cooled down it should also exhibit low strength in order to easily remove the casting from the mould [3, 14].

Besides these basic aspects, other requirements are imposed on the moulding materials for the block mould casting process [15]:

- flowability for good mould fillingprior to setting
- high reproductive accuracy
- fixation of the pattern
- low affinity to volume changes during the setting and firing processes
- no chemical reaction with the used casting material
- thermal stability
- short processing time
- cost-effective
- ecologically compatible

Block mould castings process's moulding materials consist of a refractory material, usually quartz, cristobalite, or a mixture of both, and a binder whereby gypsum, phosphate, metallic oxides or silicates are used. The choice of the moulding material for producing a block casting mould mainly results from the casting temperature of the material which has to be cast [16].

For aluminium and silver alloys as well as nickel-chrome alloys, a moulding material with 25 to 30 wt.% gypsum and 70 to 75 wt.% silicon oxide is used because gypsum-boned investments (GBI) exhibit good ability to collapse after casting [17].

In the past, the gypsum-bonded moulding material consists of a mixture of silica and plaster of Paris. Over the last few decades, this material has been modified by the addition of boric acid, pigments and reducing agents to, among other things, increase the strength [14]. Gypsum is produced from the sedimentary gypsum rock, whose prismatic crystals are bonded by water molecules. By treating in an autoclave, the gypsum is transformed into the unstable hemihydrate state, which needs water to restore the natural dehydrate state. When the GBI is mixed with water, branched gypsum crystals are formed which bond the refractory of the moulding material [4].

Although different statements are made in literatureabout the thermal stability of the binder, experts and producers of GBI agree that the gypsum's decomposition does not start in the temperature interval of 650 °C to 700 °C, which are common maximum firing temperatures for GBI [18, 19]. The GBI moulds should not be heated above 700 °C to 750 °C because the residual carbon from the pattern's wax then reacts with the gypsum from the moulding material. This reaction results in sulphur dioxide which decreases the surface quality of castings and,in the case of gold components, the mechanical properties [19].

High melting point alloys are cast in phosphate-bonded investments (PBI) consisting of 75 wt.% to 90 wt.% silica (quartz or cristobalite) and magnesium- or ammonium-magnesium-phosphate. This phosphate is formed by the reaction of magnesium with monoammonium in water during the mixing process. The firing process causes water loss, crystallisation and recrystallisation of magnesium phosphates, and forms fused glass, which lends high strength to PBI block casting moulds [12, 16].

The high strength of block casting moulds in the green and fired states is the biggest advantage of PBIs. At high temperatures, the strength and surface quality of the moulds are decreased due to a thermal decomposition of the binder, especially at casting temperatures above 1375 °C [14, 16].

Phosphate-bonded moulding materials are used for a broad spectrum of materials, e. g. gold, titanium, nickel, chrome and cobalt-chrome-alloys [20, 21].

The strength of gypsum or phosphate-bonded block casting moulds depends on a multiplicity of influencing factors, such as the binder, additives, storing and setting environment as well as the firing temperature. It has to be taken into account, that porosity in the ligated moulding material considerably influences the strength of the polycrystalline brittle block casting mould [14, 21]

2.4. Aspects

The block mould casting process exhibits several advantages which, when combined, cannot be found in other casting processes. Using the lost patterns casting process, ambitious geometries possessing undercuts can be realised with high dimensional accuracy, reproductive ability, and surface quality resulting in little finishing effort. The finished casting surface quality depends on the surface energy of the melt and on the interface energy between the melt and moulding materials. In this context, the composition and the purity of the melt as well as the reaction of the melt with the furnace chamber's atmosphere and the moulding material is important [17].

In conjunction with rapid-prototyped patterns the block mould casting process enables real-time production of metallic prototypes or small batches of castings. Using the ideal moulding material, nearly all casting materials can be processed using this casting method, whereby the process time is much shorter compared to the more time-consuming investment casting process [6, 8, 22, 23].

On utilising gypsum or phosphate-bonded investments, process specific disadvantages result. The GBIs or PBIs have not until now been in continuous use. This raises the price of the moulding material and limits the economic batch size at around 1,500 castings, depending on the size and geometry of the components [23]. Prior to their mixing with water, the moulding material exists as a powder, which promotes exposureto health problems in the workplace. The moulds of the block mould casting process exhibit low heat conductivity and high heat capacity. For this reason, the current cooling rate is low whichresults in unfavourable casting characteristics and low mechanical properties. Subsequent to their firing, block moulds exhibit a lower strength compared with other moulding materials. This limits the casting weight to approximately 10 kg. In addition to this, their low strength can lead to cracks in the moulds, decreasing the quality of the castings [23]. Regarding the economy of the block mould casting process, it can be concluded that this casting process is the more economical the more complex the geometry is and the smaller the dimensions of the potential cast components are.

3. Influence of sodium chloride and sodium fluoride on gypsum-bonded investment's green and fired strengths

The tendency of block moulds to form cracks has to be reduced to increase the casting component's quality (see chapter 2.2). This can be achieved by optimising the moulding material's processing and the firing process by using additives. A typical additive is sodium chloride to lower the thermal expansion of the moulding material [24]. Information is rarely given about the general effect of sodium chloride on the GBI's compressive strength, and especially about the interaction of the sodium chloride and sodium fluoride. Owing to this, the effect of sodium chloride and sodium fluoride on the strength of a GBI after firing was investigatedin the following sections.

3.1. Materials and methods

For the examinations, GBI specimen were produced containing 1 wt.%, 3 wt.% and 5 wt.% commercial table salt (based on the weight of the mixing water) with and without sodium fluoride. Higher salt contents were not used because they lead to foaming of the GBI, which decrease the handling of this moulding material. As a reference, GBI specimens without salt and only with sodium chloride were produced, respectively. The dimensions of the specimens conformed to the specifications in DIN EN 993-5. To produce the specimens, a silicon matrix was used. The particular amount of salt was dissolved in the mixing water for one minute prior to beginning the mixing process. After adding the moulding material, Goldstar xXx from Goldstarpowders (see table 1and figure 2 for the chemical composition), to the water, the composition was mixed for three minutes and then poured into the silicon matrix.

constituent	SiO2	Al2O3	Fe2O3	TiO2	CaO	K2O	P2O5	SO3	SrO	glowing loss
value [%]	75.27	0.74	0.04	0.03	10.22	0.02	0.01	13.62	0.05	3.94

Table 1. X-ray fluorescence spectroscopy results of the used GBI Goldstar xXx from Goldstarpowders.

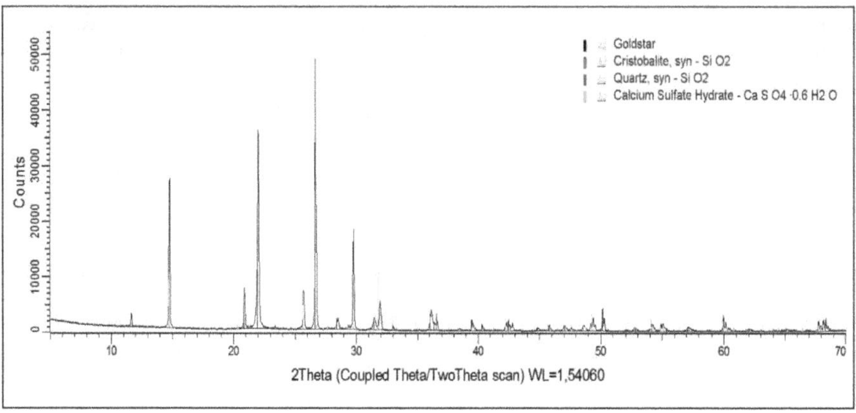

Figure 2. X-ray diffraction analysis of the used GBI Goldstar xXx from Goldstarpowders.

temperature	[°C]	135	135	720	720
acceleration time	[min]	60	---	390	---
holding time	[min]	---	180	---	240

Table 2. Furnace program of the firing program.

45 Minutes after the pouring, the specimens were removed and were fired after an additional 60 minutes. The furnace program was identical to that used for the dewaxing and firing of block moulds made from the same GBI (table 2). The fired specimens were compressed using an Instron 8033 testing machine at a cross head speed of 2 mm/ min. At least four specimens were tested for each test run.

3.2. Experimental results

The influence of sodium chloride and sodium fluoride is shown in figure 3. GBI specimens with 1 wt.% sodium chloride and without fluoride have an increased compressive strength by up to 10 % in comparison with the GBI specimen without additives. With a 1 wt.% mixture of sodium chloride and sodium fluoride, the strength can be increased by about 55 %.

With the addition of these two salts, the gypsum binder reacts with sodium chloride and sodium fluoride according to the following equations [25]:

$$CaSO_4 + 2\,NaCl = CaCl_2 + Na_2SO_4$$

$$CaSO_4 + 2\,NaF = CaF2 + Na_2SO_4$$

Because the strength in the green state of the GBI cannot be increased by a salt addition (see chapter 4), it is assumed that the effect can be attributed to the influence of the salt on the formation of sinter phases during the firing process. The decrease in strength can be explained by the increased setting time, which can be observed when GBI has a high salt content [26]. With an increased setting time, the gypsum dendrites start to coarsen due to the Ostwald maturation. This causes lesser cross-linking of the gypsum crystals. The effect of sodium fluoride cannot be explained by means of data in the literature.

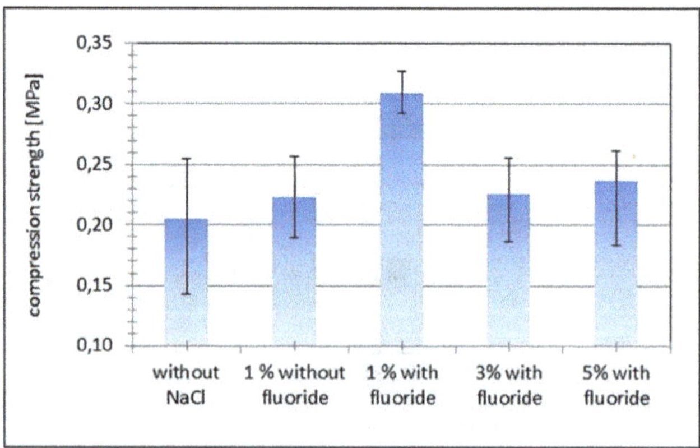

Figure 3. Influence of sodium fluoride and sodium chloride content on the gypsum-bonded investment's compression strength.

4. Influence of water temperature and content, mixing duration and quantity of salt with fluoride on the gypsum's compression strength in its green and its fired states

The quality of castings processed with the help of the block mould casting process depends to large extent on the quality of the block mould it self (see chapter 2.2 and 2.3). It is only possible to produce defect-free casting structures when the block mould exhibits strength adequate to withstand the crack initiation during the mould's production. To improve demoulding of the casting with the help of water, the moulding material should show a low green strength after casting. For this purpose, the experiments in this section were conducted to optimise the strength of the used GBI.

4.1. Materials and methods

4.1.1. Experimental design

The experiments focused on the effect of the water temperature, water content, mixing time and salt addition (sodium chloride + sodium fluoride) on the green strength and strength of a GBI. The experiments were conducted with the help of Taguchi's method. This method covers the design of experiments according to statistical aspects, the assembling of models as well as the optimisation of the process. With the help of Taguchi arrays, only a few measurements are necessary compared to a complete factorial design. To determine the effect of the focused parameters, a L9 orthogonal array was chosen. This array allows the monitoring of 4 parameters with 3 settings, to detect non-linear relationships. Each parameter exhibits two degrees of freedom with three possible settings. These lead to a total of eight [= 4 x (3-1)] degrees of freedom. This number agrees with the demand of Taguchi that the total degree of freedom of the chosen array should be larger or equal to the number of total degrees of freedom for all experiments. In table 3, the L9 array is shown with the particular test settings.

test run	water temperature [°C] (A)	water content [%] (B)	mixing time [s] (C)	salt content [%] (D)
1	1	1	1	1
2	1	2	2	2
3	1	3	3	3
4	2	1	2	3
5	2	2	3	1
6	2	3	1	2
7	3	1	3	2
8	3	2	1	3
9	3	3	2	1

Table 3. Orthogonal L9 Taguchi array.

The settings of the water, water content and mixing time parameters were specified with the help of the literature and the specifications of the GBI manufacturer. With the help of the results from chapter 3, the salt content (sodium chloride + sodium fluoride) was determined. Table 4 summarises the settings of the parameters. If the experiments were to be conducted with the help of a complete factorial array $3^4 = 81$ test runs would have to be performed.

setting	water temperature [°C]	water content [wt.%]	mixing duration [s]	salt content [wt.%]
minimum (1)	17	45	120	0
mean (2)	35	50	180	0.5
maximum (3)	55	55	240	1.0

Table 4. Parameter settings of the experiments.

The test runs are interpreted with the aid of the analysis of means (ANOM) and the analysis of variance (ANOVA). The ANOM shows the optimisation direction of the factors. The mean deviance from the total average caused by every factor level indicates the main effect of every single factor level [27, 28]:

$$\bar{\eta}_{A1} = \frac{1}{n}[(\eta_{11} + \eta_{12} + \eta_{13}) + (\eta_{21} + \eta_{22} + \eta_{23}) + (\eta_{31} + \eta_{32} + \eta_{33}) \quad (1)$$

$$\bar{\eta}_{A2} = \frac{1}{n}[(\eta_{41} + \eta_{42} + \eta_{43}) + (\eta_{51} + \eta_{52} + \eta_{53}) + (\eta_{61} + \eta_{62} + \eta_{63}) \quad (2)$$

$$\bar{\eta}_{A1} = \frac{1}{n}[(\eta_{71} + \eta_{72} + \eta_{73}) + (\eta_{81} + \eta_{82} + \eta_{83}) + (\eta_{91} + \eta_{92} + \eta_{93}) \quad (3)$$

$\bar{\eta}_{ij}$ are the measured values and n is the total number of all trials (for $L_9 = 9$).

The effects E of the level changing are for the factor A:

$E_{A1,2} = \bar{\eta}_{A2} - \bar{\eta}_{A1}$ and $E_{A1,3} = \bar{\eta}_{A3} - \bar{\eta}_{A1}$

With the aid of the ANOVA, the statistical significance of an effect of a parameter on the command variable can be evaluated. For this, the total result is partitioned into single variances. The variance expresses the squared deviance of the particular average.

The calculation of the ANOVA is performed with the help of the following equations [27, 28]:

I. Calculation of a correction factor (CF) for an easier calculation of the error:

$$SS_m = \frac{(\sum \eta_{ij})^2}{N} = CF \quad (4)$$

N is the number of all trials including all repeatings (L9 array with three repeatings per trial → N = 9 x 3 = 27)

II. Calculation of the total sum of squares

$$SS_{total} = \sum \eta_{ij}^2 - CF \tag{5}$$

III. After this the sum of the squared deviances of every factors is built, here exemplary for factor A:

$$SS_A = \frac{(\sum \bar{\eta}_{A1})^2}{N_{A1}} + \frac{(\sum \bar{\eta}_{A2})^2}{N_{A2}} + \frac{(\sum \bar{\eta}_{A2})^2}{N_{A3}} - CF \tag{6}$$

N_{A1}, $N_{A2,}N_{A3}$ are the number of trials with the parameter on level 1, 2 or 3.

IV. Error sum of squares

$$SS_{error} = SS_{total} - (SS_A + SS_B + SS_C + SS_D) \tag{7}$$

V. Degrees of freedom (f_{total} and $f_{parameter}$)

$$f_{total} = \text{total number of trials} - 1 \tag{8}$$

$$f_A = \text{number of levels of parameter A} - 1 = 3 - 1 = 2 \tag{9}$$

VI. For the evaluation of the variances for each factor (for example A)

$$V_A = \frac{SS_A}{f_A} \tag{10}$$

$$V_{error} = \frac{SS_{error}}{f_{error}} \tag{11}$$

VI. Calculation of the ratio of the variance and the error variance (for example A)

$$F_A = \frac{V_A}{V_{error}} \tag{12}$$

VIII. Verifying the significance of the factors with the help of the F-Test. For this the calculated F value is compared with a tabulated F value. If the calculated one is bigger, the observed result is statistical significant.

Within the Taguchi method the signal to noise ratio (S/N) represents the summed statistic. For this, every measured value of every test run is reweighted in a target function such that no repetition within one test run occurs and that the total degree of freedom is decreased. There are different target functions which can be chosen for the S/N-ratio depending on the quality attribute. If the aim is to minimise the number of trials (smaller-the-better-type), the S/N-ratio is calculated in the following way [27]:

$$\eta = -10x \log\left(\frac{1}{n}\sum_{i=1}^{n} y_i^2\right) \tag{13}$$

The maximisation of the S/N-ratio (lager-the-better-type) is

$$\eta = -10x \log\left(\frac{1}{n}\sum_{i=1}^{n} \frac{1}{y_i^2}\right) \tag{14}$$

Using the S/N-values, the ANOM and the ANOVA can be calculated in the same way; similar to the mean values.

Utilising the Taguchi method in this way aims at setting a process such that the target value achieves the desired maximum or minimum with low scatter. To verify the data resulting from Taguchi experiments, one should perform verifying experiments.

The value of the command variable $\eta_{optimal}$, under optimal conditions, is calculated from the optimal parameter settings [28]:

$$\eta_{optimal} = \bar{\eta} + (\bar{\eta}_{A_i} - \bar{\eta}) + (\bar{\eta}_{B_i} - \bar{\eta}) + (\bar{\eta}_{C_i} - \bar{\eta}) + (\bar{\eta}_{D_i} - \bar{\eta}) \tag{15}$$

with $\bar{\eta}$ = overall mean of the target value

The confidence interval for the approving experiment ($CI_{confirmation}$) and of the population ($CI_{population}$) are calculated by

$$CI_{confirmation} = \sqrt{F_a(1, f_e) x V_{error}\left[\frac{1}{n_{eff}}\right] + \left[\frac{1}{R}\right]} \tag{16}$$

$$CI_{population} = \sqrt{\frac{F_a(1, f_e) x V_{error}}{n_{eff}}} \tag{17}$$

$$n_{eff} = \frac{N}{1 + totalDOFassociatedintheestimateofmean} \tag{18}$$

R = sample size for confirmation experiments

$F_a(1, f_e)$= tabulated F-value

4.1.2. Production and testing of specimens

For the production of the compression specimen the corresponding amount of GBI (Goldstar xXx from Goldstarpowders, table 1 and figure 2) was mixed with max. 825 g water with the help of a drill-stirrer. If salt was used for a test run, the salt was dissolved in the mixing water for 1 minute. After mixing, the slurry was poured in a silicon matrix to produce compression specimens according to DIN EN 993-5, and after 45 minutes the specimens were removed and fired (table 2). The compression tests were conducted using an Instron 8033 at a cross head speed of 2mm/min. For every test run, three and five specimens in the green and fired states were tested, respectively.

4.2. Experimental results

Table 5 summarises the results of the compression tests and the S/N-ratio for every test run.

test run	green strength response 1	green strength response 2	green strength response 3	green strength S/N [dB] (smaller-the-better)	strength response 1	strength response 2	strength response 3	strength response 4	strength response 5	strength S/N [dB] (larger-the-better)
1	2.42	2.40	3.00	-8.37	0.20	0.24	0.18	0.29	0.21	-13.33
2	1.89	1.70	1.70	-4.93	0.24	0.22	0.25	0.34	0.24	-12.04
3	1.17	1.11	1.10	-1.03	0.25	0.23	0.24	0.31	0.24	-12.04
4	2.28	1.66	2.13	-6.19	0.38	0.51	0.37	0.40	0.51	-7.50
5	2.44	2.76	2.25	-7.93	0.20	0.24	0.20	0.22	0.21	-13.45
6	1.42	1.58	1.65	-3.82	0.18	0.16	0.16	0.16	0.18	-15.53
7	2.11	2.18	2.16	-6.64	0.20	0.23	0.24	0.22	0.21	-13.20
8	1.46	1.38	1.64	-3.50	0.25	0.23	0.27	0.26	0.24	-12.08
9	2.03	2.12	2.06	-6.32	0.18	0.18	0.20	0.20	0.13	-15.33

Table 5. Results of the compression tests and calculated S/N ratios.

The aim of the presented experiments was to maximise and minimise the GBI's strength in the fired and the green states, respectively. That is, the larger and the smaller the S/N-ratios were selected which are the better in the fired state the better in the green state, respectively.

The results of the ANOM are presented in figures 4 to 7; tables 6 and 7 present the results of the ANOVA. Figures 4a) and 5a) and table 6 show that the water temperature and the mixing time have little effect on the green compression strength of the used GBI. This contrasts with the water and salt contents (figures 4b and 5b). From the literature, it is known that the strength of GBI decreases with increasing water content.

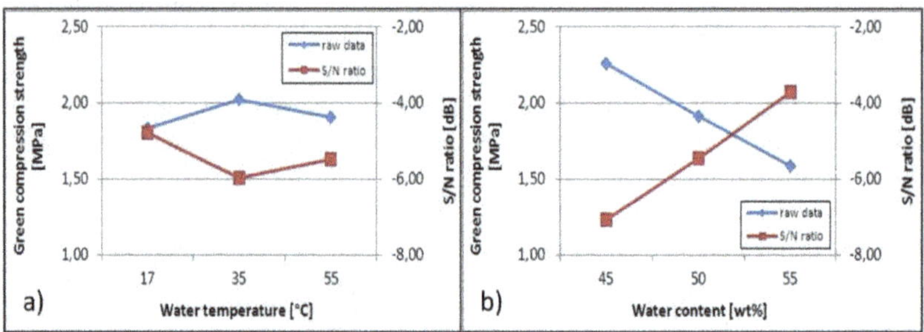

Figure 4. Influence of a) water temperature and b) water content on green compression strength of gypsum-bonded investments.

Owing to its heterogeneous nucleation, gypsum spontaneously crystallises without being in its equilibrium condition even in the first stages of the hydration process. Excess water, which is important for the flowability of the moulding material, is not bonded after the re-hydration of the gypsum and evaporates during setting leaving pores in the material. This microstructural defect considerably decreases the strength of the GBI. In addition to the pore formation, an increased water content leads to an increase in the gypsum's setting time, which produces a coarsening of the gypsum crystals due to Ostwald maturation. These crystals built a less dense network which cannot carry the magnitude of load that a better cross-linked gypsum crystal network is capable of [21, 26, 29].

On increasing the salt content from 0 wt.% to 1 wt.% severely decreases the green strength of GBI (figure 5b). An explanation of this result cannot be interpreted with the help of the literature.

In the fired state, all the varied parameters show an effect on the compression strength. The maximum strength is achieved with a mean water temperature, low water content, mean mixing time and a high salt content.

The effect of the water temperature on the GBI's compressive strength is similar in both the green and the fired states, but the impact is more pronounced in the fired state. The best water temperature for optimising the compressive strength lies in the range of 17 °C to 55 °C. A reason for this could be the change in the gypsum's solubility in water. When water reaches the temperature of approx. 27 to 35 °C, the solubility of gypsum in water is decreased resulting in a lower amount of hydrated gypsum. This gypsum is hydrated between the start of mixing and the setting. Due to this, less gypsum crystals are formed and thus the strength of the moulding material is decreased [30, 31].

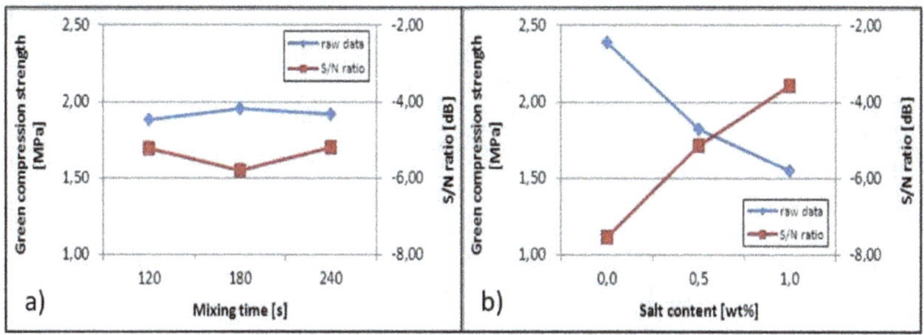

Figure 5. Influence of a) mixing time and b) salt content on green compression strength of gypsum-bonded investments.

source	SS	DOF	V	F-ratio	SS'	P (%)
water temperature [°C] (A)	0,16	2,00	0,08	pooled	---	---
water content [%] (B)	2,07	2,00	1,03	27,72	1,99	32,07
mixing time [s] (C)	0,02	2,00	0,01	pooled	---	---
salt content [%] (D)	3,29	2,00	1,65	44,17	3,22	51,82
error (pooled)	0,67	18,00	0,04	---	1,00	16,11
total (T)	6,21	26,00	---	---	6,21	100,00

Table 6. ANOVA of the green compression strength (raw data). The tabulated critical f-value is 3.55.

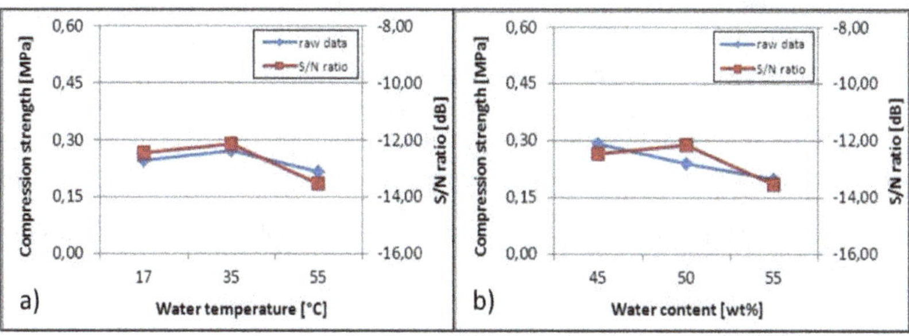

Figure 6. Influence of a) water temperature and b) water content on compression strength of fired gypsum-bonded investments.

In contrast to the green state, the effect of different water contents on the compression strength of GBI is minor. The effect of pores, which are built because of excess water, is lower because probably some sinter phases are formed during the firing process.

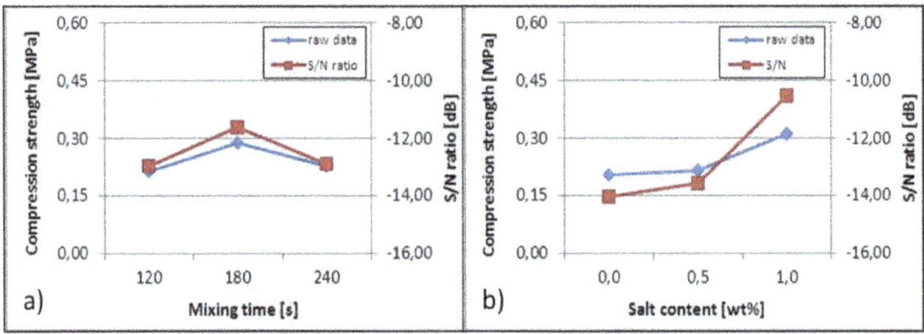

Figure 7. Influence of a) mixing time and b) salt content on compression strength of fired gypsum-bonded investments.

source	SS	DOF	V	F-ratio	P (%)
water temperature [°C] (A)	0,02	2,00	0,01	9,03	7,24
water content [%] (B)	0,06	2,00	0,03	24,83	21,49
mixing time [s] (C)	0,05	2,00	0,02	18,59	15,86
salt content [%] (D)	0,11	2,00	0,05	40,46	35,58
error (pooled)	0,05	36,00	1,30E-03	---	19,84
total (T)	0,29	44,00	---	---	100,00

Table 7. ANOVA of the compression strength (raw data). The tabulated critical f-value is 3.32.

The impact of the mixing duration on the compressive strength shows the same tendency in the green and in the fired state. The strength rises from the beginning of the mixing process up to 180 seconds. After this peak, the strength decreases with continuous mixing. 180 seconds mixing advances the hydration of the gypsum and leads to a higher strength after setting and burning, respectively, because more gypsum crystals are precipitated. However, once precipitated, the crystals are destroyed by the mixing. After mixing for the ideal duration, destruction of the crystals dominates and the resulting strength again decreases [26].

The influence of the salt content can be attributed to the formation of sinter phases during the firing process.

4.3. Confirmation tests

The specimens for the verification test runs were produced with the parameter settings summarised in table 8. In addition to this, the calculated and measured values for the reference specimens are presented.

The measured values are located within the calculated scattering bands and deviate only marginally from the calculated average values. Firstly, it can be concluded that the obtained results have a major significance. Secondly, that there is no interaction between the parameters; otherwise the calculated values would differ more strongly from the measured results since these calculations are only based on the single effect of the parameters. It is interesting that there is no interaction between the water content and the mixing duration. Although the setting process is slowedusing increased water content, a mixing duration longer than 180 seconds in combination with high water contents has no unfavourable effect on the compressive strength of GBI after either setting or firing. Since the unfavourable effect of the mixing duration can be attributed to the destruction of gypsum crystals, it follows that, independent of the water content, a nearly equal amount of gypsum crystals are present after the mixing process.

water temperature [°C]	water content [wt.%]	mixing duration [s]	salt content [wt.%]	calculated strength [MPa]	average strength [MPa]
17	45	180	1	0,41 ± 0,07	0,40 ± 0,05

Table 8. Setting of the parameters of the confirmation tests, the calculated strength for a confidence interval of 95 % and the measured strength.

5. Influence of glass fibre volume and fibre length on the strength of fired gypsum-bonded investments

When the compositions of several GBIs are examined it can be shown that some products contain glass fibres to increase the strength of the block mould. The reinforcement of GBI and the effect of glass fibres on the properties of GBI were the focus of several examinations [32-36]. The results were contradictoryregarding the effect of the glass fibres on the compressive strength of GBI. Due to this, the aim of the following analyses was to demonstrate the influence of the glass fibre volume and glass fibre length on the compressive strength of a GBI.

5.1. Materials and methods

Uncoated short glass fibres were used. Coated glass fibres could lead to both a chemical reaction and a bond between fibres and moulding material which exerts an adverse effect on the reinforcement [32]. The maximum fibre content was held constant at 1.0 wt.% based on the weight of the moulding material. Larger amount of glass fibres could not be added because the viscosity of the slurry was otherwise too low. The mean glass fibre content was 0.5 wt.%. The glass fibre lengths employed were 3 mm, 6 mm to 12 mm. For the production of the compression specimens, 1500 g of the GBI Goldstar xXx from Goldstarpowders (composition see table 1 and figure 2) was mixed with the glass fibres and then with 675 g water for three minutes. The slurry was poured into a silicon matrix to produce compression specimens according to DIN EN 993-5. The specimens were removed and fired (table 2) after 45 minutes and 60 minutes, respectively. With the help of an Instron 8033 testing machine (cross-head speed = 2mm/min), the compressive behaviour of at least four specimens were determined.

5.2. Experimental results

The results of the compression tests are shown in figure 8. In general, the reinforcement of GBI specimens using glass fibres decreases the scatter of the compression test results. The weakening effect of the pores is attenuated, which are the main reason for the deviations of the measured compressive strength, independent of the volume and length of the glass fibres.

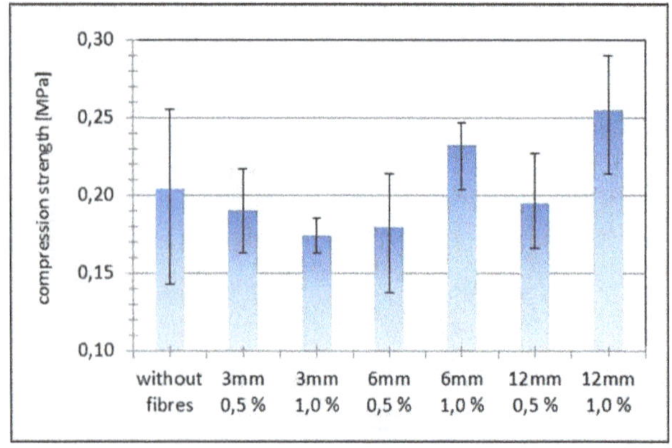

Figure 8. Influence of short fibre content and short fibre length on compression strength of fired gypsum-bonded investment.

The moulding material's compressive strength is only slightly attenuated by glass fibres with a length of 3 mm and a fibre content of 0.5 wt.%. An increase in the compressive strength of up to 25 % can be achieved by reinforcing the GBI using 1.0 wt.% of 6 mm and 12 mm glass fibres. This content and these lengths improve the resistance of the moulding material to the initiation of micro cracks and their growth. Following micro crack initiation due to loading the fibre-reinforced moulding material, the crack reaches the glass fibre and growths along the interface of the matrix and fibre; thus dissipating crack energy [32, 36].

6. Influence of metal powder in the moulding material on the GBI's setting behaviour and its compressive strength as well as on the cooling behaviour, the metallographic and mechanical properties of an A356 (AlSi7Mg0.3) alloy

Despite several advantages, the block mould casting process exhibits some unfavourable characteristics (see chapter 2.4). One of the most severe problems is the low cooling rate of this casting process during the solidification of liquid metals. For example, this cooling rate amounts to approximately 0.1 K/s during the solidification of an A356 alloy at a casting temperature of 720 °C. In contrast to this, the high pressure die casting process with a similar alloy and wall thicknesses reaches cooling rates between 50 and 85 K/s [37]. Owing to the low heat conductivity of the moulding materials used, the cast metal solidifies relatively slowly producing a coarse microstructure and a high dendrite arm spacing (DAS). These microstructural parameters lead to decreased mechanical properties of the metallic components [23, 38]. Besides the mechanical properties, the low cooling rate impairs the casting properties. Inferior casting properties lead to technical volume deficits in the casting; such as shrinkage or microporosity. Low cooling rates induce a segregation of elements in the melt adjacent to the solidification front thus promoting constitutional undercooling. This changes the solidification morphology with a tendency to produce exogenous mushy and endogenous papescent structures. These types of solidification morphologies decrease the feeding ability of the metallic melt, from which microporosity results [39]. Methods for improving the cooling rate of the block mould casting process leads directly to an improvement in the casting and mechanical properties.

The presented trials aimed at increasing the cooling rate of the block mould casting process with the help of iron powder. Besides the iron powder affecting the cooling of the metal castings in the block moulds, the impact of the iron powder on the setting and strength of the moulding material is evaluated.

6.1. Materials and methods

To evaluate the influence of iron powder, compression specimens according to DIN EN 993-5 were produced. A specific amount of iron powder was mixed with 1500 g Goldstar xXx from Goldstarpowders (composition see table 1 and figure 2) and then introduced in 50 wt.%, 55 wt.% and 60 wt.% water. After 3 minutes mixing, the liquid moulding material was poured into a silicon matrix. After 60 minutes, the GBI specimens were removed and, following another 60 minutes, the specimens were fired (table 2). At least three specimens per test run were compression tested at a cross head speed of 2 mm/min with the aid of an Instron 8033 testing machine.

To investigate the effect of applying a vacuum, some silicon matrixes, which were filled with liquid iron powder GBI mixture, were evacuated.

The effect of the metal powder in the moulding material on the cooling behaviour of an A356 alloy was measured with the aid of thermal analyses, tensile tests and metallographic sections. To enable this, a wax assembly possessing four tensile specimen patterns was produced, whereby a type K thermocouple was integratedinto the middle of one tensile specimen per assembly. With the help of preliminary tests, a maximum (75 wt.%) and a mean (37.5 wt.%) iron powder content was specified. After embedding the patterns into the GBI (Goldstar xXx from Goldstarpowders, composition see table 1 and figure 2) mixed with iron powder, the block moulds were dewaxed, fired and cooled down to 200 °C. A pre-grain refined, not modified N-degassed A356 alloy was cast into the block moulds using a casting temperature of 720 °C. After the melts had cooled down, the tensile specimens were tested. Specimens with integrated thermocouples were used to prepare metallographic sections. For this purpose, the corresponding samples were mounted using the embedding compound Araldit DBF together with the hardener Ren HY 956. The mounted samples were ground using different grades of abrasive paper and polished using VibroMet. Light microscopy images were taken at different magnifications using an Axio Imager A 1 m from the company Zeiss.

6.2. Experimental results

At the beginning of the investigations, it was attempted to determine the iron powder content at which the GBI would no longer set. It was possible to show that the setting process was never interrupted, irrespective of the iron powder content. Figures 9 and 10a) show the effect of metal iron powder on the setting time of the used GBI. In addition to the increase owing to a higher water content, the setting time increases with elevated iron powder content, yet the setting process is not interrupted. A reason for this cannot be found in literature.

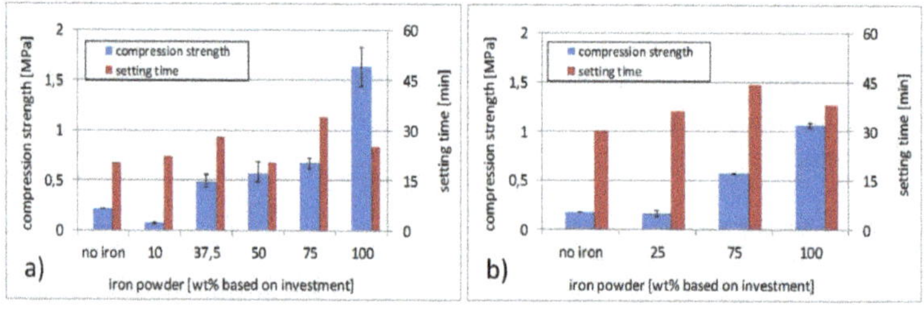

Figure 9. Influence of iron powder on the compressive strength and setting of gypsum-bonded investments with a) 55 wt.% water and b) 60 wt.% water.

Besides the setting time, the influence of the iron powder on the compressive strength of the used GBI is shown in figures 9 and 10b). From the critical value (10 wt.% to 25 wt.%), the compressive strength of the GBI is increased independent of the specimen's water content. The maximum difference in the compressive strength of specimens with 100 wt.% and without iron powder content is more than 750%. With the aid of references in the literature, it is assumed that the iron decreases the melting point of the GBI's refractory. Due to this, fewer sinter phases are formed which considerably increase the strength of the moulding material [40].

The analysis of the vacuumed GBI specimens possessing different metal powder contents indicates that the iron powder is well distributed within the fired GBI specimens (figure 10b). There is no segregation of the metal powder due to the underpressure.

Figure 11a) shows the effect of the iron powder in the moulding material on the cooling behaviour of an A356 alloy. Using the first derivative of the cooling curves, the liquidus temperature and the temperature at the end of solidification were detected. In combination with the respective times, the cooling rates achieved by the block moulds were calculated (figure 11b). The cooling rate of the block mould casting process can be increased by factor of 3.5

with the help of 37.5 wt.% iron powder in the moulding material. The addition of more iron powder exhibits no further increase.

The increased cooling rates are reflected in the dendrite arm spacing (DAS) of the A356 alloy. Owing to a 3.5 fold increase in solidification rate due to the iron powder addition, the DAS is decreased by about 33 % in comparison to specimens produced using moulds containing no iron powder (figure 11a). Since there is no changing in the amount of the cooling rate, no change in the DAS was found due to increasing the iron powder content from 37.5 wt.% to 75 wt.%.

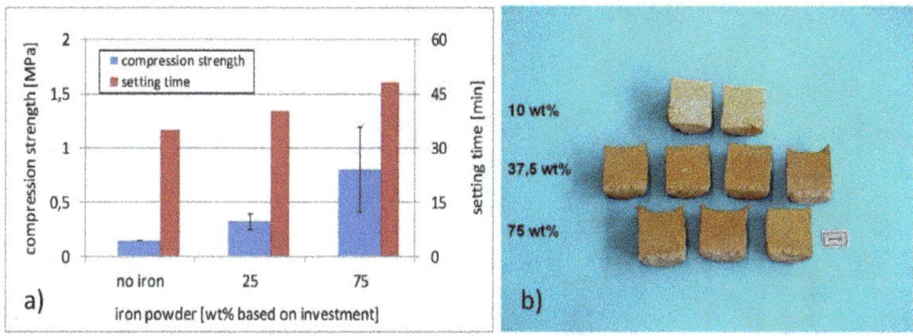

Figure 10. a) Influence of iron powder on the compressive strength and setting of gypsum-bonded investments with 65 wt.% water. b) Influence of vacuum on the distribution of iron powder in the ceramic specimens.

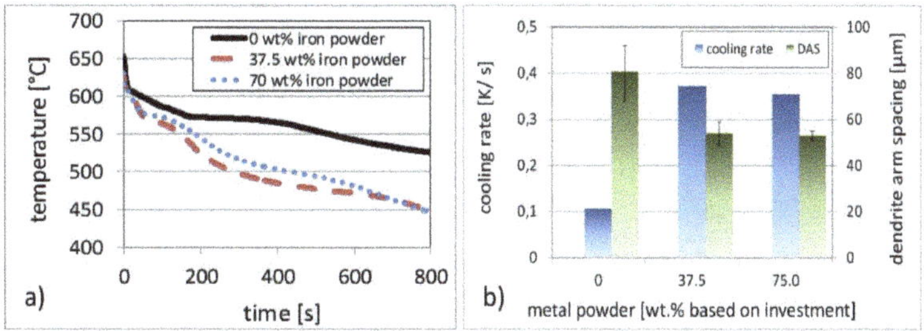

Figure 11. Influence of iron powder on the cooling rate of block moulds casted with an A356 with a casting temperature of 720 °C and a mould temperature of 200 °C. a) Cooling curves. b) Calculated cooling rates and the DAS of the specimens.

The effects of the increased cooling rate and the decreased DAS cannot be related to the mechanical properties of the A356 alloy (figure 12). According to the literature, the mechanical properties of those specimens produced with the aid of the modified moulds which promote a higher cooling rate should be better than the properties of those specimens cast using the unmodified moulds. In contrast to this, the tensile strengths and particularly the elongations to fracture of the specimens produced with the aid of the block moulds containing 37.5 wt.% iron powder are, in fact, worse compared to those of specimens cast in non-modified moulds. The tensile strengths and elongations to fracture of specimens from modified moulds can be increased if the iron powder content is raised to 75 wt.%.

The metallographic sections of the A356 alloy specimens show that these results can be attributed to iron phases in the aluminium-silicon matrix. When the mean content of the iron powder is employed in the mould, the alloy's microstructure exhibits the most iron containing phases. It appears that during the mould filling and solidification of the A356 alloy, iron was dissolved out of the mould by the aluminium melt. The solubility of iron in solidified aluminium is low, which promotes the precipitation of the iron phases during the solidification. These plate-like phases decrease the tensile strength and the elongation to fracture and also increase the yield strength (figure 12 and figure 13c).

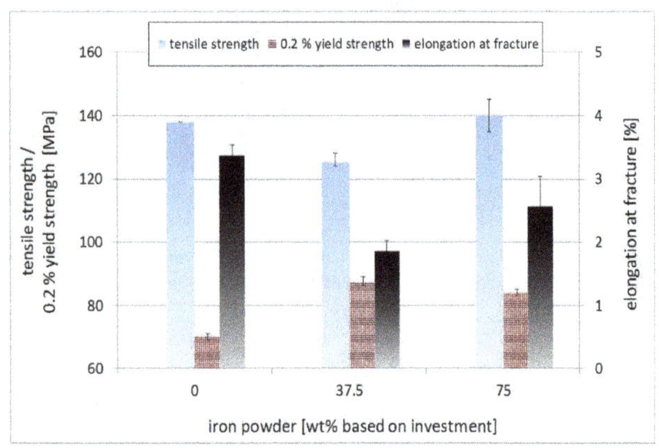

Figure 12. Influence of iron powder in the block mould on the tensile properties of an A356.

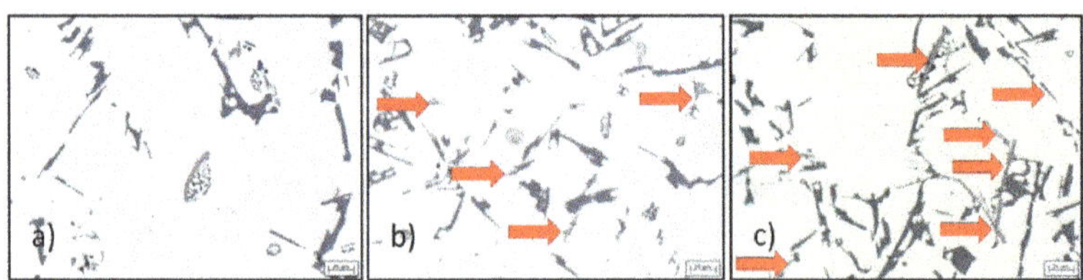

Figure 13. Metallographical sections of specimens casted in block moulds with a) 0 wt.%, b) 75 wt.% and c) 35.5 wt.% iron powder. Some iron phases are marked.

The lower yield strength and the lower fraction of iron phases in the microstructure indicate that the specimen cast in the mould containing 75 wt.% iron powder dissolved less iron in comparison to the specimen produced with the help of moulds containing 37.5 wt.% iron powder. This assumption could be explained by the higher strength of the moulds with 75 wt.% iron powder (figure 13b), since they dissolve less iron content due to mould erosion.

7. Influence of the maximum firing temperature and duration on gypsum-bonded investment's compressive strength

During the firing process, thermal stresses develop in the block mould due to the different thermal expansion coefficients of the moulding material's constituents. For example, in GBI bonded moulds the gypsum contracts during the firing process due to dehydration and the binder decomposes at elevated temperatures [14]. The dehydration takes place at temperatures of around 128 °C whereby hemihydrate is formed, in which water is removed at 163 °C. Depending on the temperature, the resulting anhydrite exhibits three polymorphic states. At 200 °C, III-$CaSO_4$ (hexagonal) is transformed into II-$CaSO_4$ (orthorhombic). Above 1200 °C, I-CaSO4 (cubic) is formed [24]. The fireproof basic material silica, mostly a mixture of cristobalite and quartz, shows a phase transformation at 220 °C to 275 °C and 573 °C, respectively. This transformation leads to an expansion of these phases. The isotropic transformation of cristobalite from the α- to β-modification takes place rapidly and causes a shearing moment between the refractory and the binder. Moreover, this moment can induce cracks in the mould. The expansion of quartz is anisotropic and can rapidly elevate the shearing moment caused by the cristobalite [14].

Independent of the crack formation due to the refractory's expansion, cracks can be initiated in the block mould by the expansion of the wax pattern during the dewaxing process or by a too rapid cooling after firing [14].

The tendency to decrease the crack initiation directly leads to a qualitative improvement of block mould cast components. For this purpose, the firing process was modified regarding the maximum firing temperaturein the followinginvestigations.

7.1. Materials and methods

The effect of the maximum firing temperature on the strength of GBI was analysed with the help of compression tests. For this, the moulding material Goldstar xXx from Goldstarpowders (composition see table 1 and figure 2) was used to produce compression specimens accordingly to DIN EN 993-5. The mixing procedure was conducted with respect to the results in section 4. Two types of specimens were produced: One with the best and one with the worst settings for the mixing parameters (table 9).

The general furnace program is summarised in table 2 with varying maximum burning temperatures. The highest furnace temperature was fixed at 950 °C to avoid a decomposition of the gypsum [14]. The compression specimens were tested using an Instron 8033 at a cross head speed of 2mm/min.

strength	water temperature [°C]	water content [wt.%]	mixing duration [s]	salt content [wt.%]
minimum	17	55	240	0
maximum	17	45	180	1

Table 9. Mixing conditions for the production of the GBI specimens.

7.2. Experimental results

Figure 14 shows the results of the firing experiments. By increasing the maximum firing temperature up to 950 °C, the compressive strength of GBI specimens is increased independent of the starting strength. A probable reason for this is that a greater quantity of sinter phases is formedat higher firing temperature which raises the strength of the moulding material. However, the scatter of the strength values is increased with elevated firing temperatures. In addition to this, there is a large difference between the strength of specimens produced using an optimised setting of the mixing parameters and that of other specimens.

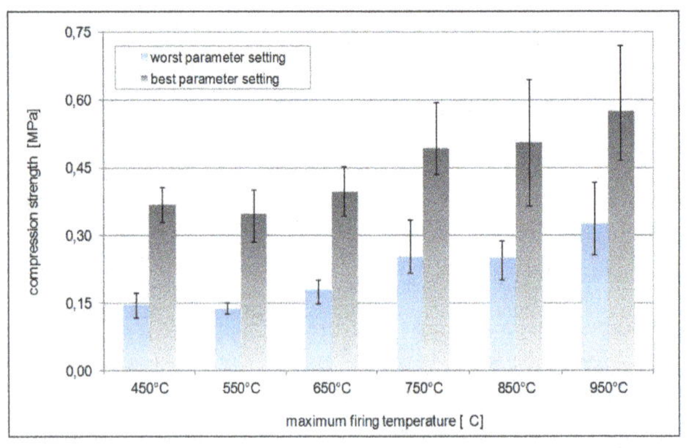

Figure 14. Influence of maximum firing temperature on the compression strength of gypsum-bonded investments.

8. Conclusions

The block mould casting process offers several attractive advantages which, when combined, only exist in this process. Besides this, high dimensional accuracy and surface quality of the potentially complicated casting result in low finishing effort. To produce high quality, block mould cast components, the tendency of the block mould to cracking has to be minimised. In the investigations presented here, different approaches were followed to modify the block moulds:

1. The addition of sodium chloride and sodium fluoride to the mixing water considerably increases the strength of the GBI block moulds. Regarding the handling of the GBI, a salt content of one weight per cent should not be exceeded.

2. In the green state, only the water and the salt content have a statistically significant influence on the strength of the GBI. Increasing the water and salt contents decreases the strength of the GBI in the green state. Here, the water temperature and the mixing duration have no significant influence.

3. The water temperature and the mixing duration exert a statistically significant effect on the fired state. The maximum strength is achieved using a water temperature of 17 °C, a low water content (here 45 wt.%), moderate mixing duration (here 180 s) and a high salt content of one weight per cent.

4. The results depicted here allow the green strength of GBI to be minimised providing a better demoulding of the block mould cast components. In addition to this, the strength in the fired state can be optimised to avoid crack initiation in the block moulds. Salt (sodium chloride + sodium fluoride) increases the green strength and decreases the fired strength.

5. One weight per cent of 12 mm glass fibres raise the strength of GBI.

6. Iron powder increases the strength and the cooling rate of GBI. Owing to the best agreement obtained between erosion resistance and cooling rate, a powder content of 75 wt. % is suggested. To avoid the assimilation of iron by the aluminium melt, coating of the wax pattern it advised.

7. With increased maximum firing temperature, the strength of GBI is raised, whereby a temperature of 950 °C should not be exceeded.

Acknowledgements

The authors would like to thank Jürgen Nominikat, Franz Ernst and Matthias Bücher for the production, testing and preparing of the ceramic and metallic specimens, and Timm Ziehm, Philipp Weiß, Johannes Brachmann, Alexander Gußfeld, Jennifer Tilli, Daria Borst, Stefan Böhnke, Sebastian Schmidtke, Anna Trentmann and Martin Schwenk for supporting them.Last but not least, the German Research Foundation (DFG) is thanked for their financial support within the scope of the Priority Program 1420.

Author details

Sebastian F. Fischer* and Andreas Bührig-Polaczek

*Address all correspondence to: s.fischer@gi.rwth-aachen.de

Foundry-Institute, RWTH Aachen University, Aachen, Germany

References

[1] Wu, M., Tinschert, J., Augthun, M., Wagner, I., Schädlich-Stubenrauch, J., Sahm, P. R., & Spiekermann, H. (2001). Application of laser measuring, numerical simulation and rapid prototyping to titanium dental castings. *Dental Materials*, 17(2), 102-108.

[2] Lautenschläger, E. P., & Monaghan, P. (1993). Titanium and titanium alloys as dental materials. *International Dental Journal*, 43(3), 245-253.

[3] Luk, H. W. K., & Darvell, B. W. (1997). Effect of burnout temperature on strength of phosphate-bonded investments. *Journal of Dentistry*, 25(2), 153-160.

[4] Guler, K. A., & Cigdem, M. (2012). Casting Quality of Gypsum Bonded Block Investment Casting Moulds. *Advanced Materials Research*, 445, 349-354.

[5] Ashby, M. F., Evans, A., Fleck, N. A., Gibson, L. J., Hutchinson, J. W., & Wadley, H. N. G. (2000). *Metal Foams- A Design Guide*, Woburn: Butterworth-Heinemann.

[6] Fischer, S. F., Schüler, P., Bayerlein, B., Fleck, C., & Bührig-Polaczek, A. (18-21 September). Paper presented at Proceedings of 7th International Conference on Porous Metals and Metallic Foams, MetFoam11, Busan, Korea. *Some aspects of the optimisation of open-pore metal foams by introducing bio-inspired hierarchical levels. In: Hur BY, Kim SE, Hyun SK. (eds.)*, BEXCO, 57-62.

[7] Bührig-Polaczek, A., Fettweis, D., & Hett, M. (12-14 October 2004). Paper presented at Proceedings of "Cellular Metals and Polymers ", Zürich, Switzerland. *Casting of metallic sponges using rapid prototyping. In: Singer RF, Körner C, Altstädt V, Münstedt H. (eds.)*, Trans Tech Publications, 65-68.

[8] Maguire, M. C., Baldwin, M. D., & Atwood, C. L. (1995). Fastcast: Integration and application of rapid prototyping and computational simulation to investment casting. . In: Society for the advancement of materials and process engineering: proceedings of 27th International Sampe Technical Conference 9- 12 October Albuquerque, New Mexico , 27, 235-244.

[9] Ivanov, T., Bührig-Polaczek, A., & Vroomen, U. (2011). Casting of microstructured shark skin surfaces and applications on aluminium casting parts. *Foundry*, 60(3), 229-233.

[10] Veeck, S., Lee, D., & Tom, T. (2002). Titanium Investment Castings. *Advanced Materials and Processes*, 160(2), 59, 62.

[11] Fischer, S. F., Thielen, M., Loprang, R. R., Seidel, R., Fleck, C., Speck, T., & Bührig-Polaczek, A. (2010). Pummelos as Concept Generators for Biomimetically-Inspired Low Weight Structures with Excellent Damping Properties. *Advanced Engineering Materials*, 12(12), 658-663.

[12] Lacy, A. M., Mora, A., & Boonsiri, I. (1985). Incidence of bubbles on samples cast in a phosphate-bonded investment. *Journal of Prosthetic Dentistry*, 54(3), 367-369.

[13] Chandler, H. T., Fisher, W. T., Brudvik, J. S., & Bottiger, G. (1973). Vacuum-air pressure investing. *Journal of Prosthetic Dentistry*, 29(21), 225-227.

[14] Luk, H. W. K., & Darvell, B. W. (2003). Effect of burnout temperature on strength of gypsum-bonded investments. *Dental Materials*, 19(6), 552-557.

[15] Phillips, R. W. (1947). Relative merits of vacuum investing of small castings as compared to conventional methods. *Journal of Dental Research*, 26(5), 343-352.

[16] Chew, C. L., Land, M. F., Thomas, C. C., & Norman, R. D. (1999). Investment strength as a function of time and temperature. *Journal of Dentistry*, 27(4), 297-302.

[17] Ott, D., & Raub, C. J. (1985). Investment casting of gold jewellery. . Gold Bulletin; , 18(4), 140-143.

[18] O'Brien, W. J., & Nielsen, J. P. (1959). Decomposition of Gypsum Investment in the Presence of Carbon. *Journal of Dental Research*, 38(3), 541-547.

[19] Matsuya, S., & Yamane, M. (1981). Decomposition of Gypsum Bonded Investments. *Journal of Dental Research*, 60(8), 1418-1423.

[20] Luk, H. W. K., & Darvell, B. W. (1997). Effect of burnout temperatures on strength of phosphate-bonded investments- Part II: effect of metal temperature. *Journal of Dentistry*, 25(5), 423-430.

[21] Juszczyk, A. S., Radford, D. R., & Curtis, R. V. (2000). The influence of handling technique on the strength of phosphate-bonded investments. *Dental Materials*, 16(1), 26-32.

[22] Bonilla, W., Masood, S. H., & Iovenitti, P. (2001). An Investigation of Wax Patterns for Accuracy Improvement in Investment Cast Parts. *International Journal of Advanced Manufacturing Technology*, 18(5), 348-356.

[23] Brevick, J. R., Davis, J. W., & Dincher, C. (1991). Towards improving the properties of plaster moulds and castings. *Journal of Engineering Manufacture*, 205(4), 265-269.

[24] Mori, T. (1986). Thermal Behavior of the Gypsum Binder in Dental Casting Investments. *Journal of Dental Research*, 65(6), 877-884.

[25] Gryzlova, E. S., & Kozyreva, N. A. (2009). Displacement of Chemical Equilibria in 12-Salt Six-Component Reciprocal Systems in Melt. *Doklady Chemistry*, 424(1), 19-22.

[26] Eichner, K., & Kappert, H. F. (1995). *Dental materials and their processing*, Stuttgart: Thieme; (in german).

[27] Klein, B. (2007). *Design of experiments-DoE: Introduction to Taguchi/Shainin*, München: Oldenburg Wissenschaftsverlag; (in german).

[28] Kumar, S., Kumar, P., & Shan, H. S. (2008). Optimization of tensile properties of evaporative pattern casting process through Taguchi's method. *Journal of Materials Processing Technology*, 204(1-3), 59-69.

[29] Amathieu, L., & Boistelle, R. (1988). Crystallization Kinetics of Gypsum from dense Suspension of Hemihydrate in Water. *Journal of Crystal Growth*, 88(2), 183-192.

[30] Singh, M., & Garg, M. (1992). Glass Fibre Reinforced Water-Resistant Gypsum-Based Composites. *Cement & Concrete Composites*, 14(1), 23-32.

[31] Klepetsanis, P. G., Dalas, E., & Koutsoukos, P. G. (1999). Role of Temperature in the Spontaneous Precipitation of Calcium Sulfate Dihydrate. *Langmuir*, 15(4), 1534-1540.

[32] Ali, M. A., & Grimer, F. J. (1969). Mechanical Properties of Glass Fibre-Reinforced Gypsum. *Journal of Material Science*, 4(5), 389-395.

[33] Liu, K., Wu, Y. F., & Jiang, X. L. (2008). Shear strength of concrete filled glass fiber reinforced gypsum walls. *Materials Structure*, 41(4), 649-662.

[34] Wertz, S. (1990). Innovative Use of Glass Reinforced Gypsum. *The Construction Specifier*, 43, 76.

[35] Heckl, J., & Krcmar, W. (2006). Improving the stability of gypsum mould by fibre reinforcement. *Ziegelindustrie International- Brick and Tile Industry International*, 59(12), 44-45.

[36] Majumdar & A. J. (1970). Glass Fibre Reinforced Cement and Gypsum Products. *Mathematical Physical and Engineering Science*, 319(1536), 69-78.

[37] Majumdar & A. J. (1974). The role of the interface in glass fibre reinforced cement. *Cement and Concrete Research*, 4(2), 247-266.

[38] Yamagata, H., Kasprzak, W., Aniolek, M., Kurita, H., & Sokolowski, J. H. (2008). The effect of average cooling rates on the microstructure of the Al-20% Si high pressure die casting alloy used for monolithic cylinder blocks. *Journal of Materials Processing Technology*, 203(1-3), 333-341.

[39] Borreguero, A. M., Carmona, M., Sanchez, M. L., Valverde, J. L., & Rodriguez, J. F. (2010). Improvement of the thermal behaviour of gypsum blocks by the incorporationof microcapsules containing PCMS obtained by suspension polymerization with an optimal core/coating mass ratio. *Applied Thermal Engineering*, 30, 1164-1169.

[40] Engler, S. (1972). Solidification morphology and casting properties. *International Journal of Materials Research*, 63(7), 375-379, (in german).

[41] Mabie, C. P. (1973). Petrographic Study of the Refractory Performance of High-Fusing Dental Alloy Investments: I. High-Fired, Phosphate-Bonded Investments. *Journal of Dental Research*, 52(1), 96-110.

Chapter 4

Progress in Investment Castings

Ram Prasad

Additional information is available at the end of the chapter

1. Introduction

Certain basic elements in the earlier development of investment casting process, as well as many current advancements made to the process, some due to the rapid growth in technology, are described in this chapter.

Investment casting is often called 'lost wax' casting, and it is based on one of the oldest metal forming processes. The Egyptians used the process, some 5000 years ago, to make gold jewelry, as exact replica of many intricate shapes, cast in gold from artfully created beeswax patterns. In the investment casting process, a ceramic slurry is applied, or 'invested', around a disposable pattern, usually wax, and is allowed to harden to form a disposable casting mold. The wax pattern is 'lost', when it is melted out from the 'disposable' ceramic mold, which is later destroyed to recover the casting.

In investment casting, the ceramic molds are made by two different methods: the solid mold process and the ceramic shell process. The solid mold process is mainly used for dental and jewelry castings, currently has only a small role in engineering applications, and as such will not be covered in this chapter. The ceramic shell process has become the predominant technique for a majority of engineering applications, displacing the solid mold process.

The ceramic shell process is a precision casting process, uniquely developed and adapted to produce complex-shaped castings, to near-net-shape, and in numerous alloys. Continued advancements in materials and techniques used in the process, are driven and supported by R&D on many fronts, both in the industry as well as in many schools for foundry metallurgy. For instance, earlier research, funded by Rolls Royce Limited, UK, at the University of Birmingham, UK, investigating the feeding behavior of high temperature alloys has assisted in the development of optimal gating and feeding systems for investment castings, [1-6]. The influence of alloy and process variables on producing sound investment castings is detailed later in the chapter, under the section on the design of gating and feeding system.

The ceramic shell process, however, requires careful control during many steps or operations. The basic steps in the process, involving both materials and techniques are presented here, in the sequence illustrated in Fig.1.

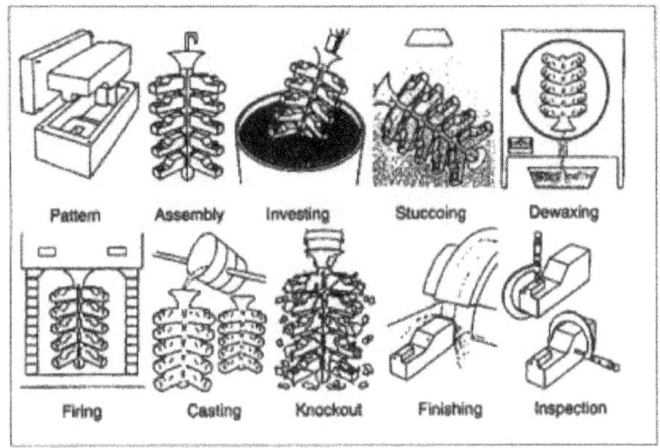

Figure 1. Showing sequence of the steps in the investment casting process.

2. Pattern Materials

Pattern materials currently in use are waxes, and plastics, while other pattern materials are used sometimes, and for specific applications. Waxes, blended and developed with different compositions, are more commonly used, while use of plastic patterns, generally polystyrene, may sometimes be required, to produce thin- walled, complex -shaped castings, such as in aerospace integrally cast turbine wheels and nozzles.

2.1. Pattern Waxes

Waxes are mostly the preferred material for patterns, and are normally used, modified and blended with additive materials such as plastics, resins, fillers, antioxidants, and dyes, in order to improve their properties, [7].

Paraffins and microcrystalline waxes are the most widely used waxes, and are often used in combination, because their properties tend to be complementary.

Paraffin waxes are available in many controlled grades, with melting points ranging from 52 to 68 °C (126 to 156 °F). They are readily available in different grades, have low cost, high lubricity and low melt viscosity. Their usage is, however, limited because of high shrinkage and brittleness.

Microcrystalline waxes tend to be highly plastic and provide toughness to wax blends. Available in both hard, nontacky grades as well as soft, adhesive grades, they have higher melting points, and are often used in combination with paraffin.

Other waxes used include: Candelilla, a vegetable wax, which is moderately hard and slightly tacky. Carnauba wax is a vegetable wax with higher melting point, low coefficient of thermal expansion, and is very hard, nontacky and brittle. Beeswax is a natural wax, widely used for modeling, and in pattern blends, provides properties similar to microcrystalline waxes.

Fischer-Tropsch waxes are synthetic hydrocarbon waxes resembling paraffins, but are available in harder grades, with higher melting points. Ozocerite is a mineral wax sometimes used in combination with paraffin.

Waxes, in general, are moderately priced, and can easily be blended to suit different requirements. Waxes have low melting points and low melt viscosities, which make them easy to blend, inject, assemble into tree- or cluster-assemblies, and melt out with out cracking the thin ceramic shell molds.

2.1.1. Additives to Pattern Waxes

Waxes with their many useful properties are, however, deficient in two practically important areas:

(a) Strength and rigidity especially required to make fragile patterns; and (b) Dimensional control, especially in limiting surface cavitation due to solidification shrinkage, during and after pattern injection. Additives are made to waxes to cause improvements needed in these two deficient areas.

The strength and toughness of waxes are improved by the addition, in required volumes, of plastics such as polyethylene, nylon, ethyl cellulose, ethylene vinyl acetate and ethylene vinyl acrylate.

Solidification shrinkage causing surface cavitation in waxes, is reduced to some extent by adding plastics, but is reduced to a greater extent by adding resins and fillers.

Resins suitable for this are: coal tar resins, various rosin derivatives, hydrocarbon resins from petroleum and tree-derived resins such as dammar, Burgundy Pitch, and the terpene resins. These resins have a wide range of softening points and varying viscosity at different temperatures. These factors must be considered while blending and using resins in pattern waxes.

Fillers are powdered solid materials, and are used more selectively in waxes than resins. This leads to the description of pattern waxes as being either filled or unfilled. Fillers have higher melting point and are insoluble in the base wax, thereby contributing to reduced solidification shrinkage of the mixture, in proportion to the amount used. Fillers that have been developed and used in pattern waxes include: spherical polystyrene, hollow carbon microspheres, and spherical particles of thermosetting plastic.

Several other additives can be used in pattern waxes to obtain additional properties. Antioxidants can be used to protect waxes and resins subject to thermal deterioration. Colors in the form of dyes are used to enhance appearance, to provide identification, and to facilitate inspection of injected patterns.

- Typical composition of unfilled waxes, with these additives, is in the following ranges:

Waxes: 30-70%; Resins: 20-60%; Plastic: 0-20%; Other additives: 0-5%.

Filled waxes have similar base composition, and are normally added with 15 to 45% filler.

2.1.2. Factors for Pattern Wax Selection

Process factors while selecting and formulating wax pattern materials, that must be addressed are listed below, grouped with the material properties required or to be considered:

- Injection: Freezing range, softening point, ability to duplicate detail, setup time.
- Removal, handling, and assembly: Strength, hardness, rigidity, impact resistance, weldability.
- Dimensional control: Solidification shrinkage, thermal expansion, cavitation tendency.
- Shell mold making: Strength, wettability, and resistance to binders and solvents.
- Dewaxing and burnout: Softening point, viscosity, thermal expansion, and ash content.
- Miscellaneous: Availability, cost, ease of recycling, toxicity, and environmental factors.

2.2. Plastics

Plastic is the most widely used pattern material, next to wax. Polystyrene is usually used, because it is economical, very stable, can be molded at high production rates on automatic equipment, and has high resistance to handling damage, even in extremely thin sections.

Use of polystyrene is however limited, because of its tendency to cause shell mold cracking during pattern removal, and it requires more expensive tooling and injection equipment than for wax.

However, the most important application for polystyrene is for delicate airfoils, used in composite wax-plastic integral rotor and nozzle patterns, assembled using wax for the rest of the assembly.

2.3. Other Pattern Materials

Foamed Polystyrene has long been used for gating system components. It is also used as patterns with thin ceramic shell molds in a separate casting process known as Replicast Process.

Urea-based patterns, developed in Europe, have properties similar to plastics; they are very hard, strong and require high-pressure injection machines. Urea patterns have an advantage over plastics: they can easily be removed, without stressing the ceramic shell, by simply dissolving in water, or an aqueous solution.

3. Production of Patterns

Patterns are usually produced by injecting pattern material in to metal dies, made with one or more cavities of the desired shape, in each die. Different equipments, with different operating parameters, have been developed to suit different pattern materials.

Wax patterns are injected at lower temperatures, (110 to 170 °F), and pressures, (40 to 1500 psi), in split dies using specially designed equipment. The wax injection equipment ranges from simple pneumatic units, to complex hydraulic machines, which can accommodate large dies, and at high injection pressures.

Polystyrene patterns are injected at higher temperatures, (350 to 500 °F), and pressures, (4 to 20 ksi), in hydraulic machines, normally equipped with water cooled platens that carry the die halves.

Advanced techniques have been developed currently to produce prototype, or experimental patterns, when only a few patterns are required. For such limited and/ or temporary usage of patterns, 'Rapid prototype patterns' are being produced in machines, with some utilizing advanced techniques such as SLS (selective laser sintering) or SLA (stereo lithography) and with special polymer material called photopolymers [8]. These techniques known alternatively as '3D-printing' or 'Additive- manufacturing', produce prototype patterns after building parts by depositing fine layers of various materials and using lasers only where necessary to achieve the finished shape, as defined by CAD. These patterns have been found to have favorable dewaxing response, resulting in substantially improved surface quality for investment cast prototypes in many alloys.

3.1. Pattern Dies

Various pattern tooling options are available for waxes because of their low melting point and good fluidity. Many die materials are used, including: rubber, plastic, plaster, metal-filled plastic, soft lead-bismuth tin alloys, aluminum, brass, bronze, beryllium copper, steel or a combination of these. The selection is based on considerations of cost, tool life, delivery time, pattern quality, and production efficacy in available patternmaking equipment.

Plastic patterns usually require steel or beryllium copper tooling. Pattern dies made by machining use CNC (computer numerical controlled) machine tools and electric discharge machining. Alternatively, cast tooling made in aluminum, steel or beryllium copper is also used effectively. Wax can be cast against a master model to produce a pattern, which is then used to make an investment cast cavity for this type of cast tooling.

4. Pattern Assembly

Patterns for investment casting produced in dies are prepared for assembly in different ways. Large patterns are set-up and are processed individually, while small to medium size patterns are usually assembled into clusters for economy in processing. For example, pattern clusters of aircraft turbine blades may range from 6 to 30 parts. For small hardware parts, patterns set in clusters may range from tens to hundreds. Most patterns are injected with the gates.

However, large or complex parts are injected in segments, which are assembled into final form. The capacity of injection machines and the cost of tooling are important considerations. Gating components, including pour cups, gating and runner components forming trees or clusters are produced separately, and patterns assembled with these to produce the wax-tree or pattern cluster. Standard extruded wax shapes are often used for gating, especially for mock-up work. Preformed ceramic pour cups are often used in place of wax pour cups. Most assembly is done manually, with skilled personnel.

Wax components are assembled by wax welding, using hot iron or spatula, or a small gas flame. Wax at the interface between two components is quickly melted, and the components are pressed together until the wax solidifies. The joint is then smoothed over. A hot melt adhesive can be used instead of wax welding. Currently, laser welding units have been developed to provide improvements in assembling of wax components. Fixtures are essential to ensure accurate alignment in assembling patterns. Joints must be strong, and completely sealed with no undercuts. Care also must be taken to avoid damaging patterns or splattering drops of molten wax over the patterns being assembled.

Polystyrene pattern segments are assembled by solvent welding. The plastic at the interface is softened with solvent, and the parts are pressed together until bonded. However, polystyrene becomes very tacky when wet with solvent, and readily adheres to itself. Frequently, only one of the two halves needs to be wet. The assembly of polystyrene to wax is done by welding, with only the wax being melted.

Most assembly and setup operations are performed manually, but some automation is currently being introduced in some investment casting foundries. In one application, a robot is used to apply sealing compound in the assembly of patterns for different integrally cast nozzles, with each nozzle having, from 52 to 120 airfoils apiece.

4.1. Design of Pattern Tree or Cluster

The following preliminary requirements are considered essential:

- Providing a tree or cluster design that is properly sized and mechanically strong enough to be handled through the process
- Meeting all metallurgical requirements
- Providing test specimens for chemical or mechanical testing, when required.

Once these essentials are satisfied, other factors are adjusted to maximize profitability. Since the process is very flexible, foundries approach this goal in various ways. Some foundries prefer cluster design tailored to each individual part to maximize parts per cluster and metal usage. Others adopt standardized trees, or clusters, to facilitate handling and processing. When close control of grain is required, such as for equiaxed, directionally solidified columnar, or single crystal casting, circular clusters are often used to provide thermal uniformity during solidification in the casting process.

The design of the pattern tree or cluster is however, critical and important, since it can affect every aspect of the investment casting process. The design of the assembled pattern tree, or cluster critically impacts various stages of the investment casting process, as well as, in effectively meeting all the quality and metallurgical requirements of the final product.

As such, it is presented here in three parts, namely: (a) Basic design requirements,(b) Design of gating and feeding system, and (c) Use of computer solidification simulation software.

4.1.1. Basic Design Requirements

Contribution towards the final casting quality due to any specific design of the pattern tree, or pattern cluster needs to be carefully evaluated. Factors to be considered in the basic design of wax tree or cluster assembly include:

number of pieces processed at a time, ratio of metal poured to castings shipped, number of pieces assembled in each tree or cluster, ease of assembly, handling strength, ease of dipping or mold forming and drying, wax removal, shell removal, ease of cut off and finishing, and available equipment and processes at all stages.

Additional factors affecting metallurgical casting quality include: liquid metal flow in terms of tranquil, laminar flow or turbulence in flow, top fill versus bottom fill, gas or air entrainment, filling of thin sections, control of grain size and shape (when specified), effect on inducing favorable melt- temperature gradients, efficacy in feeding of shrinkage, ceramic bridging at added joints aggravating shrinkage, or the propensity for hot tears and cracks in casting sections.

4.1.2. Design of Gating and Feeding System

The critical aspects of tree/ cluster design are gating and risering, or feeding. Basic concepts of feeding sand castings, such as progressive solidification toward the riser or feeder, Chvorinov's rule and its extensions, solidification mode and feeding distance as a function of alloy, and section size also have been found to apply to investment casting, [1-3], [9]. The step-by-step procedure towards designing gating and feeding system is described, with two practical examples, at the Appendix. Feeding distances in hot investment molds are generally found to be longer than in sand molds. In investment castings, while separate feeders or risers are used sometimes, more often the gating system also performs the risering or feeding function. This applies specifically to numerous small parts that are commonly investment cast.

The use of wax trees or clusters permits great flexibility in the design of feeding systems. Wax clusters for process development are readily mocked up for trial. Extruded wax shapes are easily bent into feeders that can be attached to any isolated sections of the part that are prone for shrinkage. Once proved, they can be incorporated into tooling, if this is cost-effective. If not, they can be applied manually during tree or cluster assembly. This capability makes it practical to cast very complex parts with high quality. It also makes it feasible to convert fabrications assembled from large numbers of individual components into single-piece investment castings at substantial cost-savings.

4.1.3. Use of Computer Solidification Simulation Software

Considerable development efforts have been made to provide many solidification simulation models of value in investment casting production. Currently, alternative computer simulation software systems are available applying heat transfer models, based either on finite element or finite difference methods. These are being utilized on the shop floor in many larger foundries, especially in the design of gating and feeding systems, to determine effect of solidification conditions on alloy microstructure, and for accurate predictions of tooling dimensions. The use of simulation models plays a major role in the development of investment casting process for gas turbine blades, specified with equiaxed grains, DS (directionally solidified) columnar grains, or with single crystal, in many super alloys. Additionally, rapid advancements in the solidification software show continual improvement in the ability to predict accurately many grain defects that can occur in the production of directionally solidified, DS, or single crystal components.

5. Production of Ceramic Shell Molds

Investment shell molds are made by applying a series of ceramic coatings to the pattern tree assemblies or pattern clusters. Each coating consists of a fine ceramic layer, with coarse ceramic 'stucco' particles embedded in its outer surface. The tree assembly or cluster is first dipped into a ceramic slurry bath, then withdrawn from the slurry, and manipulated to drain off excess slurry, and to produce a uniform layer. The wet layer is immediately stuccoed with coarser ceramic particles, either by immersing it into a fluidized bed of the particles, or by sprinkling or 'raining' on it the stucco particles from above.

The fine ceramic layer forms the inner face of the mold, and reproduces every detail, including the smooth surface of the pattern. It also contains the bonding agent, which provides strength to the structure. The coarse stucco particles serve to arrest further runoff of the slurry, help to prevent it from cracking or pulling away, provide keying or bonding between individual coating layers, and build up shell thickness faster.

Each coating is allowed to harden or set before the next one is applied. This is accomplished by drying, chemical gelling, or a combination of these. The operations of coating, stuccoing, and hardening are repeated a number of times, until the required shell thickness is achieved.

The final coat, often called a seal coat, is left unstuccoed, in order to avoid the occurrence of loose particles on the shell mold surface.

Various ingredients and preparation efforts required to make ceramic shell molds are described in the following section.

5.1. Refractories

Silica, zircon, alumina and various aluminum silicates are commonly used refractories for both slurry and stucco in making ceramic shell molds. Alumina is expensive, and as such used selectively, such as in directional solidification processes. Other refractories, such as graphite, zirconia and Yttria have been used with reactive alloys. Yttria is used in prime coats for casting titanium. Typical properties of refractories are listed in Table 1 [11].

Silica is generally used in the form of fused silica (silica glass). Fused silica is made by melting natural quartz sand and then solidifying it to form a glass, which is crushed and screened to produce stucco particles, and it is ground to a powder for use in slurries. The extremely low coefficient of thermal expansion of fused silica, Fig. 2, imparts thermal shock resistance to molds. Its ready solubility in molten caustic solutions provides a means of chemically removing shell material from areas of castings that are difficult to clean by other methods. Silica is sometimes used as naturally occurring quartz, expense of which is very low. However, its utility is limited because of its high coefficient of thermal expansion and by the high, abrupt expansion at 573 °C (1063 °F) accompanying its α-to-β-phase transition, causing excessive cracking of shell mold, if the mold is not fired slowly.

Table 1 Nominal compositions and typical properties of common refractories for investment casting

Data are for comparison only, are not specifications, and may not describe commercial products.

Material	Nominal composition %	Crystalline-form	Approximate theoretical density g/cm³	lb/in.³	Relative leachability (a)	Approximate melting point ·C	·F	PCE temperature (b) ·C	·F	pH	Color
Aluminosilicates											
42%...	Al_2O_3-53 SiO_2	Mixture	2.4-2.5	0.086 - 0.090	Poor	...	...	1750	3180	6.5-7.8	Gray to tan
47%...	Al_2O_3-49 SiO_2	Mixture	2.5-2.6	0.090 - 0.094	Poor	...	...	1760	3200	6.5-7.8	Gray to tan
60%	Al_2O_3-36 SiO_2	Mixture	2.7-2.8	0.097 - 0.10	Poor	...	...	1820	3310	6.5-7.8	Gray to tan
70%...	Al_2O_3-25 SiO_2	Mixture	2.8-2.9	0.10 - 0.104	Poor	...	...	1865	3390	6.5-7.8	Gray to tan
73%...	Al_2O_3-22 SiO_2	Mixture	2.8-2.9	0.10 - 0.104	Poor	...	...	1820	3310	6.5-7.8	Gray to tan
Alumina	99% + Al_2O_3	Trigonal	4.0	0.144	Poor	2040	3700	...	...	8.5-8.9	White
Fused silca	99.5% SiO_2	Typically 97% + amorphous	2.2	0.079	Good	1710	3110	...	...	6.0-7.5	White
Silica-quartz	99.5% SiO_2	Hexagonal	2.6	0.094	Good	1710	3110	...	...	6.4-7.5	White to tan
Zircon	97% + $ZrSiO_4$	Tetragonal	4.5	0.162	Moderate	2550	4620	...	...	4.7-7.0	White to tan

(a) **Poor:** slight reaction in hot concentrated alkali; **Good:** soluble to very soluble in hot concentrated alkali or hydrofluoric acid; **Moderate:** reacts with hot concentrate
(b) **PCE**, Pyrometric cone equivalent. Source: "Ceramic Test Procedures" Investment Casting Institute, 1979 [Ref 11]

Table 1. Normal compositions and typical propeties of common refractories for investment casting.

Zircon occurs naturally as a sand, and used in this form as a stucco. Its primary advantages are high refractoriness, resistance to wetting by molten metals, and round particle shape. Use of zircon is generally limited with prime coats, as it does not occur in sizes coarse enough for stuccoing backup coats. It is ground to powder for use in slurries, often in conjunction with fused silica and aluminosilicates.

Aluminum silicates are generally composed of stable compound, mullite ($Al_2O_{3.}2SiO_2$) with some free silica, which is usually in the form of silica glass. They are made by calcining fireclays, to produce different levels of mullite, (which contains 72% alumina) and free silica. Refractoriness and cost increase with alumina content. Fired pellets are crushed or ground and carefully sized to produce a range of powder sizes for use in slurries, and granular materials for use as stuccos.

Alumina, produced from bauxite ore by the Bayer process, is more refractory than silica or mullite, and is less reactive toward many alloys. However, its use is primarily confined to superalloy casting.

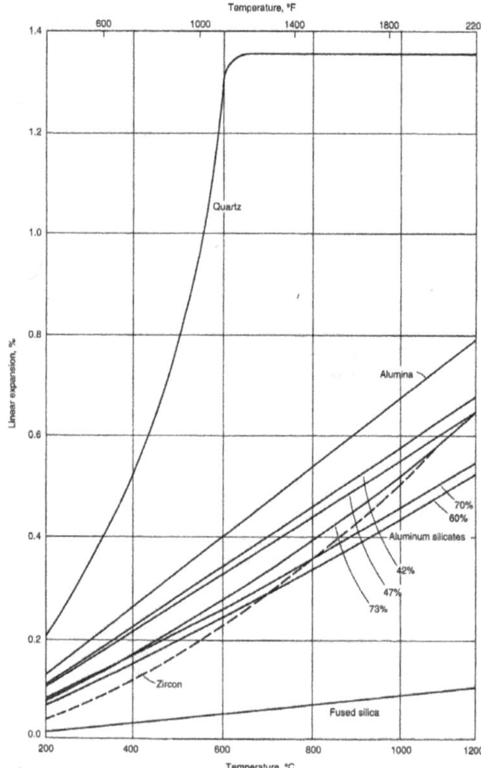

Figure 2. Linear thermal expansion of refractories commonly used for investment casting [11].

5.2. Binders

The commonly used binders include colloidal silica, hydrolyzed ethyl silicate and sodium silicate.

Colloidal silica is most widely used. It consists of a colloidal dispersion of spherical silica particles in water. It is an excellent general purpose binder. Its main disadvantage is that its water base makes it slow drying, especially in inaccessible pockets or cores.

Ethyl silicate which has no bonding properties is converted to ethyl silicate binder by hydrolysis, a reaction with water carried out in ethyl alcohol, using acid catalyst such as hydrochloric acid.

Ethyl silicate, with its alcohol base, dries much faster than colloidal silica. Ethyl silicate slurries are readily gelled after rapid dripping cycles, by exposure to an ammonia atmosphere. Use of this binder is however, rapidly declining, since it is much more expensive, and poses fire and environmental hazards.

Liquid sodium silicate solutions are sometimes used where a very inexpensive binder is desired. They have poor refractoriness, which limits their application.

Other Binders: The operation of directional solidification and single crystal processes, which subject the mold to high temperatures, for longer times, along with the introduction of more reactive superalloys, has led to the development of more refractory binders, such as colloidal alumina and colloidal zirconia binders. Both these binders, however, are inferior to colloidal silica in room temperature bonding properties.

5.3. Other Ceramic Shell Constituents

Wetting Agents. Slurries generally contain wetting agents, in addition to the refractory and the binder, to promote wetting of the pattern or prior slurry coats. Wetting agents such as sodium alkyl sulfates, sodium alkyl aryl sulfonates, or octyl- phenoxy polyethoxy ethanol, are generally used in amounts of 0.03 to 0.3% by weight of the liquid. Wetting agents are sometimes omitted from ethyl silicate alcohol slurries and from water based back-up slurries.

Antifoam Compounds. Where wetting agents are used, especially in prime coats, an antifoam compound is included to suppress foam formation and to permit air bubbles to escape. Commonly used defoamers are aqueous silicone emulsions and liquid fatty alcohols such as *n*-octyl alcohol. These are effective in low concentrations of 0.002 to 0.10%, based on the liquid weight.

Other Constituents. Nucleating agents, or grain refiners, which are refractory cobalt compounds such as aluminates, silicates, and oxides are added to the prime slurry (in amounts from 0.5 to 10% by weight of the slurry) for equiaxed superalloy castings, where close control of grain size is required. Organic film formers are sometimes used to improve green strength. Small additions of clay have been used to promote coating characteristics.

5.4. Slurry Preparation

Compositions of the slurry, which are usually proprietary, are based on the particular refractory powder and the type of binder. Slurry composition is generally in the following broad range:

- Binder solids: 5 -10%
- Liquid (from binder or added) : 15 – 30%
- Refractory powder: 60-80%

Slurries are prepared by adding refractory powder to binder liquid, using agitation to break up agglomerates, remove any air entrainment. Stirring is continued until viscosity falls to its final level before the slurry is put to use. Continued stirring is also required in production to keep the powder from settling out of suspension. Either rotating tanks with baffles or propeller mixers are used for this purpose.

Control procedures for slurries vary considerably among foundries. The most prevalent controls are the measurement of the initial ingredients, slurry temperature, density, pH and viscosity. Viscosity is measured with a No.4 or 5 Zahn cup, or a Brookfield type rotating viscometer. Properties of the finished ceramic shells that are monitored include: weight, modulus of rupture (green and fired), and permeability.

5.5. Pattern Tree or Cluster Preparation

Before dipping, pattern trees or clusters are usually cleaned to remove injection lubricant, loose pieces of wax, or dirt. Cleaning is accomplished by rinsing the pattern clusters in solution of wetting agent, or a suitable solvent that does not attack the wax. The trees, or clusters, are usually allowed to return to room temperature and dry, before dipping.

5.6. Coating and Drying

Dipping, draining, and stuccoing of clusters are carried out manually, robotically, or mechanically. Foundries are increasingly using robots in order to heighten productivity, to process larger parts and clusters, and to produce more uniform coatings. When robots are introduced, they are often programmed to reproduce actions of skilled operators. Dedicated mechanical equipment can sometimes operate faster, especially with standardized clusters.

Most dipping is done in air, but dipping under vacuum has been found, in some limited applications, very effective for coating narrow passageways and for eliminating air bubbles.

The cleaned wax cluster is dipped into the prime slurry and rotated. It is then withdrawn and drained over the slurry tank with suitable manipulation to produce a uniform coating. Next the stucco particles are applied by placing the cluster in a stream of particles falling from an overhead screen in a rainfall sander, or by plunging the cluster into a fluidized bed of the particles. In the fluidized bed, the particles behave as a boiling liquid, because of the action of pressurized air passing through a porous plate in the bottom of the bed.

Generally, prime slurries contain finer refractory powder, are used at a higher viscosity, and are stuccoed with finer particles than the backup coats. These characteristics provide a smooth surfaced mold, capable of resisting metal penetration.

Backup coats are formulated to coat readily over the prime coats (which may be somewhat porous and absorbent), to provide high strength, and to build up the required thickness with a minimum number of coats. The number of coats required is related to the size of the clusters and the metal weight to be poured. It may range from 5 for small clusters, to 15 or more for large ones. For most applications, the number ranges from 6 to 9.

Between coats, the slurries are hardened by drying or gelling. Air drying at room temperature with circulating air of controlled temperature and humidity is the most common method. Drying is usually carried out on open racks or conveyors, but cabinets or tunnels are sometimes used.

Drying is complicated by the high thermal expansion and contraction characteristics of waxes. If drying is too rapid, the chilling effect causes the pattern to contract, while the coating is still wet and unbonded. Then, as the coating is developing strength and even shrinking, the wax begins to expand, as the drying rate declines and it regains temperature. This can actually crack the coating. Therefore, to prevent this, relative humidity is normally kept above 40%, usually at a recommended value of 50%.

6. Ceramic Cores

Ceramic cores are widely used in investment castings to produce internal passageways in castings; and cores are either self-formed or preformed.

a. *Self-formed cores* are produced during the mold building, with the wax patterns already having corresponding openings. Metal pull cores in pattern tooling are used for simple shapes, while soluble cores for other shapes are made and placed in the pattern tooling, and the pattern injected around them. The soluble core is then dissolved out in a solution that does not affect the wax pattern, such as an aqueous acid. Citric acid is used commonly.

b. *Preformed cores* are required when self-formed cores can not be used, and are produced by a number of ceramic forming processes. Simple tubes and rods are commonly extruded from silica glass. Many cores are made by injection molding of fine ceramic powder with a suitable organic binder into steel dies and subjecting the cores to a two stage heat treatment. In the second stage, the core is sintered to its final strength and dimensions.

Preformed cores are normally used by placing in pattern die and injecting wax around them. Since cores expand differently than the shell molds, due to their differences in composition, cores must be provided with slip joints in the mold [12].

7. Removal of Pattern

Pattern removal is the operation that subjects the shell mold to the most stresses, since the thermal expansion of waxes are many times those of refractories used for molds. When the mold is heated to liquefy the wax, this expansion differential leads to enormous pressure that is capable of cracking, or even destroying the mold. In practice, this problem is effectively circumvented by heating the mold extremely rapidly from the outside in. This causes the surface layers of wax to melt very quickly, before the rest of the pattern can heat up appreciably. This molten wax layer either melts out of the mold or soaks into it, thus providing the space to accommodate the expansion as the remainder of the wax is heated. Melt-out tips are sometimes provided, or holes are drilled in the shell to relieve wax pressure.

Even with these techniques, the shell is subject to high stress. To get the shell as strong as possible, it should be thoroughly dried before dewaxing. Shells are subject to 16 to 48 h of extended drying after the last coat, sometimes enhanced by the application of vacuum or extremely low humidity. Two methods have been developed to implement the surface melting concept: autoclave dewaxing and high temperature flash dewaxing.

- *Autoclave dewaxing* is the most widely used method. Saturated steam is used in a jacketed vessel, with a steam accumulator to ensure rapid pressurization. Autoclaves are equipped with a sliding tray to accommodate a number of molds, a fast acting door with a safety lock, and an automatic wax drain valve. Operating pressures of approximately 550 to 620 kPa (80 to 90 psig) are reached in 4 to 7 s. Molds are dewaxed in approximately 15 min. or less. Wax recovery is good. Polystyrene patterns cannot be melted out in the autoclave, but require flash dewaxing.

- *Flash dewaxing* is carried out by inserting the shell into a hot furnace at 870 to 1095 °C (1600 to 2000 °F). The furnace is equipped with an open bottom so that wax can fall out of the furnace as soon as it melts. Some of the wax begins to burn as it falls, and even though it is quickly extinguished, there is greater potential for deterioration than with an autoclave. Nevertheless, wax from this operation can be reclaimed satisfactorily. Flash dewaxing furnaces must be equipped with an afterburner in the flue or some other means to prevent atmospheric pollution.

Polystyrene patterns are readily burned out in flash dewaxing. However, polystyrene can cause extensive mold cracking, unless it is embedded in wax in the pattern (as in integral nozzle patterns), or unless the polystyrene patterns are very small.

- *Hot liquid dewaxing* has found some use among smaller companies seeking to minimize capital investment. Hot wax at 177 °C (350 °F) is often used as the medium, while other liquids can also be used. Cycles are longer than for autoclave and flash dewaxing, and there is potential fire hazard.

8. Mold Firing and Burnout

Ceramic shell materials are fired to remove moisture (free and chemically combined), to burn off residual pattern material and any organics used in the shell slurry, to sinter the ceramic, and to preheat the mold to the temperature required for casting. In some cases, these are accomplished in a single firing. Other times, preheating is performed in a second heating, after the mold is cooled down, inspected and repaired if necessary. Cracked molds can be repaired with ceramic slurry or special cements. Many molds are wrapped with a ceramic-fiber blanket at this time to minimize the temperature drop that occurs between the preheat furnace and the casting operation, or to provide better feeding by insulating selected areas of the mold. Gas fired furnaces are used for mold firing and preheating, except for molds for directional solidification processes, which are preheated in the casting furnace with induction or resistance heating. Batch and continuous pusher-type furnaces are most common, but some rotary furnaces are also in use.

Burnout furnaces operate with temperatures between 870 and 1095 °C (1600 and 2000 °F), and have some 10% excess air provided to ensure complete combustion of organic materials. Preheat temperatures vary depending on part configuration and the alloy to be cast. Common ranges are: 150 to 540 °C (300 to 1000 °F) for aluminum alloys, 425 to 870 °C (800 to 1600 °F) for many copper base alloys, and 870 to 1095 °C (1600 to 2000 °F) for steels and superalloys. Molds for the directional solidification process are preheated above the liquidus temperature of the alloy being cast.

9. Melting and Casting

Different types of equipment are currently in use for melting, and support different casting methods adopted.

9.1. Melting Equipment

Coreless type Induction furnaces are used with capacities ranging from 15 to 750 lb., with normal melting rates of 3 lb/min. They are usually tilting models, and can be employed for melting in air, inert atmosphere or vacuum. They are extensively used for melting steel, iron, cobalt and nickel alloys, and sometimes copper and aluminum alloys. *Gas- fired crucible furnaces* are used for aluminum and copper alloy castings, while *electrical resistance furnaces* are sometimes preferred for aluminum casting, since they help reduce hydrogen porosity. The crucibles typically used are magnesia, alumina and zirconia, which are made by slip casting, thixotropic casting, dry pressing, or isostatic pressing. Magnesium alloys can be melted in gas-fired furnaces using low- carbon steel crucibles.

Consumable-electrode vacuum arc skull furnaces are used for melting and casting titanium.

Electron beam melting has been used in Europe as an alternative to vacuum arc melting for casting titanium [13], and in the United States, for melting superalloys for directionally solidified and single crystal casting [14].

9.2. Casting Methods

Both air and vacuum casting methods are used in investment casting. There is some use of rammed graphite molds in vacuum arc furnaces for casting titanium. Most castings are gravity poured.

- *Air casting* is used for many investment-cast alloys, including aluminum, magnesium, copper, gold, silver, platinum, all types of steel, ductile iron, most cobalt alloys, and nickel-base alloys that do not contain reactive elements. Zinc alloys, gray iron and malleable iron are usually not investment cast for economic reasons.
- *Vacuum casting* provides cleaner metals with superior properties and is used for alloys that can not be cast in air, such as the γ'-strengthened nickel base alloys, some cobalt alloys, titanium and the refractory metals. Batch and semicontinuous interlock furnaces are normally used.

A major advantage of investment casting is its ability to cast very thin walls, due to the use of hot mold. This advantage is further enhanced by specific casting methods, such as vacuum-assist casting, pressurized casting, centrifugal casting and countergravity casting.

- In *vacuum-assist* casting, the mold is placed inside an open chamber, which is then sealed with a plate and gaskets, leaving only the mold opening exposed to the atmosphere. A partial vacuum is drawn within the chamber and around the mold. The metal is poured into the exposed mold opening, and the vacuum serves to evacuate air through the porous mold wall and to create a pressure differential on the molten metal, both of which help to fill delicate detail and thin sections.
- In *pressurized casting,* rollover furnaces are pressurized for the same purpose. The hot mold is clamped to the furnace-top with its opening in register with the furnace opening, and the furnace is quickly inverted to dump the metal into the mold, while pressure is applied using compressed air or inert gas.
- *Centrifugal casting* uses the centrifugal forces generated by rotating the mold to propel the metal and to facilitate filling. Vacuum arc skull furnaces discharge titanium alloy at a temperature just above its melting point, and the centrifugal casting is usually needed to ensure good filling. Dental and jewelry casting use centrifugal casting to fill thin sections and fine detail.
- *Countergravity casting* assists in filling thin sections, by applying a differential pressure between molten metal and the mold. This technique developed for over 30 years, works effectively in air or under vacuum, for air melted and vacuum melted alloys, to produce castings in aluminum and nonferrous alloys, many types of steels and superalloys, in weights from a few grams to 20 kg (44 lb) [15].

- *Countergravity Low-Pressure Air (CLA) Process* has the preheated shell mold, with an extended sprue, placed in a chamber above the melt surface of an air melted alloy, as described in Fig. 3. The sprue is lowered to below the melt surface, vacuum applied to mold chamber to cause controlled filling of the mold. The vacuum is released after castings and in-gates solidify, causing molten metal in the central sprue to return to the melt crucible, for use in the next cycle. Besides substantial savings in alloy usage and improved gating efficiency, the other benefits from the process include improved casting quality with reduced dross and slag inclusions.

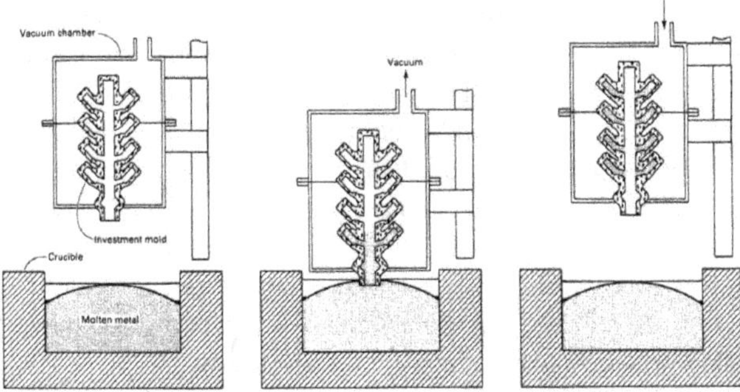

Figure 3. Schematic of the operations in the CLA process. (a) Preheated investment shell mold placed in the casting chamber. (b) Mold is lowered to the filling position, and vacuum is applied. (c) After casting and in-gates have solidified, vacuum is released, and all of the central sprue flows back into the melt.

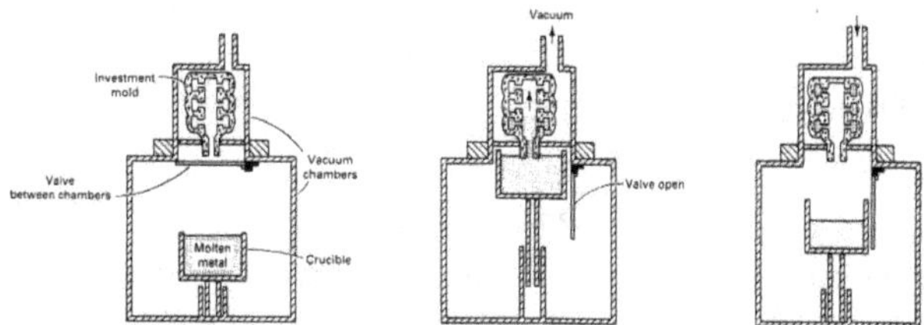

Figure 4. Schematic showing steps in the CLV process. (a) Metal is melted in vacuum, hot mold introduced into upper chamber, which is then evacuated. (b) Both chambers are flooded with argon, valve between the chambers is opened, and fill pipe enters the melt. Vacuum is the applied to the upper chamber to draw metal upward. (c) The vacuum is released after the parts are solidified, and the remaining molten metal in the gating system returns to the crucible.

- *Countergravity Low-Pressure Vacuum (CLV)* Process is similar to CLA process, and as described in Fig. 4, has the crucible in a vacuum chamber for vacuum melted alloys such as in nickel-base and cobalt-base superalloys.

10. Postcasting Operations

Post casting operations represent a significant portion, often 40 to 60%, of the cost of producing investment castings. A standard shop routing is provided for each part, and large savings can be realized by specifying the most cost-efficient routing. For example, it is often cost-effective to scrap early to avoid wasting finishing time, even if this means including an extra inspection operation. Alternative methods may be available for performing the same operation, and the most efficient one should be selected. The actual sequence in which operations are performed can be important. Some specifications require verification of alloy type, and this is done before parts are removed from the cluster.

a. *Knockout.* Some shell material may spall off during cooling, but a good portion usually remains on the casting and is knocked off with a vibrating pneumatic hammer or by hand. Brittle alloys require special attention. Part of the prime coat sometimes remains adhered to the casting surface, and the bulk shell material may remain lodged in pockets or between parts. This is removed in a separate operation, usually shotblasting. Clusters are hung on a spinner hanger inside a blasting cabinet, or are placed on a blasting table. If cores are to be removed in a molten caustic bath, the entire cluster can be hung in the bath, and the remaining refractory can be removed along with the cores. High-pressure water (6 to 10 ksi) is sometimes used instead of mechanical knock out, especially for aluminum and other nonferrous alloy parts.

b. *Cutoff.* Aluminum, magnesium, and some copper alloys are cut off with band saws. Other copper alloys, steel, ductile iron, and superalloys are cut off with abrasive wheels operating at about 3500 rpm. Torch cutting is sometimes used for gates that are inaccessible to the cutting wheel. Some brittle alloys can be readily tapped off with a mallet. Some steel and ductile iron parts can be cutoff after soaking parts in frozen liquid nitrogen (-320 °F). Gates are to be properly notched for these two cutoff techniques. Shear dies have also been used to remove castings from standardized clusters. Following cutoff, gate stubs are ground flush and smooth using abrasive wheels, and small hand grinders equipped with mounted stones.

c. *Core Removal.* Cores can be removed by abrasive or water blasting. If blasting can not be used, the cores can be dissolved out, using molten caustic bath (sodium hydroxide) at 900 to 1000 °F, or a boiling solution of 20 to 30% sodium hydroxide or potassium hydroxide in an open pot, or in high-pressure autoclave.

d. *Heat Treatment.* Air or vacuum heat treatments are performed extensively as needed to meet property requirements. Before they are heat treated, single-crystal castings must be handled very carefully to avoid recrystallization during subsequent heat treatment.

e. *Abrasive Cleaning.* Blast cleaning is used to remove scale resulting from core removal or heat treatment, using pneumatic and centrifugal blasting machines. Steel or iron grit or shot, and silica, or alumina sand are commonly used.

f. *Other Postcasting operations*. Hot isostatic pressing (HIP) is being increasingly adopted to eliminate porosity, and to improve properties especially for titanium, and used selectively for steel, aluminum, and superalloys.

Machining is often performed on investment castings, normally confined to selected areas requiring closer dimensions. Broaching, coining and abrasive grinding are used to improve dimensional accuracy. Straightening of investment castings, heated usually, is performed when required, either manually or using hydraulic presses, with suitable fixtures.

Chemical finishing treatments are also used, in applications such as: acid pickling to remove scale, passivation treatment for stainless steel, chemical milling to remove α-case on titanium, and chemical treatment to apply a satin finish to aluminum or to polish stainless steel.

11. Inspection and Testing

(a) *Alloy Type Test*. This test is often an alloy verification test, and is conducted on the tree or cluster before cutoff. A spectrometer or X-ray analyzer is used to verify that the correct alloy has in fact been poured.

(b) *Visual inspection*. An early visual inspection is essential so that obvious scrap does not get passed on to expensive finishing or inspection operations. Some commercial parts require only visual inspection.

(c) *Fluorescent Penetrant Inspection* is extensively used for nonmagnetic alloys. It can also be used for magnetic alloys, but a magnetic penetrant is generally specified instead. This test detects defects on, or open to the surface, such as porosity, shrinkage, cold shuts, inclusions, dross, and cracks of any origin (hot tears, knockout, grinding, heat treat, straightening). In this technique, the surface is cleaned and a liquid penetrant with a low surface tension and low viscosity is applied, and is drawn into the defects by capillary action. Excess liquid is wiped away, a developer is applied that functions as a blotter to draw the liquid out, and the area is examined visually in a dark enclosure under black (ultraviolet) light, which reveals defects that cannot be detected visually.

(d) *Magnetic Particle Inspection* detects similar defects, but is preferred for use on ferromagnetic alloys. The method involves surface preparation, magnetization of the casting, and application of either a liquid suspension of magnetic particles (wet method) or fine magnetic particles (dry method). The presence of defects causes a leakage field that attracts the magnetic particles and causes them to cling to the defect and define its outline. Colored particles and fluorescent particles (for viewing under black light) are available and can be used as required.

(e) *X-ray Radiography* is used to detect such internal defects as shrinkage, gas porosity, dross, inclusions, broken cores, and core shift. Many parts receive 100% inspection, while others are inspected according to a specified sampling plan. Even when not specified, X-ray inspection is used as a foundry control tool to monitor process reliability, to establish a satisfactory gating and feeding system, and to troubleshoot foundry problems. It is sometimes used to examine wax patterns containing delicate ceramic cores, to ensure that the cores were not broken during the pattern injection operation.

(f) *Miscellaneous Inspection Methods*:

Hardness testing is widely used to verify response of castings to heat treatment. Chemical analysis is generally controlled through the use of certified master heats. Mechanical properties are determined on separately cast test bars or test specimens mounted on production clusters. Specimens machined from castings are used for process development and periodic audits.

Grain size is regularly checked on many equiaxed castings, following chemical or electrolytic etching, and often as a part of process development effort. Electrolytic etching is also used for examining and detecting grain defects in directionally solidified and single crystal castings.

The orientation of single crystal castings is determined by *Laue back- reflection x-ray diffraction.*

Pressure tightness tests on investment castings are conducted for a variety of engineering and other applications.

Dimensional inspection ranges from manual checks with a micrometer or simple go/no-go gages, to the use of coordinate measuring machines (CMM) and automatic three dimensional inspection stations capable of checking a sculptured surface in one continuous sweep. *Wall thickness* on many cored turbine blade castings is determined ultrasonically. Nodularity is checked metallographically from the first and the last tree or cluster poured from heats of ductile iron. *Metallography* is an essential part of process development for high performance castings.

12. Design Advantages of Investment Castings

Designing for investment castings primarily aims to make full use of the enormous capability and flexibility inherent in the process to produce parts that are truly functional, cost effective and often more aesthetically pleasing. The principal advantages are described below.

a. *Casting Complexity.* Almost any degree of external complexity as well as internal complexity (which is limited only by the state of the art in ceramic core manufacturing), can be achieved. As a result, numerous parts previously manufactured by assembling many separate, individual components are currently being made as integral castings, at much lower costs and often with improved functionality.

b. *Wide Alloy Selection.* Any castable alloy can be used, including ones that are impossible to forge, or are too difficult to machine.

c. *Close Dimensional Tolerances.* The absence of parting lines and the elimination of substantial amounts of machining by producing parts very close to final size give investment casting an enormous advantage over sand casting and conventional forging.

d. *Prototype Tooling.* The availability of temporary and prototype tooling is a major advantage in the design and evaluation of parts. Quick and low expense tooling methods such as SLS, SLA available in rapid prototype production, described earlier, facilitate timely collaboration between designer and the foundry to produce parts that are functional and manufacturable. This capability is not found in such competitive processes as die casting, powder metallurgy or forging.

e. *Reliability.* The long-standing use of investment castings in the aircraft engines for the most demanding applications has fully demonstrated their ability to be manufactured to the highest standards.

f. *Wide Range of Applications. In* addition to complex parts and parts that meet the most severe requirements, investment casting also produces numerous simple parts competitively, due to the low tooling costs involved. Investment castings are competitively produced in sizes ranging from a few grams to more than 300 kg (660 lb).

13. Design Recommendations

The following recommendations provide a guide to the design of investment castings [16]:

- Focus on final component cost rather than cost of the casting.
- Design parts to eliminate unnecessary hot spots through changes in section sizes, use of uniform sections, location of intersections, and judicious use of fillets, radii, and ribs.
- Use prototype castings to resolve questions of functionality, producibility, and cost.
- Do not overspecify; permit broader than usual tolerances wherever possible.
- Indicate datum planes and tooling points on drawings; follow ANSI Y 14.5M for dimensioning and tolerancing.

14. Applications

Applications for investment castings exist in numerous manufacturing industries. A partial list is given in Table 2. The largest applications are in the aircraft and aerospace industries, especially turbine blades and vanes cast in cobalt- and nickel-base superalloys, as well as structural components cast in superalloys, titanium, and 17-4 PH stainless steel. Examples of applications with castings in ceramic shells using wax patterns, ceramic and soluble cores, are shown in Fig. 5 to Fig. 7.

Table 2. A partial list of applications of Investment Casting	
Aircraft engines, air frames, fuel systems	Machine tools
Aerospace, missiles, ground support systems	Materials handling equipment
Agricultural equipment	Metalworking equipment
Automotive	Oil well drilling and auxiliary equipment
Bailing and strapping equipment	Optical equipment
Bicycles and motorcycles	Packaging equipment
Cameras	Pneumatic and hydraulic systems
Computers and data processing	Prosthetic appliances
Communications	Pumps
Construction equipment	Sports gear and recreational equipment
Dentistry and dental tools	Stationary turbines
Electrical equipment	Textile equipment
Electronics, radar	Transportation, diesel engines
Guns and small armaments	Valves
Hand tools	Wire processing equipment
Jewelry	

Table 2. A partial list of applications of Investment Casting

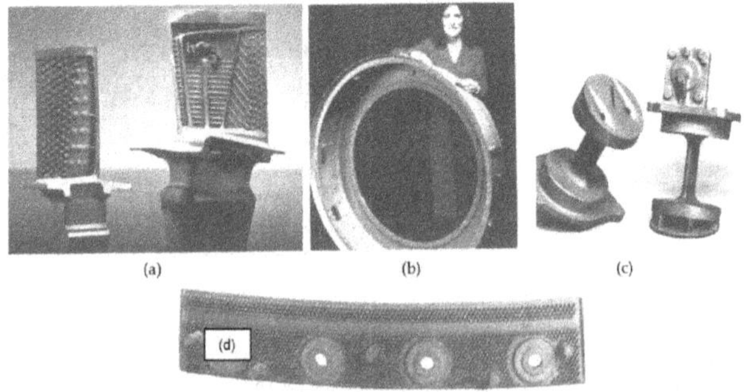

Figure 5. Aircraft and aerospace applications: (a) Single crystal turbine blades investment cast using complex ceramic cores. (b) 17-4 PH stainless steel fan exit case; weight: 96 kg (212 lb). (c) Aircraft fuel sensor strut cast in 17-4 PH stain-

less steel, and (d) Aircraft combustion chamber floatwall, with numerous small posts on the wall. Courtesy of: Pratt Whitney Aircraft [7].

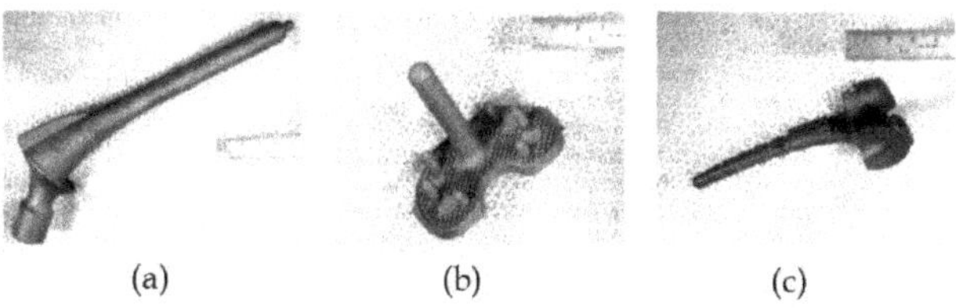

Figure 6. Biomedical applications for investment castings. (a) Hip-femoral prosthesis. (b) Knee-tibial base. (c) Elbow-humeral prosthesis. All cast in ASTM F75 cobalt-chromium-molybdenum alloy. [7].

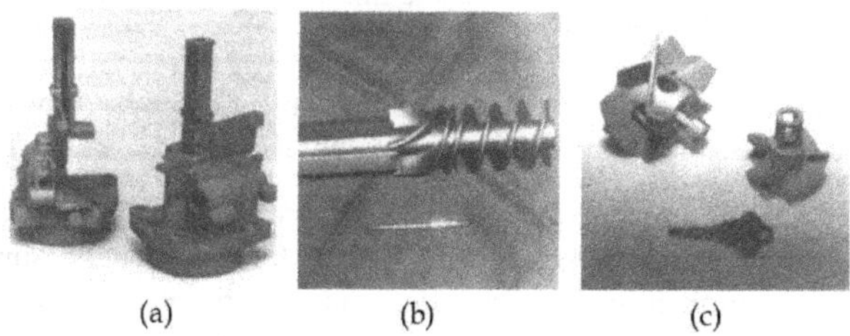

Figure 7. Examples of other applications: (a) Nosepiece for nailgun cast in 8620 alloy steel. (b) Ni-resist Type II cast iron inducer for deep oil drilling. (c) Small 17-4 PH turbine vanes; small vane weighs 71 g (2.5 oz), larger vane 185 g (6.5 oz). [7].

Appendix to: Progress in Investment Castings

Guidelines for Designing Gating and Feeding Systems

1. Introduction

In any casting process, when an alloy is poured into a mold, it starts to shrink or contract in volume as it cools down to liquidus temperature, and subsequently solidifies. Foundries compensate for these two stages of volumetric contractions by providing reservoirs – feeders - as parts of the mold cavity design. (The term "feeder" used here has the same meaning as "riser" used in sand- casting). The feeders freeze slower than the casting, thus allowing some of the liquid in the feeders to flow into the casting toward the areas where the shrinkage occurs.

Note: The third stage of volumetric alloy contraction or shrinkage, from solidus temperature to room temperature, generally called *'patternmaker's shrinkage'*, is compensated for in the design of mold cavity and included in pattern tooling.

The feeding behavior of selected investment cast alloys was studied in a research carried out at University of Birmingham, UK, funded by Rolls Royce Limited, UK. [1-4]. The solution for a complete feeding system design includes the dimension, shape, location, method of attachment of feeders to the casting and the materials used. Similar approaches to what are used for sand casting can also be applied to investment casting, as was established in the above study.

However, unlike most sand casting applications, in investment casting, many gating and feeding elements perform dual or combined functions / roles: many elements adjacent to the cast component, initially assist or govern the mold filling operation, guide the molten metal as it flows in to the component cavity, and later during solidification, guide the flow of feeding liquid into this cavity to compensate for the first two stages of alloy contraction described earlier.

Additionally, as is the case with many small size investment castings, some of these gating-feeding elements are required to provide portions of feed liquid to many parts joined together to these elements, specifically for such a purpose.

In general, variations in the design of feeding systems for investment casting components arise because of differences in melting, molding and pouring methods used in foundries. Nevertheless, some general guidelines in procedure for feeding system design can be followed.

An approach is presented here for combining the feeder dimensioning concepts based on Chvorinov's rules, [9] with the experimental, theoretical and practical data available on feeding systems used for investment castings. This approach has been found effective in actual usage at many investment casting foundries, and with a wide range of cast alloys. To show how the procedure is used in practical applications, two examples of investment castings are included later in this appendix.

The feeding concepts for investment castings generated from this study have also been expanded to a PC-based computer program, *The AFS Investment Casting Feeding System*, distributed earlier by American Foundry Society [AFS].

2. Effect of Process Variables on Feeding:

A summary of findings from the above study, primarily on the effects of process variables on feeding investment castings, (some of which show similarities, while some show marked differences with sand castings), are described here. The feeding concepts developed from these findings, as well as rules for dimensioning feeding elements and the procedure for feeding system design are described in subsequent sections: 3.0, 4.0 and 5.0.

2.1. Casting Geometry: Length, Width, Thickness, Taper, Cross-section Shape

(a) Length

i. Casting soundness improves, invariably, with reduced casting length, due to the increase in longitudinal temperature gradients.

ii. Feeding range or feeding distance, f_d, of a single feeder was generally found to be (10 x t), where t is the casting section thickness, under normal casting conditions.

iii. Feeding distance was found to increase by varying levels, with specific changes in process conditions, causing measured increases in longitudinal temperature gradients, such as with increased casting taper, freezing ratio, or with applied mold temperature gradients.

(b) Width

Increasing casting width, up to a certain level, improves casting soundness, which was found to be due to the availability of wider channels for the feeding liquid.

(c) Thickness

Thicker castings are easier to feed and have improved soundness owing to wider spacings between dendrites, both primary and secondary, which allow correspondingly wider feeding flow channels.

(d) Taper

An increase in taper for both 'plate' (rectangular cross-section) and 'blade' (nearly trapezoidal cross-section) geometries used in this study, substantially improves casting soundness, due to improvement in longitudinal temperature gradients.

(e) Cross-section Shape

'Plate' shapes with rectangular cross-section are easier to feed than 'blade' shapes, with identical overall length, width, and thickness, for similar conditions of casting, namely, casting and mold temperatures, feeding system and pressure head on the feeder. Wider feed flow channels available with 'plate' primarily cause this improvement.

2.2. Freezing Ratio, k

Freezing ratio, k, defined as the ratio of modulus of feeder to that of the casting (M_f / M_c), is described more in section 4.0.

(a) Increasing the freezing ratio, k, in the range from 1.0 to 2.3, consistently improves casting soundness of 'plate' and 'blade' specimens. Increased longitudinal gradients, with increased primary and secondary dendrite arm spacings, cause wider feed flow channels, all of which promote strong directional and progressive solidification towards the feeder.

(b) The minimum value of freezing ratio, k, and volume ratio, V_f / V_c required to achieve radiographic soundness were obtained as follows:

Casting geometry	k	V_f / V_c
• 4 mm thick 'plates'	1.7	0.8
• 4 mm thick 'blades'	2.0	0.8
• 8 mm thick 'plates'	1.2	0.4
• 8 mm thick 'blades'	1.7	0.6
• Aerofoil blade	1.76	0.8

(c) The minimum limits of freezing ratio, k, and volume ratio, V_f / V_c can however be reduced appreciably by: the application of casting taper, or longitudinal gradients on the shell, or an increase in pouring temperature (from 1430 ºC to 1500 ºC for IN 100), or a suitable change in the gating design.

2.3. Mold Temperature and Applied Temperature Gradients

i. Uniform mold temperature allows shallow temperature gradients to develop and persist, especially with increased casting length, causing increased shrinkage porosity.

ii. Applying longitudinal temperature gradients to the mold substantially improves casting soundness.

iii. To obtain improved mold filling and avoid casting misrun, it is advantageous to adjust (increase) mold temperature, while maintaining pour temperature, dependent on the casting geometry and the gating design.

2.4. Pouring Temperature

i. Increase in pouring temperature (in the specified range of 1430 °C to 1500 °C for nickel base alloy IN 100) causes considerable improvement in casting soundness, for studied casting thicknesses.

ii. However, as is widely acknowledged, lower pour temperature generally gives improved soundness in massive or thick casting sections, due to improved temperature gradients sustained during solidification; increased pour temperature creates shallow gradients in later stages of solidification in these thick sections, enhancing shrinkage pores.

iii. On the other hand, with extended, thin sections, increased pour temperature, besides allowing easy mold fill (preventing any misrun), provides widened interdendritic channels, with adequate gradients for effective feeding.

2.5. Pressure: Metallostatic pressure, Atmospheric or Ambient Pressure

(a) Metallostatic Pressure

i. An increase (of 60 mm) in metallostatic head, h, of 'plates' and 'blades' makes a substantial improvement in casting soundness, since it is a major propulsive force for feed liquid.

ii. An increase of about 9% in metallostatic pressure (defined as $P_h = \varrho_L \times g \times h$) in nickel base alloy MAR M002, contributes partly to its improved feeding behavior compared to IN 100 (for the same casting conditions), because of the increased density in the liquid state, ϱ_L.

iii. The availability of metallostatic pressure prevailing at the feeding source can become restricted, or curtailed, due to premature lateral solidification, locally, at thinner sections in the running system of many bottom gated designs. This can result in macro-cavities or extensively interconnected micro-pores.

iv. For casting under vacuum, metallostatic pressure is the main force propelling feed liquid to the growing solid-liquid interface both by macro-flow and by micro-flow.

(b) Atmospheric or Ambient Pressure

i. The application of atmospheric pressure on top of the pouring basin or cup creates a substantial increase in the pressure head acting on the feed liquid.

ii. However, a continuous liquid column from the region where atmospheric or any ambient pressure is applied, to the main source of feed supply, namely the feeder, is essential for adequate and continuous feeding to occur.

iii. In all applications of pressure, the moduli of the pouring and of the running system are required to be larger than that of the feeder to prevent any premature lateral solidification cutting off the transmission of the applied pressure to the feeding liquid.

2.6. Dissolved Gas

i. Increased levels of dissolved gases (for e.g., dissolved oxygen and nitrogen in excess of 15 ppm in nickel base alloy IN 100 bar stock) are found to substantially augment shrinkage pores, resulting in macro-cavities in many thicker sections in the casting.

ii. Pores in fully- fed* casting sections (*as determined from separate experiments) are uniquely created by dissolved gases, and can be clearly identified by metallography due to their rounded shape, or the spherical surfaces.

iii. Pores caused by dissolved gas can be reduced partially or locally in some casting sections by making changes to the gating design; however, separate degassing techniques are required to fully eliminate gas pores from the casting.

2.7. Gating and Feeding System

i. Top- gated and side-gated feeding systems show consistently improved feeding efficiency as compared to bottom gated designs.

ii. Top- gated and side-gated feeding systems show increased levels of longitudinal temperature gradient which promotes directional freezing, and hence assist effective feeding.

iii. Top- gated and side-gated feeding systems show continued availability of metallostatic pressure on the feeder during solidification in all sections of the casting, as against the possible cut-off mechanism due to premature lateral solidification, in the bottom- gated design, which interrupts feed flow by curtailing metallostatic pressure from the feeder.

2.8. Alloy : Alloy Constitution, Structural Parameters

The feeding characteristics of the following three groups of metals and alloys are classified and identified, based on their solidification behavior, namely:

i. alloys freezing with marked 'skin' formation, normally short freezing range alloys;

ii. alloys showing 'pasty' or 'mushy' zone during freezing, normally longer freezing range alloys*;

iii. grey cast irons, which show little shrinkage during solidification, since the formation of graphite flakes involves an expansion, and requires rigid molds to inhibit casting expansion.

*NOTE: Some alloys, generally with the intermediate freezing range, show 'mushy-skin' type of solidification. These alloys are found sensitive to rate of freezing: an alloy showing mushy-skin freezing in a sand casting may behave like a skin forming alloy when chill cast.

Freezing range of an alloy is defined as the difference between liquidus and the solidus temperature.

Alloy Constitution

(a) Specimens cast in nickel base alloy MAR M002 have consistently improved soundness compared to identical specimens cast in IN 100. The wider freezing temperature range of MAR M002, as compared to that of IN 100, is observed to have negligible detrimental effect on the process of feeding in MAR M002.

(b) The improved feeding behavior of MAR M002 compared to IN 100 is attributed to the following six factors:

i. Consistently larger induced temperature gradients in MAR M002 specimens, compared to identical specimens in IN 100.

ii. A progressive decrease in the rate of solidification from specimen ends to feeder, enabling the ease in macro- and micro –flow of feed metal, similar to IN 100 specimens cast in identical conditions.

iii. Higher γ-γ' eutectic content, of the order of 10-20% in MAR M002, than that in identical specimens in IN 100 with 1.0- 4.0% eutectic, contribute to the feed supply at the later stages of solidification.

iv. The chain-like morphology of eutectic in MAR M002 suggests a more favorable morphology of feed channels for both macro-flow and micro-flow, than the kidney shaped eutectic observed in IN 100. This is consistently substantiated by the similarity in morphology of micropores, and γ-γ' eutectic grains in MAR M002, as well as in IN 100 alloy. In addition, the micropores are observed to occur frequently adjacent to the eutectic nodules in both the alloys.

v. Variations in the thermal properties of the two alloys, mainly specific heat and thermal conductivity in the liquid and the constituent solid phases, may also alter the morphology of feed channels during solidification and hence feeding characteristics of these alloys, and

vi. An increase of about 9% in metallostatic pressure in MAR M002, for the same metallostatic head as available with IN 100 specimens, because of the increase in density of MAR M002 liquid.

Structural Parameters

(a) Matrix Morphology and Grain Size

i. Chill grains observed at surfaces of IN 100 specimens, especially at a lower pour temperature, do not influence the feeding process, i.e., have neither beneficial nor detrimental effect on feeding.

ii. The morphology of columnar γ – grains in both IN 100 and MAR M002, clearly indicate the relative levels of heat abstraction from solidifying micro-regions through different parts of the section, the possible directions of growth and identifies the 'hot spot' region, especially in a complex geometry such as a 'blade' section, the thermal center for the section and hence freezes last. An examination of microstructure thus leads to an explanation for the morphology and distribution of shrinkage pores, resulting from inadequate feeding.

iii. Measured width of columnar grains is directly related to primary dendrite arm spacing. While a wider grain leaves wider feed passages for micro-flow, a steep linear increase in arm spacing from specimen end to the feeder, which is inherent with steep longitudinal gradients, allows optimum size of channels for macro-flow.

(b) Dendrite Arm Spacing

i. Measured values of secondary dendrite arm spacing can be directly related to rate and time of local solidification in sections at various distances from feeder, to indicate the progress of solidification as well as the morphology of feed channels available to feeding as affected by variations in casting parameters.

ii. For the various casting parameters studied, including specimens with out taper (with constant section thickness), cast at uniform mold temperature (zero applied gradient), and at the lower pour temperature, the dendrite arm spacing increases from specimen end to the feeder. However, any variation from linearity in this increase leads to simultaneous lateral solidification over longer regions of the casting, and cut-off mechanisms operating on macro- and micro-flow, resulting in shrinkage porosity.

iii. Application of taper on casting section thickness gives a steeper decrease in dendrite arm spacing with distance from feeder. An increase in thickness, or pouring temperature, or freezing ratio, or the application of gradients on the shell results, consistently, in wider arm spacing towards the feeder, and hence provide wider feed channels.

3. Feeding Concept for Investment Castings

Primarily based on the above findings on effect of process variables on feeding, the following components of the general concept have been developed on the feeding process in investment castings.

3.1. Feeding Conditions

Conditions, as specified by rules on 'thermal and volumetric criteria' (described in section 4.0) for dimensioning the feeders, followed by appropriate spacing of feeders to meet the 'feeding range demand' in complex castings, are sufficient to ensure radiographic soundness in skin freezing alloys [1]. However, in mushy-skin and mushy freezing alloys, the following additional conditions are required to be met in the design of feeding systems, particularly to obtain HDR, High Definition Radiography-soundness, Fig.11, in investment casting of long and thin components similar to airfoil blades:

a. a strongly directional and progressive freezing from the end/s of the casting towards the feeder/s. This directional progress of solidification is governed mainly by three factors, namely: (i) thermal gradients in both longitudinal and transverse directions, (ii) the local rate of lateral solidification, and (iii) the local time of lateral solidification;

b. an adequate pressure head on the liquid in the main source of liquid-feed supply, namely the feeder/s, to sustain propulsion of feed liquid to the growing solid-liquid interface/s.

3.2. Feeding High Temperature Alloys and Other Investment Cast Alloys

(a) The metallographic examination of specimens, cast under different process parameters, lead to the conclusion that two of the alloys studied, IN 100 and MAR M002 can be classed with 'mushy-skin' freezing, and identified with intermediate freezing range alloys.

(b) From the observed morphology of the matrix and interdendritic spaces or feeding channels in these two alloys, the mechanisms operative in feeding can be grouped into: 'macro-feeding' and 'micro-feeding'. Macro-feeding through longitudinal channels is the main source of supply for lateral micro-flow, which feeds the solid-liquid interface through interdendritic channels. The feeder is the main source of supply for macro-feeding, which in turn, terminates in micro-feeding.

(c) While, in mushy freezing alloys, liquid feeding and mass feeding mechanisms operate during macro-feeding, in skin freezing alloys, liquid feeding is the main source for macro-flow. Interdendritic feeding is the mechanism operating in micro-flow [1-2].

(d) The driving force propelling both the macro-flow and the micro-flow is the pressure on the liquid in the feeder, mainly metallostatic and ambient pressure, and the negative contraction pressures developed with the growth of the solid-liquid interface.

(e) The major resistance to the flow of feed liquid, by both macro-flow and micro-flow is the viscosity of the liquid which increases as solidification progresses locally, the flow channels get narrowed, the liquid temperature drops and intermediate phases are precipitated in the liquid.

(f) Continuity in the supply of feed liquid by macro-flow can be cut-off by premature lateral solidification upstream, mainly due to shallow longitudinal gradients. Macro-flow mechanism through spaces between dendrite branches can be interrupted by growth cut off in flow direction caused by shallow longitudinal and transverse temperature gradients.

3.3. *Temperature Gradients*

(a) Shallow longitudinal and transverse gradients lead to non-directional freezing, while steep gradients in the requisite directions, result in directional solidification. The flow distances required both in macro-flow and micro-flow get shortened in direct proportion to the increase in these gradients.

(b) Primarily poor thermal conductivity and high level of solidification temperatures, allow such alloys as steels, nickel base and cobalt base alloys generally to have steep temperature gradients. In addition, large chilling capacity of zircon normally used in the ceramic shell for these alloys, also contributes appreciably to the magnitude of the gradients. However, sustaining beneficial gradients through solidification for effective feeding requires controlling and monitoring other process factors identified in this study.

3.3.1. *Longitudinal Gradients*

(a) Longitudinal gradients are composed of three distinct components in the long plates and blades samples studied, namely: at the casting ends, at mid-sections, and near the feeder. While, the mid-component of the gradient is found to be shallow with non-directional freezing, it is made steep by strongly directional solidification.

(b) Application of longitudinal gradient on the shell mold, or a taper in the casting thickness, contributes selectively to one gradient component in the casting, mainly to increase the magnitude of the mid-section component.

(c) The following seven factors have been found to contribute, in varying levels, to some, or all three components, described above, in the longitudinal temperature gradient: (i) the casting 'end effect', (ii) loss of melt temperature as liquid metal fills in the mold, (iii) convection in the bulk liquid, (iv) casting geometry, (v) feeder contribution, (vi) variations in the chilling capacity of the mold, and (vii) gating method / design.

3.3.2. Transverse Gradients

As determined in the above investigation, transverse temperature gradients are mainly dependent on the following factors: (i) the difference between pouring temperature and the mold temperature, (ii) the thermal conductivity of the melt, and (iii) the mold chilling capacity [1-2].

3.4. Rate and Time of Solidification

(a) Spacings between dendrite arms, d, both primary stems and secondary branches are related to the rate and time of local solidification, and hence, indicate the progress of lateral solidification with distance (from feeder) and time (from start of freezing). Measured 'd'-values illustrate the advance or growth of solid-liquid interface and the consequent morphology of feed channels available for macro-flow and micro-flow.

(b) Wide spacings between primary and secondary dendrites promote an appreciable improvement in the efficiency of both macro-feeding and micro-feeding.

(c) For optimum conditions of feeding, dendrite arm spacing increases linearly from casting / specimen end to the feeder, with as wide spacing as compatible, with the local cooling rate, which is mainly dependent on: casting and shell mold geometry, casting and mold temperatures, and heat transfer properties of alloy melt and mold.

3.5. Feeding Mechanisms in High Temperature Alloys and Other Cast Alloys

(a) Liquid feeding mechanism is the main source of supply for macro-flow which, in turn, feeds numerous micro-flow paths. A steep thermal gradient, both in longitudinal and transverse directions, is mainly responsible for the continuity in the supply of liquid feed.

(b) Observations of cast structure of IN 100 and MAR M002 suggest that primary γ- constituent does not participate in the mass flow mechanism of feeding. However, residual liquid streams may carry primary carbides or other intermediate phases formed near the liquidus temperature, early in the solidification. This could be to the detriment of end-feeding in the micro casting regions where the carried primary constituents may block or interrupt liquid flow into interdendritic spaces.

(c) Interdendritic feeding occurs by micro-flow in spaces between primary and sequential dendrite arms to compensate the shrinkage due to the growth of dendrites and interdendritic regions, continuously within the mushy zone, under optimum conditions of thermal gradient and freezing rate.

(d) A solid skin on castings forms at early stages of solidification, restricting the mechanism of solid feeding in all the casting geometries studied, namely: 'plates' (rectangular cross-section),' blades' (nearly trapezoidal cross-section) and airfoils.

4. Dimensioning Feeding Elements

The optimum size of a feeder is obviously that which will ensure the required soundness of the casting, with an adequate safety factor. For skin-freezing alloys the minimum feeder size to eliminate macroshrinkage is also adequate for satisfactory feeding. Larger feeder sizes do no harm apart from the possible danger of the feeder pipe extending into the casting, and being uneconomical.

Mushy freezing alloys, on the other hand, are more sensitive to the temperature gradients produced by the feeder. Feeder volume below the optimum size results in macropores as well as in distributed microporosity. Larger feeder heads are self-defeating since they delay freezing, diminish the temperature gradients and increase the amount of interdendritic porosity near the feeder.

4.1. Modulus Method

Based on the original concepts by Chvorinov and subsequently elaborated by Wlodawer [9-10], rules have been formulated to determine the adequate feeder size for castings, Fig.8 to Fig.11.

Modulus is defined as the ratio of the volume to the cooling surface area of the casting (or a part of the casting) or the feeder.

$$M = V / CS \tag{1}$$

Where, V is the volume of casting, or feeder, and CS is the cooling surface area.

Modulus, also alternatively termed 'cooling modulus', or 'freezing modulus', relates amount of heat stored in casting or feeder (the volume component) to the rate of heat dissipation (via the cooling surface area). Many experimental observations have found the average solidification time in castings of numerous alloys, to be proportional to the square of the modulus.

In the modulus method, two rules have been postulated to calculate the adequate feeder size:

(a)The first rule ensures that the feeder is 'thermally adequate', that it freezes adequately later than all the casting sections being fed:

$$M_f = k \times M_c \tag{2}$$

where, k is identified as the freezing ratio, and M_f and M_c are moduli of feeder and casting respectively. The freezing ratio 'k' is initially set at a minimum of 1.2, for both short and long freezing range alloys.

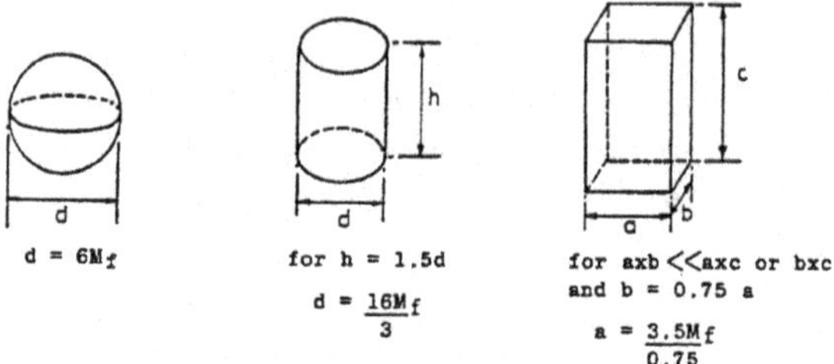

Figure 8. Feeder dimensions calculated from modulus value.

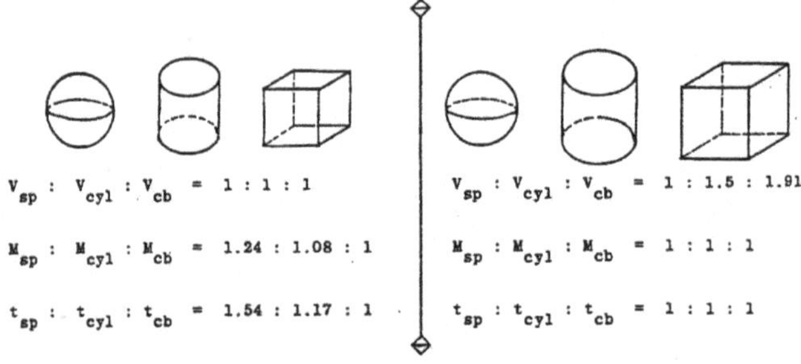

Figure 9. a) Comparing moduli and solidification times of a sphere, cylinder and cube of the same volume; (b) comparing volume of these shapes for the same solidification time.

Figure 10. Examples of casting designs requiring separate feeding of sections with different moduli, or to meet feeding distance requirements.

(b) The second rule ensures that the feeder is 'volumetrically adequate', that feeder has adequate volume of feed liquid to supply to casting sections being fed:

$$V_f = V_c x \{\beta / (\eta - \beta)\} \quad (3)$$

where, V_f and V_c are the volumes of feeder and the casting sections being fed, β is alloy solidification shrinkage, and η is the available feeding liquid factor, relating to the feeder efficiency and is empirically determined.

Note: (i) β is composed of the first two stages of alloy contraction, namely, $\beta = \beta_L + \beta_F$, where β_L, the liquid contraction volume, is of the order of 1% of the total liquid volume, per 180 °F (100 °C), and β_F is the alloy freezing contraction, varies from 0 to 6.5%, and (ii) η is found empirically to be about 18% under normal casting conditions; and can be substantially increased (thereby, correspondingly reducing required feeder size) with specific changes in process conditions made to enhance local temperature gradients, such as with casting taper, applied mold gradients, or selective mold ceramic insulation.

4.2. *Number of Feeders and Positioning of Feeders*

(a) Feeders are placed in direct contact with the heavier sections of the casting, as this enables directional solidification to be maintained through out freezing. While castings of simple shapes can be fed by a single feeder, in many castings with varying section thicknesses, the concept of feeder/s location divides itself into natural zones of feeding, each feeder centered on a heavy section, separated from the others by a thinner section, Fig.8 to Fig.10.

Each zone is fed by a separate feeder and hence the number of such zones determines the number of feeders required.

(b) In the case of rangy castings (mainly with long, thin sections), the 'feeding range' or 'feeding distance' of the feeder, rather than the constriction (or thin sections between thick ones), is the factor limiting the operation of each feeder. This is due to the fact that appreciable and sustained temperature gradients are required to feed extended thin sections.

(c) To achieve maximum metallostatic pressure, feeders should be positioned as high as possible relative to the casting. This is particularly important in mushy freezing alloys, since metallostatic head is often the only effective source of feed pressure; owing to the absence of an intact skin, the casting sections as well as the feeder remain accessible to atmospheric pressure. In skin forming alloys, atmospheric pressure is an important factor in feeding.

4.3. *Attachment of Feeders to Castings*

(a) Partly because of the economic importance of reducing costs in removal of feeders from castings and partly to control temperature gradients, feeders are connected to castings by 'feeder necks' with reduced cross sections.

(b) For satisfactory completion of feeding, the neck should solidify after the casting, but before the final solidification of the feeder. Should neck freeze prematurely, the casting will contain shrinkage, no matter how large the feeder itself is.

(c) The necessary minimum neck dimensions vary with the casting design and method, but as a first approximation the neck requires a cross sectional area equal to, or greater than, the section it is designed to feed.

(d) Based on the modulus concept, modulus of feeder neck, M_n, is determined from:

$$M_n = k_n x\, M_c \quad (4)$$

where k_n is the freezing ratio of feeder neck. It is recommended that initially k_n,is set at 1.1, and the neck shape and dimensions are such that the moduli of casting, neck and the feeder are in the ratio (1: 1.1: 1.2).

5. Procedure for Feeding System Design for Investment Castings

As mentioned earlier, a complete feeding system design involves, to varying levels, each of the different stages of the investment casting process. A step- by- step procedure is described here, and includes designing the optimum feeder system based on the modulus method, after modifying all other process factors towards a common goal of superior casting quality.

{References and links in these instructions are made to the serial number attached to each step, such as: 1, 2....16, 17}

Casting Design

Step 1. *Design Modifications:* Examine casting geometry to verify if freezing of casting sections provides directional cooling gradients towards the feeder(s). If needed, obtain customer's agreement on design modifications to improve temperature gradients for directional solidification, for attaching feeders and for quality control purposes.

Step 2. *Alloy Selection:* Alloy is normally selected by the designer from standard specifications and alloy handbooks. To meet process requirements, consultation between the designer and the foundryman is necessary. If several alloys are feasible, carry out a value analysis.

Step 3. *Feeding System:* Using the modulus method, identify from the casting design the number and location of feeders. Select an appropriate gating and feeding system from a standard grid (or cluster, or tree), modified grid, or a specially designed feeder system. Generally such a cluster or grid consists of a framework of vertical and horizontal members attached to a pouring basin (or cup) and patterns. Detailed adjustment to feeder(s) location is made later.

Step 4. *Cutoff & Cleaning:* Check on the suitability of proposed feeding system for available ways of cutting off the feeders.

Alloy Properties

Step 5. *Feeding Properties*: In consultation with designer, make necessary adjustments to the alloy composition (small additions of / or dilution in alloying elements) of the selected alloy to improve feeding characteristics, such as freezing range, eutectic volume and casting fluidity.

Step 6. *Melt Treatments:* Feeding behavior of many alloys can also be enhanced by influencing the freezing nucleation and growth behavior, by various melt controls. Check on the specified testing procedures for alloy melt quality controls: composition, grain size, microstructure, residual gases, inclusions, superheat times and temperatures, pouring temperatures, and for obtaining test bars for mechanical property tests.

Step 7. *Product Quality:* Final casting quality is influenced by mold phenomena such as mold atmosphere, flow of metal, cooling directions and rates due to mold materials. The production of porosity-free castings depends on the success of the whole feeding system. Check on the specified product quality control tests (types of procedures for nondestructive or destructive tests in relation to the casting design, the alloy used and product applications).

Mold Construction

Step 8. *Patternmaking and Assembly:* Examine the effects of gating and feeding system on making pattern dies and pattern assembly. Explore alternative methods of clustering patterns on the tree, or grid, (location and orientation) or in the mold to produce favorable feeding temperature gradients during mold preheating and alloy pouring, while ensuring adequate cluster rigidity. Remember, although each casting is freezing on its own, the whole grid acts as a single cooling system.

Step 9. *Gating:* Select the gating system to optimize flow for favorable feeding temperature gradients, which minimize flow turbulence and maximize feeding pressure. Explore the feasibility of horizontal, vertical and bottom gating. Review the feeding requirements in relation to pattern assembly while optimizing gating system functions. Design the gating and feeding systems by considering jointly both the flow and feeding requirements. For grid systems, identify parts of the grid that act at the same time as feeding and gating elements.

Step 10. *Mold Materials:* Examine whether standard mold, feeder wall materials, mold making and preheating techniques should be altered (insulation, chilling) to improve feeding temperature gradients.

Step 11. *Feeding Pressure:* The total feeding pressure acting on the feeding liquid flow is important in the last stages of freezing when flow channels become narrow and fine. Obtain the fullest possible benefit of atmospheric and metallostatic pressures on the feeding liquid, and reduce, separately, dissolved gases and inclusions in the alloy to a minimum.

Dimensioning Feeding Elements

Step 12. *Number of Feeders:* Establish from casting design, step (2) and steps (3), (6), (8), (9), (10), and (11), whether directional feeding of the whole casting could be obtained from one single feeder. Calculate freezing moduli (V/ CS) of section(s) that have to be fed separately. Exclude the calculation of all "parasite" sections (thinner casting sections-such as bosses, fed by thicker adjacent sections) attached to the feeding parts. Exclude noncooling surfaces (casting surface not facing the mold) in modulus calculation. For complex sections, apply the shape substitution principle.

Step 13. *Feeder(s) Shape and Locations:* Select feeder shape and location from steps (2), (4), (8), (9), (10), (11) and (12). For grid systems, verify that the elements of the gating system that take part in the feeding flow also obey the freezing modulus rules, in step (14).

Step 14. *Feeder Volume from Freezing Time:* Calculate the feeder volume from: $M_f = k \times M_{c'}$, where the freezing moduli of casting, $M_{c'}$, are obtained from step (12), while the value of 'k' is selected from steps (1), (8), (10), and (11), and feeder shape from step (13).

Step 15. *Feeder Volume from Volumetric Criterion*: Compare feeder volume obtained from step (14) with that derived from: $V_f = V_c \times (\beta / \eta - \beta)$. where V_c is the total volume of all parts, including parasite sections, which are fed from a single feeder, β is total (liquid and freezing) contraction percent and η is the available feeding liquid factor, e.g., feeding volume delivery percent of a feeder arising under conditions noted in steps (8) and (13).

Step 16. *Feeder Neck:* Calculate the feeder neck dimensions from: $M_n = k_n \times M_c$, where k_n value and neck shape are selected from steps (1), (3), (4), (6), (8) and (9). Note: k_n can initially be set at 1.1.

Step 17. *Feeding Distance:* Finally, calculate the feeding distance, f_d of each feeder from the equation, $f_d = 10 \times t$. Verify from steps (3), (8), (10), (11) and (12) that required temperature gradients would be obtained to meet the necessary feeding distance to completely feed all casting sections, Fig. 11.

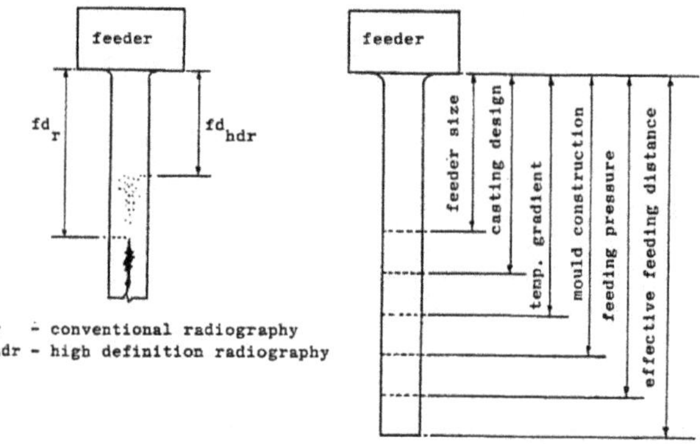

Figure 11. Feeding distance sensitivity to various alloy, mold and feeding system parameters studied.

The design of feeding system based on the above procedure for two practical investment casting samples are described below:

Example 1

The engineering drawing, Fig 12, gives essential dimensions and other requirements for the selected casting "Clamping Arm".

Gunmetal, the customer specified alloy has the composition: Cu 88%, Sn 8%, Zn 4%. Alloy shrinkage, β, is estimated at 6%. At a volume feed delivery, η of 18%, its volume ratio derived from equation-3, $VR = (V_f / V_c) = 0.50$ (or 50%). This criterion sets the minima for feeder sizes required to ensure volumetric adequacy in the supply of feed liquid.

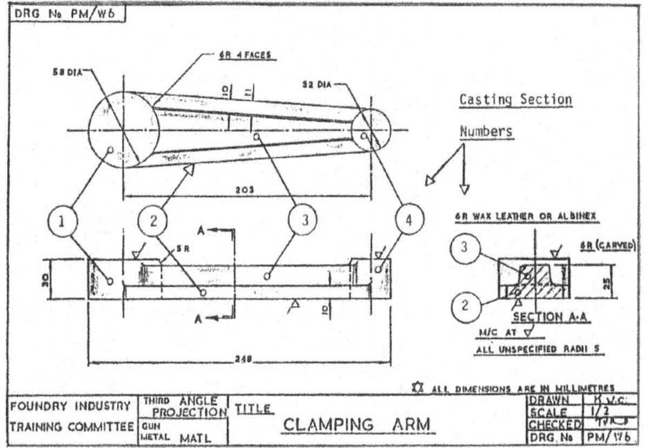

Figure 12. Engineering drawing of 'Clamping Arm' casting with casting section numbers for modulus calculation.

Casting Section	V (mm^3)	CS (mm^2)	M = V/CS (mm)	Remarks
1. 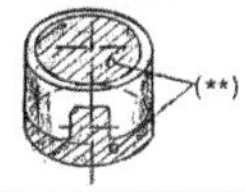 (**)	79,262.4	6,717.2	11.8	o Slowest to freeze. o Requires a separate and adjacent feeder.
2. 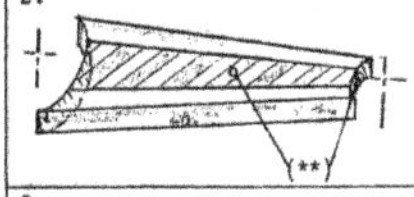 (**)	74,118.4	13,984.6	5.3	o First portion to freeze.
3. 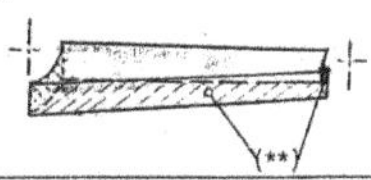 (**)	66,360.0	9,216.7	7.2	o Second portion to freeze.
4. 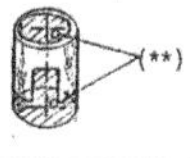 (**)	24,127.4	3,093.3	7.8	o Freezes later than arms #2 and #3. o Requires a separate and adjacent feeder.

Total volume = 243,868.2 (**Hatched portions are non-cooling surfaces.)

Figure 13. shows modulus calculations for the Clamping Arm.

The casting is divided into four sections, and modulus for each calculated, Fig. 13. From their modulus values, it is seen that the arm (#2) and rib (#3) solidify ahead of the smaller cylinder (#4). The big cylinder (#1) is the last section to freeze. This shows that a *feeder #1* is required to feed the big cylinder (#1), which in turn can feed the arm (#2) and the rib (#3).

However a separate feeder #2 is required to feed the small cylinder (#4), since thinner sections cut off feed supply to it from *feeder #1*.

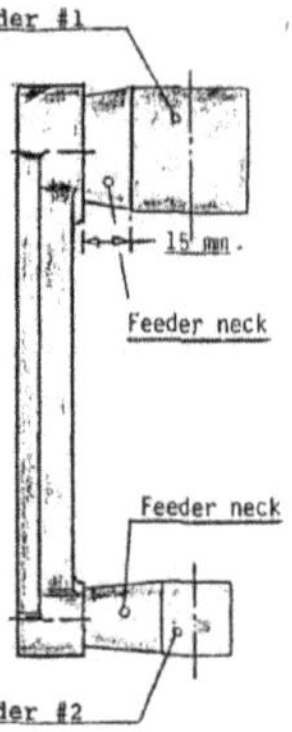

Feeder Details	$\sum V_{C(i)}$* (mm³)	$\hat{M}$** (mm)	$M_f = k \times \hat{M}$ (mm)	Feeder Size
• Feeder #1 15 mm. Feeder neck	V_{C1} + V_{C2} + V_{C3} = 219,740.8	11.8	14.2 (k=1.2)	• Side of square feeder #1 d = 4 x M_f = 57.0mm • At 70mm length, V_f = 227,430mm³ • It is volumetrically capable of feeding two castings on either side, with VR = 0.52
	o Feeder Neck Minimum size = 4 x 1.1 x $\hat{M}$ = 52.0mm dia. This neck leaves 3mm. wide annulus called 'Witness,' which provides reference surface for snagging and other finishing operations after cut-off.			
Feeder neck	V_{C4} = 24,127.4	7.8	9.4 (k=1.2)	• Side of square feeder #2 = 4 x M_f = 38.0mm • At 38mm length, V_f = 54,872mm³
• Feeder #2	• Feeder Neck Size = 4x1.1x7.8 = 34.3mm			• On feeding #4 on either side VR = 1.14

Note: *Total volume of sections fed; for e.g. volume of sections #1, #2, #3 for feeder #1.
**Highest modulus $\hat{M}$ amongst fed casting sections; for e.g. $\hat{M} = M_1$ for feeder #1.

Figure 14. Feeder Dimensions for the Clamping Arm.

In order to take advantage of the substantial taper in the arm (#2) and the rib (#3), this component can be cast in a vertical orientation, Fig. 14. In this orientation, gravity provides necessary feed pressure to feeder #1 (and the cylinder #1) to feed the long arm (#2) and the rib (#3).

Feeder #1 is designed to feed three sections, namely, the big cylinder (#1), the arm (#2) and the rib (#3).

Feeder #1 modulus M_f =k x M_c, from equation -2. With freezing ratio, k, at 1.2, and M_c, the highest modulus among fed sections as that of section #1, M_f = 14.2

Modulus of a square bar feeder of side d, M_f = d/4, when feeder is considered part of a grid and two side surfaces are noncooling, Fig 14. Hence, the side of *feeder #1*, d = 4 x M_f = 57.0 mm.

At 70 mm length*, V_f = 227, 430 mm³. This feeder #1 has a volume ratio, VR = V_f / V_c = 1.04, if attached to one casting. It is hence advantageous to attach a second casting on the other side of the same feeder.

This results in, VR= 0.52, which is > 0.50, and hence satisfies the volumetric criterion.

*NOTE: The length of the two feeders # 1 and # 2 in this grid, effective in feeding is assumed to be ~ 120% of the larger dimension, e.g., diameter of attached casting section (# 1 and # 4, respectively).

Feeder neck sizes are calculated from equation-4: $M_n = k_n \times M_c$

Minimum neck size (frustrum of cone) = (4 x 1.1 x 11.8) = 52 mm diameter. This allows a 3 mm wide annulus on the casting surface called 'witness', which provides a reference surface for finish operations. The length of the feeder neck, shown as 15 mm, is to allow for easy fettling or cutoff.

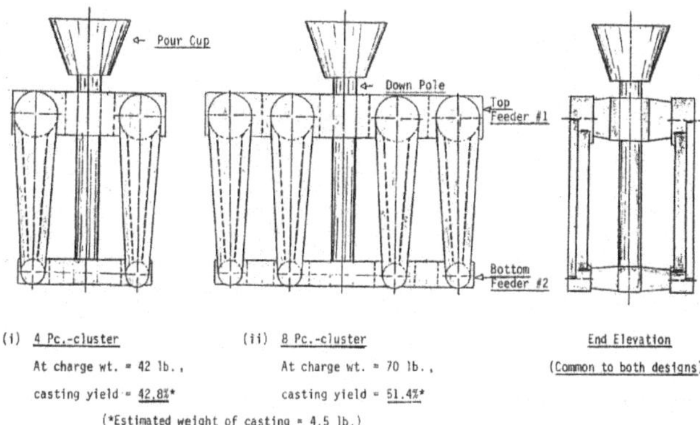

Figure 15. Alternative designs of feeding systems.

Minimal size for the second *feeder #2* is determined in a similar way: $M_f = k \times M_c$, where M_c is the modulus of cylinder #4. Side, d, of a square bar feeder is found to be 38 mm.

To check if *feeder #2* has sufficient volume at a length of 38 mm * (see note above), V_f is found to be 54,872 mm^3 and volume ratio VR = 1.14, on feeding cylinder #4 on either side. Thus, it easily satisfies the volume criterion of VR ≥ 0.50.

Feeder neck size for this feeder is found similarly. Minimum neck size is found to be: (4 x 1.1x.7.8) = 34.3 mm.

To verify the feeding distance, $f_{d,}$ of the feeders: $f_d = 10 \times t = 250$ mm, using the combined thickness (t= 25 mm) of arm and rib (section AA in the drawing Fig. 12). This shows that *feeder #1* can feed the cylinder #1 effectively, as well as the complete length of arm (#2) and the rib (#3). The second *feeder #2* can feed the adjacent section #4 effectively.

Two alternative designs for feeding grids are given in examples in Fig. 15, namely: a 4-piece cluster that can be cast successfully with an alloy charge weight of 42 lb (19 kg); this gives a casting yield of 42.8%; 8- piece cluster with an alloy charge weight of 70 lb (32 kg) giving a casting yield of 51.4%.

Other alternative designs, such as 6-piece clusters, are possible, and are again fitted with optimal sizes of feeders. An optimum feeding grid can be selected based on available foundry facilities, such as in shell- making or melting. For instance, the 8-piece cluster mold would require robotic dipping for making the shell, while manual dipping is adequate in making the 4-piece cluster.

Example 2

The engineering drawing in Fig. 16 shows casting section numbers (S/N) for a typical aero-engine turbine rotor blade.

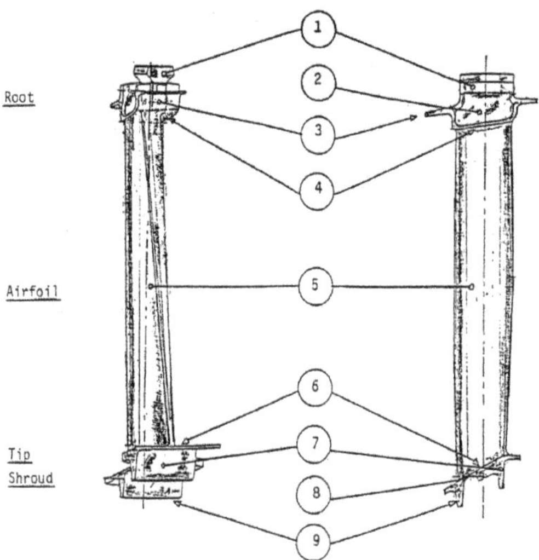

Figure 16. Depicting casting section numbers for turbine blade.

Nickel-base alloy IN-713C selected by the customer for this blade has been found to have good castability. Alloy shrinkage, β, is estimated at 6%, and volume feed delivery, η of 18% to give a minimum volume ratio, VR= $V_f / V_c = 0.50$ (or 50%), as derived from equation-3.

Calculated modulus values for various sections of the blade are shown in Fig. 17. Figure 18 shows two examples of calculating modulus for complex and irregular shapes.

To calculate the modulus for airfoil section #5, the end profiles (usually supplied by the customer at magnifications of 10X or 20X) are imprinted on suitable graph paper. Area for each of these profiles is obtained by counting the squares (A_n) and adjusting the magnification ($A_n \div 10^2$ or 20^2). The perimeter for each profile is obtained by measuring the curved lines (P_m) with a bow compass(set at constant 10 mm in this case) and adjusting for the magnification ($P_m \div 10$ or 20).

Modulus for the airfoil section #5 is given by the following equation- 5.

$$V = L_{AV} \times \frac{A_T + A_B}{2},\ CS = L_{AV} \times \frac{P_T + P_B}{2},\ M = V \big/ CS \tag{5}$$

where, L_{av} is the average length of airfoil; (A_T, A_B) and (P_T, P_B) are area and perimeter at sections TT and BB, Fig. 18 (a)

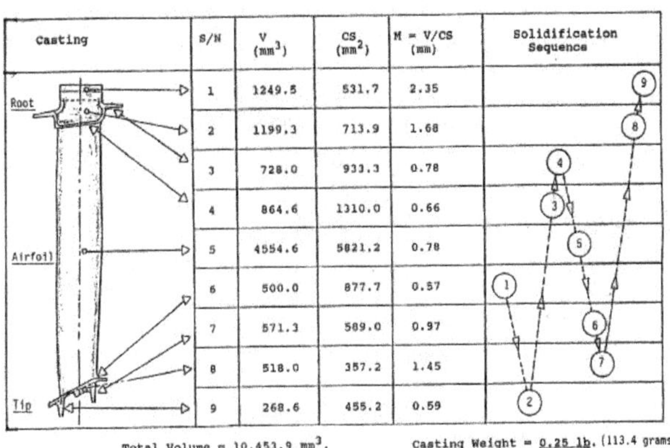

Casting	S/N	V (mm³)	CS (mm²)	M = V/CS (mm)	Solidification Sequence
Root	1	1249.5	531.7	2.35	
	2	1199.3	713.9	1.68	
	3	728.0	933.3	0.78	
	4	864.6	1310.0	0.66	
Airfoil	5	4554.6	5821.2	0.78	
	6	500.0	877.7	0.57	
	7	571.3	589.0	0.97	
	8	518.0	357.2	1.45	
Tip	9	268.6	455.2	0.59	

Total Volume = 10,453.9 mm³. Casting Weight = 0.25 lb. (113.4 grams)

Figure 17. Calculated modulus values for turbine rotor blade.

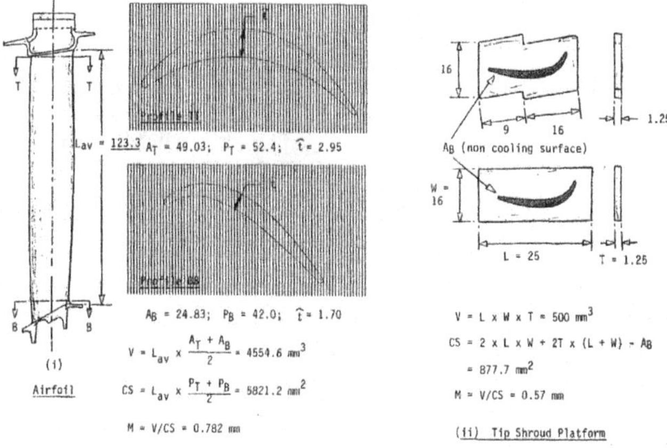

Figure 18. Modulus calculations using: (a) magnified airfoil profile, (b) the ' substitution principle.

Note: Using a digitizer assists greatly in these measurements. Additional airfoil profiles at intermediate locations can be measured to improve on the accuracy or to analyze long airfoils with large variations in the profiles. However, with current advancements in CAD and a variety of computer simulation software, the above computed modulus values are readily displayed for analysis and design of feeders by the foundry.

An irregular shape, like the Z-shaped shroud platform, is substituted by a rectangular plate as shown in Fig. 18 (b). Here, area A_B of the airfoil profile, where the top joins this shroud, is a noncooling surface and, hence, is deducted from the total surface area of the rectangular plate. Individual expertise in usage of this technique is attained easily by working out the differences between actual shape and the substituted shape for the first few castings.

The sequence of solidification can be inferred from modulus for the sections, Fig.17. Solidification is observed to commence at the parasite sections #6, 9, 4 and 3. The airfoil section #5 freezes next, followed by the tip sections #7 and 8 and root sections #2 and 1. This progress of solidification clearly calls for two feeders #1 and 2, with feeder # 1 feeding root sections and the airfoil, while feeder #2 feeds the tip sections.

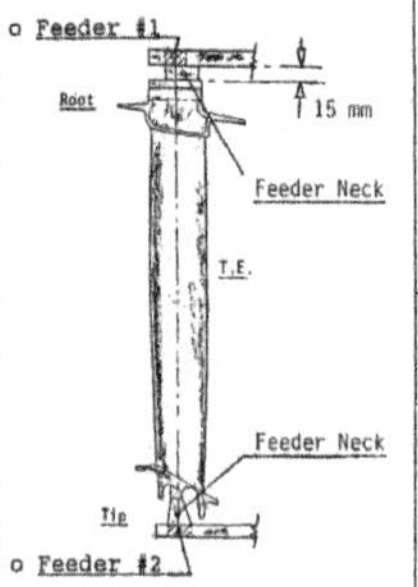

Feeder Details	$\sum V_{C(i)}$* (mm³)	$\widehat{M}$** (mm)	$M_f = k \times \widehat{M}$ (mm)	Feeder Size
o Feeder #1 Root 15 mm Feeder Neck T.E.	$\sum_{i=1}^{5} V_{C(i)}$ = 8596.0	M_1 = 2.35	2.82 (k=1.2)	• Side of square feeder #1 d = 4 x M_f = 11.5mm • But, to be volumetrically adequate at L_f = 25mm , d = 13.1mm
	• Feeder Neck: a rectangular wedge-shaped bar from the feeder, with min. size = 4 x 1.1 x M = 9.4mm			
Feeder Neck Tip o Feeder #2	$\sum_{i=6}^{9} V_{C(i)}$ = 1857.9	M_8 = 1.45	1.74 (k=1.2)	• Side of square feeder #2 d = 4 x M_f = 7.0mm • But, to meet volumetric contraction , d = 7.9mm
	• Feeder Neck: a rectangular wedge from the square bar feeder, with min. size = 4 x 1.1 x M = 6.4mm			

Note: *Total volume of Sections fed; for e.g. $\sum$ volume of sections #1, #2, #3, #4, #5 for feeder #1.
**Highest modulus $\widehat{M}$ amongst fed casting sections; for e.g. $\widehat{M} = M_8$ for feeder #2.

Figure 19. Feeder Dimensions for: turbine rotor blade.

Calculations for feeder dimensions are shown in Fig.19. Modulus, M_f for feeder#1 = k x M_1 = 2.82; modulus of root section #1, M_1 being the highest. Side of a square bar feeder, d=4 x M_f = 11.5 mm.; V_f = 3306.3 mm ^{3} at a feeder length of 25 mm. But VR = V_f / V_c = 3306.3/ 8596. VR = 0.38 is below the allowed minimum volume ratio. To adequately feed the volumetric contraction of these sections, required V_f = 0.5 x 8596 = L_f x d ^{2} ; Feeder length, L_f = 25 mm.

Required side, d, of square bar feeder is given by: d= √171.9 = 13.1 mm.

For this feeder, the minimum feeder neck size = 1.1x 4 x M_1 = 9.4 mm. To ensure ease in cut off operation, a neck height of 15 mm is provided.

Feeder #2 is designed to feed tip sections # 6, 7, 8 and 9 with the highest modulus M_8 = 1.45.

$M_f = 1.2 \times 1.45 = 1.74$. Side of bar feeder= $4 \times M_f = 7.0$; $V_f = 735\ mm^3$ at $L_f = 15$ mm. But this gives VR= 0.39. To be sufficient volumetrically, $V_f = 929 = 15 \times d^2$. Required side, d, of square bar feeder #2, is given by: $d = \sqrt{61.9} = 7.9$ mm.

Minimum feeder neck size for this feeder = $4 \times 1.1 \times M_8 = 6.4$ mm. This can be a rectangular wedge, tapering from the square feeder #2 toward rail section # 8, Fig. 19.

To check out the efficacy of these feeders, especially over the length of the airfoils, the feeding distance for each feeder is computed as follows: for feeder # 1, $f_d = 10 \times t = 10 \times 2.95 = 29.5$ mm; and for feeder #2, $f_d = 10 \times t = 10 \times 1.70 = 17.0$ mm.

The combined feeding distance is only 46.5 (= 29.5+17.0) mm. But this analysis ignores the sharp taper of 0.62 º in the airfoil from tip to the root (2.95 mm steeply tapering to 1.70 mm over the airfoil length of 123.3 mm), which allows for certain "auto feeding" along the length of the airfoil.

However, this casting-taper-directed help towards feeding can be obstructed if convergence of liquid metal flow through bottom and top runners occurs at mid-airfoil. When this happens, adverse temperature gradients develop in the bottom half of the airfoil, which work against feeding due to the sharp taper. To prevent this and to effectively feed the length of the airfoil, it is necessary to create beneficial gradients by techniques such as the following:

i. An alumina / ceramic ball (~ 30 mm diameter) properly seated on a downpole, D/P (~ 25 mm diameter) is used to deflect the pour stream; the liquid metal is allowed to flow through the "top star" runners only, Fig. 20. This uni-directional fill induces temperature gradients that support the self-feeding actions of the casting taper. This "ceramic-ball-valve" technique is dependent on the location and size of the ball, D/P, and the ball seating for functioning efficiently/ effectively. This ball placed on a spherical or tapered seating stays locked in place (submerged) when the melt stream hits it. It blocks the flow through D/P completely and allows flow only laterally through the top runners. When all the component cavities are filled up, the flow surges up through the D/P, "popping-up" the ball into the pour cup (P/C), Fig. 20. Evidence that the ball has performed as expected is obtained when the ball is seen lodged on top of the P/C after pouring / casting.

ii. Selective use of insulating ceramic kaowool wraps around the mold (wrapped prior to preheating and maintained during casting) has been found effective in feeding long airfoil sections, with reduced number of intermediate feeders to airfoil midsections. Normally, this selective- wrapping involves having about ½" thick kaowool fully around the mold (leaving Pour cup top open) including the mid-airfoil sections, and placing gradually increased layers of kaowool, (each layer with ~ 3/8" to ½" additional blanket) towards the top as well as the bottom feeders, to promote the required gradients towards the top as well as the bottom feeders. The required temperature gradients are developed both during the preheating of mold as well as selective heat dissipation during solidification. This technique, however, may require iterative trials to obtain the required level of soundness in airfoil castings.

iii. A combination of the above two techniques can be adopted to promote beneficial gradients in long airfoil sections. Other innovative techniques to create required mold temperature gradients have been effectively developed [4-6] for long, thin airfoil investment castings, required to be cast sound, specifically with out intermediate feeders (to attain the required higher, specific mechanical properties in these castings). The mold gradients in such cases have been created and maintained, using graphite susceptor in Interlock furnaces with multi-coil induction heating capabilities (conventionally used for the production of D.S. and single crystal investment castings) and with the selective use of selected mold materials, with varying chilling capacities.

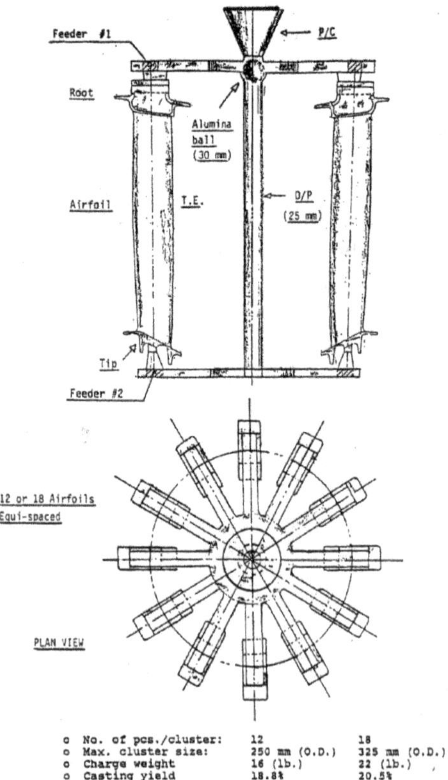

Figure 20. A comparison of alternative feeding systems.

A comparison of two alternative designs of gating-cum-feeding systems is shown in Fig. 20; both are circular clusters, each fitted with optimal feeder sizes: (i) a 12-piece cluster, 250 mm O.D., with a casting yield of 18.8%, and (ii) an 18-piece cluster, 325 mm O.D. with a casting yield of 20.5%. Where foundry facilities permit, it is advantageous to use the larger size cluster with a higher casting yield.

Acknowledgements

Late Professor Voya Kondic, University of Birmingham, UK, for his inspiring guidance during author's research on feeding investment castings, research supported by Rolls-Royce Limited, UK.

R.A. Horton, with thanks for permitting use of figures and tables from his article on 'investment casting', ASM Handbook, Volume15, 2008

Author details

Ram Prasad*

Address all correspondence to: rprasad@aerometals.com

Aero Metals Inc., USA

References

[1] Prasad, Rama T. V. (1980). University of Birmingham, UK, * Note: the author Ram Prasad's name before US Citizenship on September 16, 2011

[2] Prasad, Rama T. V., & Kondic, V. (1994). Basic elements of Feeding Investment Castings. *American Foundry Society*.

[3] Prasad, Rama T. V., & Kondic, V. (1994). Designing Feeding Systems for Investment Castings AFS Modern Casting, September

[4] Prasad, Rama T. V. (1989). Method of Casting a Metal Article. , U.S. Patent 4,809,764,

[5] Prasad, Rama T. V. (1990). Method of Casting a Metal Article,. *U.S. Patent*, 4,905,752,.

[6] Prasad, Rama T. V. (1991). Method and Apparatus for Casting a Metal Article,. *U.S. Patent*, 5,072,771,.

[7] Horton, R. A. (2008). Investment Casting. Casting, ASM Handbook,, 15

[8] Brooks, R. (2012). Still Rapid, but Much More than Prototyping. Foundry Management & Technology March, 16.

[9] Chvorinov, N. (1940). *Giesserei,*, 27, 177.

[10] Wlodawer, R. (1966). Directional Solidification of Steel Castings,. Pergamon London,

[11] Anonymous. (1979). *Ceramic Test Procedures,*, Investment Casting Institute.

[12] Uram, S. (1978). Commercial Applications of Ceramic Cores, Paper presented at the 26th Annual Meeting, Oct 1978, Investment Casting Institute

[13] Magnier, P. (1973). Titanium Investment Castings for Aircraft Applications. *May, British Investment Casters Technical Association.*

[14] Jackson, J. D., & Fassler, M. H. (1985). *Developments in Investment Casting,,* Investment Casting Institute.

[15] Mikkola, P., & Scholl, G. (2008). Low- Pressure Countergravity Casting. Casting, ASM Handbook,, 15

[16] Bidwell, H. T. (1969). Design and Application of Investment Castings. *Investment Casting,,* Machinery Publishing,.

Section 2

Reusable Mold Castings

Chapter 5

Mould Fluxes in the Steel Continuous Casting Process

Elena Brandaleze, Gustavo Di Gresia,
Leandro Santini, Alejandro Martín and
Edgardo Benavidez

Additional information is available at the end of the chapter

1. Introduction

During the last decades, the continuous casting process has made enormous advances and more than 90% of the world steel production is now continuously cast [1]. In this process, the liquid steel is poured into a water-cooled copper mould through a submerged entry nozzle (SEN), see Figure 1 [2]. At this stage the solidification process begins. In this way semi-finished products with specific characteristics such as slabs and billets are obtained. During this process the mould fluxes perform several critical functions to obtain products with the quality required.

The mould fluxes are synthetic slags constituted by a complex mix of oxides, minerals and carbonaceous materials. The main oxides are silica (SiO_2), calcium oxide (CaO), sodium oxide (Na_2O), aluminum oxide (Al_2O_3) and magnesium oxide (MgO). The (CaO/SiO_2) ratios are 0.7 to 1.3 with fluorite (F_2Ca) and carbonaceous materials additions in their compositions. The compounds content ranges and their effects on mould fluxes behaviour at process conditions are summarized in Table 1.

These fluxes can be added through the top of the mould on the liquid steel, manually or automatically, the second way being the one that offers greater stability and constancy of the required properties.

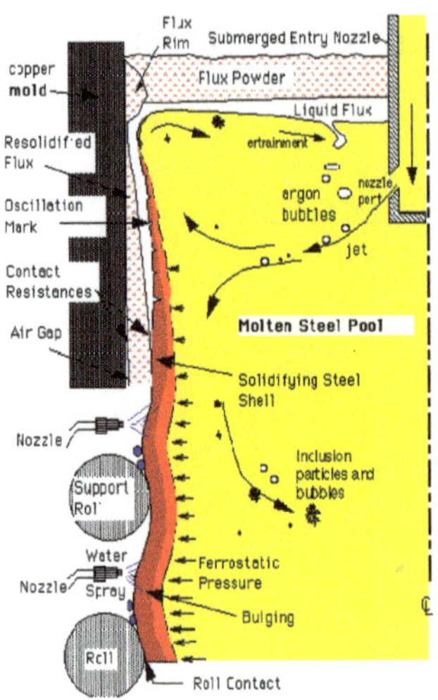

Figure 1. Schematic drawing of the continuous casting process [2].

Glass formers	SiO_2	17 – 56 %
	Al_2O_3	0 – 13 %
	B_2O_3	0 – 19 %
	Fe_2O_3	0 – 6 %
Basic oxides or modifiers	CaO	22 – 45 %
	MgO	0 – 10 %
	BaO	0 – 10 %
	SrO	0 – 5 %
Alkalis	Na_2O	0 – 25 %
	Li_2O	0 – 5 %
	K_2O	0 – 2 %
Fluidizing	F	2 – 15 %
	MnO	0 – 5 %
Melting control	C	2 – 20 %

Table 1. Typical composition of mould fluxes (wt %).

2. Mould fluxes functions

The continuous casting process is a very complex one which involves many variables: casting speed, mould oscillation characteristics, steel grade, mould dimensions and metal flow. All these variables need to be optimized but this is very difficult because it is not possible to see what is occurring inside the mould. In general, it is important to collect information on: analysis of plant data, simulations of different phenomena and measurements of different specific physical properties of the fluxes.

The additions of mould fluxes on the free liquid steel surface form different layers that are described in Figure 2. Each layer in isolation or combined with another one, provides the required functions of the powder.

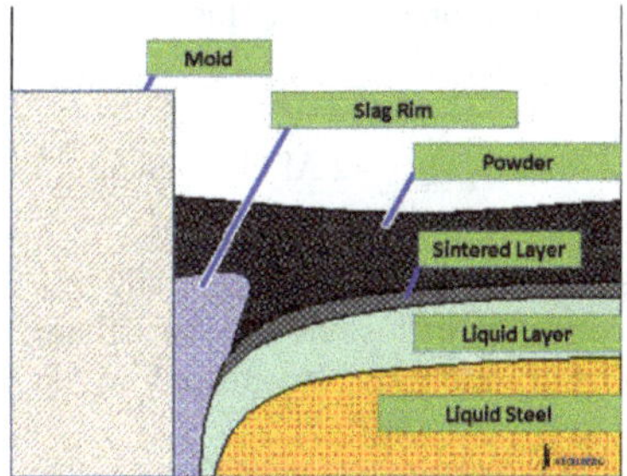

Figure 2. Different layers formed by the mould flux on the liquid steel.

The functions of the mould fluxes can be divided into two types, depending on the specific contact zone:

i) Zone of contact with the liquid steel

2.1. Thermal insulation

In this case, the objective is avoiding heat loss that could cause the premature solidification of the liquid steel in the meniscus zone. The properties of the mould fluxes that control these functions are:

- The mould flux density
- The thickness of the flux layer
- The carbon content
- The particle size distribution in the material

A bad thermal insulation in the meniscus promotes operation problems such as breakouts and could also cause surface defects in the products, such as cracks and oscillation deep marks.

2.2. Prevention of reoxidation

The liquid slag constitutes a barrier to avoid steel reoxidation by contact with air and the entrapment of other gases, such as nitrogen.

The steel reoxidation in the surface promotes oxide generation that could be incorporated as inclusions into the liquid steel (i.e. Al_2O_3) or into slag, changing its physical properties.

2.3. Inclusions entrapment

Mould fluxes are also designed to have the capacity to absorb or entrap inclusions in the interface of liquid slag–metal. In this way, it is possible to improve the cleanliness of the steel within certain operation parameters and depending on the process conditions. One of the important conditions is the depth of the liquid pool of slag [1].

The control of alumina (Al_2O_3) pickup in the liquid slag, during a certain period of time gives information of the slag absorption capacity. This oxide is produced by the reaction between the metal and the slag (Eq. 1):

$$4\ Al + 3\ SiO_2\ _{(slag)} = 2\ Al_2O_3 + 3\ SiO_2 \qquad (1)$$

The large particles can cross the slag/metal interface easily but the smaller inclusions need more time to do it. Absorption of inclusions can be enhanced using fluxes with high (CaO/SiO_2) ratios, high Na_2O, Li_2O and CaF_2 contents or low Al_2O_3, TiO_2 contents.

ii) Zone of contact with solidified steel

2.4. Lubrication between the solidified steel shell and the mould

Good lubrication is the most important function of the mould fluxes. The lubrication capacity of the liquid slag is related to the viscosity and the solidification temperature. For this reason, it is important to establish the viscosity values at operation temperatures by experimental tests or applying theoretical models. The lubrication is indirectly influenced by process conditions such as casting speed, superheat temperature and submerged nozzle (SEN) design. When the liquid slag layer is interrupted for any reason, sticker breakouts or cracks could occur. Surface cracks in slabs are also promoted by bad lubrication.

2.5. Heat transfer control

Heat transfer in the mould can be divided into horizontal and vertical heat transfer. The horizontal heat transfer has the more significant effect on the surface quality of the product. Nevertheless, the control of vertical heat flux permits to overcome problems such as pinholes and deep oscillation marks [1].

The heat transfer in the continuous casting mould is largely controlled by the film generated in the gap between the steel shell and the mould, due to the solid and liquid proportion characteristics of the slag. These characteristics are associated with the high or low crystallization tendency of the mould flux, because in this way a greater or lesser heat extraction can be controlled. For this reason, the mould flux has to be specifically selected for each steel grade.

3. Operation problems and product defects associated with mould fluxes

3.1. Sticking

Sticker breakouts occur when the solidified shell is broken in or out the mould and, as a consequence, the liquid steel can not be contained by the solidified shell. Figure 3 describes a normal shell formation and a distorted shell produced by a sticking.

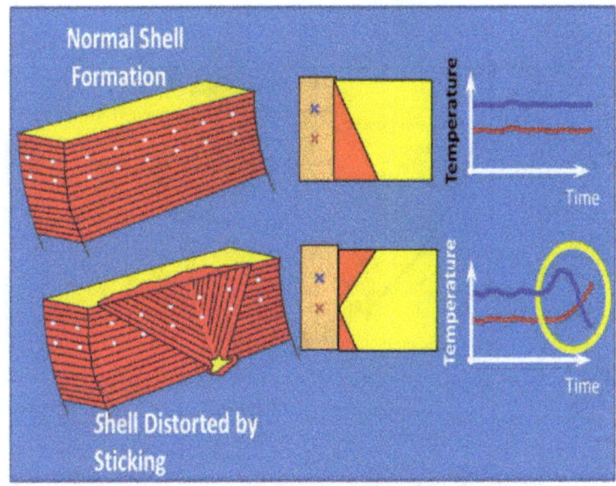

Figure 3. A normal steel shell formation and a distorted shell produced by sticking.

As it was mentioned, the mould fluxes are the responsible for providing a continuous lubrication between the mould and the strand This continuous lubrication has to be guaranteed because if it is interrupted, the steel sticks on the mould wall. This fact promotes considerable stresses due to the friction, increasing the risk of breakout.

In Figure 3, two solidification patterns are visualized: a normal solidification pattern and the solidification pattern when sticking occurs. In the right part of the figure, the graphics (temperature–time) show the behaviour of thermocouples during both mentioned situations.

The control system with thermocouples represents an important and effective tool in order to prevent damage of the steel shell by sticking. Another relevant application is to avoid the equipment detriment caused by the liquid steel leak. Possible causes of sticking problem to consider are:

- **a.** Changes of the slag viscosity due to Al_2O_3 enrichment
- **b.** Important variation of the liquid steel level
- **c.** Oscillating system in poor conditions (change in the oscillation curve)
- **d.** Interrupted lubrication by deficient mould flux supply
- **e.** Freezing of the meniscus by poor insulation or by altering the flow pattern inside the mold.

3.2. Mould flux consumption

It is important to consider that the mould flux consumption gives information about the liquid slag infiltration between the mould and the steel shell, thus estimating the present lubrication. The consumption of mould powder depends on the process conditions and the materials characteristics. Shin et al. [3] reported results on the influence of the casting speed on mould powder consumption. Figure 4 show that when the casting speed increases, the mould powder consumption decreases.

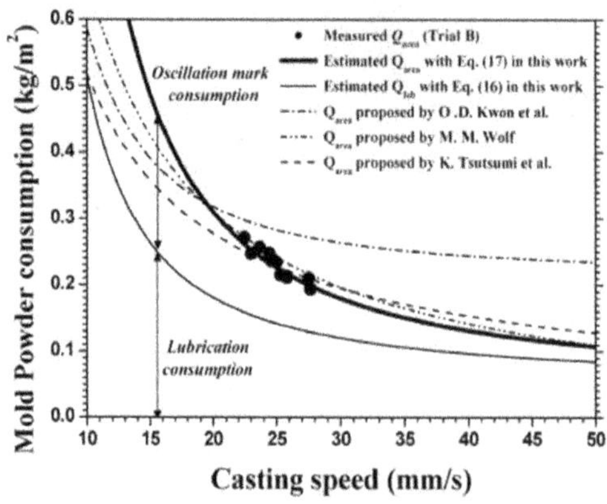

Figure 4. Influence of the casting speed on mould powder consumption [4].

Meng and Thomas [4] studied the influence of the mould oscillation parameters on the flux consumption. The authors concluded that oscillation frequency decrease implies lower powder consumption but higher oscillation amplitude or the increase in positive and negative strip increases the flux consumption.

3.3. Surface and subsurface defects

i) Slag entrapment

This type of defect may be associated with flow conditions and the physical properties of the liquid slag at steel meniscus level (see Figure 5). The main causes of this defect are associated with a high flow speed of the liquid steel at meniscus level. These conditions generate important forces that promote the entrapment of slag drops in the liquid steel. The viscosity and surface tension of the liquid slag constitute the primary physical properties related with the phenomenon of slag entrapment. Another reason for this type of entrapment be promoted, in the subsurface of the product, is the excessive changes of level in the mould. When the slag entrapments are large, they can interfere in the normal heat flow producing a thinner (and weaker) steel shell. As a consequence, the risk of breakout increases when the product leaves the mould.

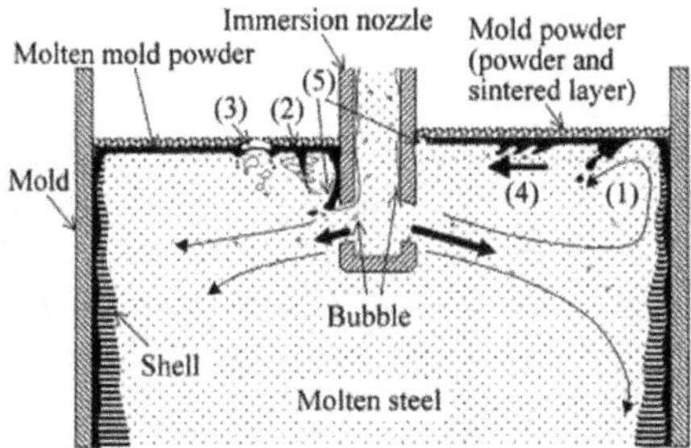

Figure 5. Mechanisms of slag entrapment indicated as 1, 2, 3, 4 and 5.

ii) Longitudinal crack

Steel grades with a chemical composition similar to peritectic steel are susceptible to develop longitudinal cracks. The origin of the problem involves the differences of the contraction coefficient between δ and γ iron that result in an irregular shell. As a consequence of stress concentration, the mentioned cracks are generated. Wolf [5] proposed the use of the carbon equivalent calculation in order to predict the longitudinal crack susceptibility. For example in the case of low alloy steel it is possible to use Eq.2 and Eq.3:

$$FP=2.5\cdot\left(0.5-[\%C_P]\right) \tag{2}$$

$$\begin{aligned} Cp &= [\%C] + 0.04\,[\%\,Mn] + 0.10\,[\%\,Ni] + 0.70\,[\%\,N] - \\ &-0.14\,[\%\,Si] - 0.04\,[\%\,Cr] - 0.10\,[\%\,Mo] - 0.24\,[\%\,Ti] \end{aligned} \tag{3}$$

where FP is the ferrite potential and C_p is carbon equivalent for peritectic transformation. Here FP > 1 signifies a fully ferritic structure and FP < 0 means fully austenitic structure. This leads to classify the steel grades in two groups: type A with high depression tendency and type B with tendency to sticking and solidification cracking. FP criterion also predicts inner crack sensitivity.

The strategy to avoid the longitudinal crack is to obtain a homogeneous shell through a uniform heat extraction. The mould powder is the tool which permits to minimize the crack tendency and these tendencies decrease at higher powder consumption because the film thickness increases. All longitudinal cracks are formed near the meniscus zone.

4. Physicochemical properties and structure of mould fluxes

The knowledge of physicochemical properties of mould powders is necessary to solve problems in industry and to develop mathematical models of the process. Generally, the determination of these properties is very complex due to the high temperatures involved (usually higher than 1000°C) and the reactions with the containers of mould powders. Besides, it is necessary to know a large number of properties such as density, thermal conductivity, viscosity, melting temperatures, surface tension, etc. Due to the complexities of these measurements, mathematical models are often used. Since, the chemical composition is information available through the suppliers; this information is used to estimate the values of physicochemical properties at high temperature.

In order to estimate these properties more complex models that make use of the structure of the molten mould powders, phase equilibrium diagrams, thermodynamic data, and neural network based models, have been used. In all cases, it should be noted that the model results are compared to experimental data, which in turn, possess a certain degree of error. Accordingly, the accuracy of the results obtained by means of the models can not be greater than those obtained experimentally.

To compute the properties of the mould fluxes several models have been used. They can be classified in [6]: (i) numerical adjustments, (ii) neural networks, (iii) structure based models and (iv) thermodynamic models.

The structure of the mould fluxes is based on silicate chains of the silicon oxide (SiO_2), where each Si^{4+} ion is surrounded by four O^{2-} (tetrahedral structure $SiO_4{}^{4-}$). Each of the anion O^{2-} is connected to two others O^{2-} (called bridging oxygen) forming a three dimensional network. This network is broken when entering the cations type Na^{+} or Ca^{2+}. These cations break silicate chains forming non-bridging oxygens O^{-} and free oxygens O^{2-} that are not bound to cations Si^{4+} but to the network breakers: Na^{+}, Ca^{2+}, Mg^{2+}, etc. [7].

Cations of type Al^{3+} can enter the polymer chain but they should be located close to others cations such Na^{+} (or Ca^{2+}) to maintain the local charge balance. Cations Fe^{3+}, in low concentrations, act as network modifiers, while in greater proportions, may be incorporated into the chain silicate similarly to Al^{3+}.

Thus, the properties of the mould powders are affected by the composition and arrangement of the individual compounds. Namely, they depend on the concentration of network formers (SiO_2, Al_2O_3) and network modifiers (Na_2O, Li_2O, CaO, MgO, K_2O).

4.1. Viscosity

The viscosity (η) expresses the difficulty with which a layer of liquid moves over another. Thus, when the length of the chains Si-O increases, this difficulty also increases. Therefore, a higher viscosity is associated with a higher degree of polymerization (higher content of network formers).

On the one hand, mould powders called "glassy", present a smooth change in viscosity versus the temperature curve when during cooling the material changes from liquid to supercooled liquid at the glass transition temperature (T_g). This temperature is associated to a viscosity of $10^{13,4}$ Pa (Figure 6a). On the other hand, for mould powders called "crystalline" the curve logη vs 1/T presents a significant change in slope at the temperature at which crystallization begins (Figure 6b). This temperature is called "break temperature" (T_{br}).

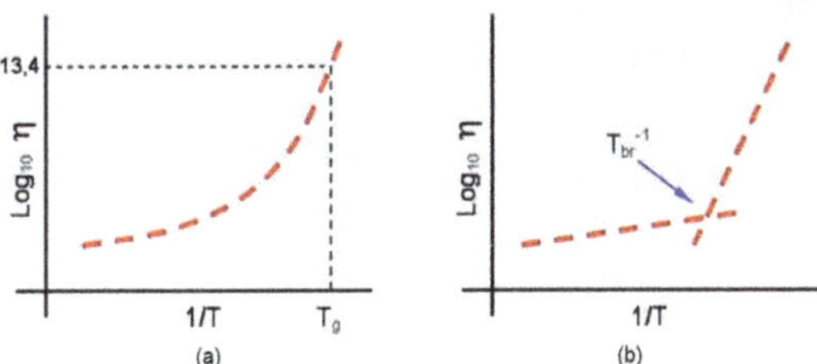

Figure 6. Plots of logη vs. 1/T of (a) glassy and (b) crystalline mould fluxes.

The viscosity of the molten material presents a significant dependence on temperature. This dependence is expressed by an equation of type Arrhenius (Eq. 4):

$$\eta = A_A.\exp(-E_A / R.T) \tag{4}$$

Or type Weymann (Eq. 5):

$$\eta = A_W.\exp(-E_W / R.T) \tag{5}$$

Where A_A, A_W are constants, R is the gas constant, and E_A, E_W are the activation energies for viscous flow.

Viscosity models of mould powders have been developed on a large amount of experimental data. A review of models based on the chemical composition [8] showed that the minor differences between the estimated values and those determined experimentally were presented by both the Iida and Riboud models. The greatest differences were within 30%.

The Riboud model [9] uses the following expression to compute the viscosity (Eq. 6):

$$\eta = A.T.\exp\left(\frac{B}{T}\right) \tag{6}$$

where T is temperature in Kelvin, and A and B are parameters obtained by means of the mould powder composition. On the other hand, in the Iida model [10] the expression for calculating the viscosity is (Eq. 7):

$$\eta = A.\eta_0 \exp\left(\frac{E}{B_i}\right) \quad (7)$$

where A and E are parameters set by adjustments to experimental data, η_o is the viscosity of the melted components not forming network and Bi is the modified basicity index.

An alternative method to compute the viscosity of mould fluxes was used by Brandaleze et al. [11]. This method is based on the model presented by Moynihan [12] that uses the width of the glass transition, which can be determined by DTA or DSC. According to this model, the viscosity can be calculated using the following equation (Eq. 8):

$$\log\eta = -5 + \frac{14.2}{[0.147(T - T'g)/(T'g)^2 \Delta(1/Tg)] + 1} \quad (8)$$

where η is the viscosity in Pa.s, T is the temperature in K, T_g is the glass transition temperature, T'_g is the end point of the glass transition and $\Delta(1/T'_g) = 1/T_g - 1/T'_g$.

Using Eq.8 the viscosity of two mould powders (10F and PC) between 1200-1450°C was estimated (Figure 7). Powder PC is of commercial origin and 10F was prepared in laboratory, both containing fluorine (for chemical composition see Table 2).

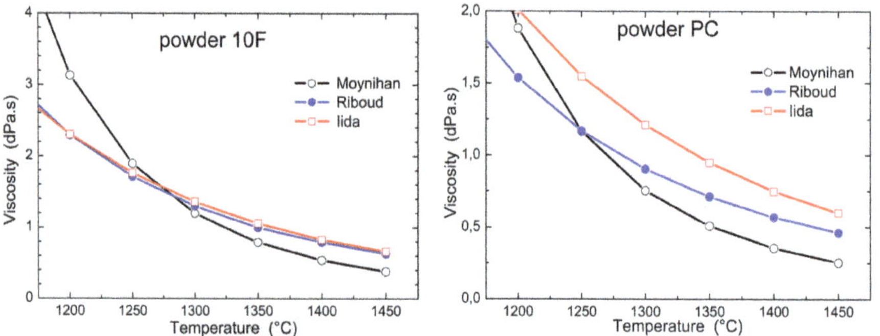

Figure 7. Viscosity values estimated from different methods.

The η-values calculated by this method were compared with those calculated for the Iida and Riboud models. The differences between the values of viscosity obtained by the Moynihan model with respect to these two traditional models were within 33%.

4.2. Thermal conductivity

The thermal conductivity of the liquid slag tends to increase as the SiO_2 content increases. This behaviour can be attributed to a better thermal conduction along the polymer chains. This transport is hindered by the presence of non-bridging oxygen (O^-) and cation breakers at the ends of the polymer chains. This interpretation has been experimentally supported by Eriksson et al. [13] in a work on liquid slags in the system $CaO\text{-}Al_2O_3\text{-}SiO_2$.

On the other hand the thermal conductivity seems to be affected by the nature of the cations modifiers, according to the following relationship $k_{LiO2}> k_{Na2O}> k_{K2O}$ [14].

Thus, when the content of network formers increases, the higher is the thermal conductivity. Therefore, an increase in the thermal conductivity may be associated with an increase of the viscosity [14]. However, this behaviour is interpreted based on the heat conduction on the network (lattice) k_L, but contributions of the heat transfer by convection (k_C) in the liquid layer and radiation (k_R) are unknown.

When the melted mould flux layer solidifies, forming either crystalline or amorphous structures, it should be noted that heat transfer by radiation in a crystalline solid decreases due to scattering of radiation by the crystal, grain boundaries and pores. Thus, an amorphous solid (glass) has greater heat conduction by radiation than a crystalline one. When comparing a glassy mould flux with a crystalline one, below the onset of crystallisation temperature (T_{br}) the conduction by radiation will be lower in the crystalline solid.

The low reliability in measurements of thermal conductivity impacts in obtaining a reliable database to develop a secure model to estimate k based on temperature and chemical composition.

Furthermore, it should be noted that, during the continuous casting process, the first layer of slag that forms against the copper mould is glassy (because of high cooling rate). But then with time, it tends to crystallize. When this layer crystallizes, it contracts (high density) and generates pores, near the crystals, and a rough surface at the interface mould/slag which is equivalent to an air gap in this interface. This air gap is represented by an interfacial resistance $R_{Cu/sl}$. For example, Hanao and Kawamoto [15] calculated an interfacial resistance $R_{Cu/sl}$ = 0.2 10^{-3} $m^2.W^{-1}.K^{-1}$, while Brandaleze et al. [16] measured $R_{Cu/sl}$ = 1.9 x 10^{-3} $m^2.W^{-1}.K^{-1}$. Thus, if the crystallization occurs with the time, this leads to a reduced flow of heat from the steel to the copper mould.

The relationship among k_R and k_L with the degree of crystallization was studied by Ozawa et al. [17]. They observed that: (i) k_L tends to increase with the degree of crystallization and (ii) k_R decreases until reaching a constant value when the fraction of crystals exceeds 15%. Meanwhile, Nakada et al. [18] studied the heat transfer through a mould flux layer and concluded that the k_R constitutes less than 20% of the total heat flow. The authors noted that the extraction of heat was very sensitive to both the thickness and the emissivity of the mould flux layer. So, if two mould powders that do not tend to crystallize on cooling (glassy) are compared, the one presenting a higher viscosity tends to generate a thicker layer of molten powder. This results in a lower extraction of heat in the hottest zone (top) of the mould.

4.3. Surface tension

This property is affected primarily by constituents who have the lowest values of surface tension (surfactants), which tend to occupy the surface layer of the liquid. The surface concentration depends on the surface tension (γ) and the activity of the components.

To estimate the surface tension different models have been used, being the simplest method that which uses partial molar fractions (X_i) of components [19]. In this model the components are divided into two classes: (i) oxides with high surface tension and (ii) components of lower surface tension or surfactants (such as B_2O_3, CaF_2, Na_2O, K_2O, Fe_2O_3) according to Eq. 9.

$$\gamma = X_1 . \gamma_1 + X_2 . \gamma_2 + X_3 . \gamma_3 + ... \tag{9}$$

The uncertainties of this model are within ± 10%.

Another model [20] also used the molar volume of components and the ionic radii.

4.4. Liquidus and break temperatures

The liquidus temperature (T_{liq}) can be determined by DTA or DSC tests (melting endotherm) or by Hot Stage Microscopy (HSM). In the latter case T_{liq} must be associated to fluidity temperature (T_F). Due to the different components of these materials, the first occurrence of liquid is detected at a temperature lower to melt flow (T_F). The fluidity temperature is considered as one in which the material reaches a viscosity apt to flow into the mould-steel gap. Models to calculate T_{liq} based on chemical composition have a high degree of uncertainty.

Moreover, the break or crystallisation temperature (T_{br}) is usually between 1100-1200 °C. A numerical model to calculate T_{br} (within an error of ± 30°C) has been used [21]. These method estimates break temperature according to the following equation (Eq.10):

$$\begin{aligned} T_{br}(K) = 1393 - 8.4\%Al_2O_3 - 3.3\%SiO_2 + 8.65\%CaO - 13.86\%MgO - 18.4\%Fe_2O_3 \\ - 3.2\%MnO - 9.2\%TiO_2 - 2.2\%K_2O - 3.2\%Na_2O - 6.47\%F \end{aligned} \tag{10}$$

4.5. Melting rate

Although the melting rate depends on process parameters such as casting speed, it is also influenced by the quality and content of free carbon [22]. The melting rate decreases with increasing carbon content and/or its particle size decreases, and increases when the reactivity of the carbonaceous material is larger [23]. An estimation of the reactivity of the carbonaceous material can be performed if decomposition kinetics is known.

Benavidez et al. [24] conducted a study on the kinetics of decomposition of two carbonaceous materials: petroleum coke (sample C) and synthetic graphite (sample G). Both materials are often used to include free carbon in mould powders composition.

The activation energy (Ea) associated with the decomposition of carbonaceous materials was calculated using four methods applied to non-isothermal thermogravimetric curves (TG) performed at different heating rates. An average value of Ea ≈ 48 kJ/mol for the powder with 15 wt% of coke, and Ea ≈ 67 kJ/mol for the powder with 15 wt% of graphite was obtained from the different methods. The lower activation energy of the decomposition process of the coke is associated with increased reactivity of this carbonaceous material relative to the graphite. This behaviour is in agreement with the higher degree of crystallinity observed in the synthetic graphite, since the greater amount of crystals results in the need of a greater amount of energy (heat) to decompose the carbonaceous material (low reactivity).

4.6. Density and molar heat capacity

Because of the strong covalent type bonds that presents SiO_2, its coefficient of linear thermal expansion (α) is very low. Thus, as the value of α is proportional to the change of density with temperature (dϱ / dT), then the density is slightly affected by the temperature. According to this, the value of α increases when the percentage of cations network modifiers increases. It is also observed that the coefficient of thermal expansion increases to a greater extent for M_2O monovalent oxides than for MO bivalent ones. In both cases, the coefficient of thermal expansion increases according to the following cation size relationship: K > Na > Li (oxides M_2O) and Ba > Ca > Mg (oxides MO). The density of the slags can be estimated using thermodynamic models [25]. However, considering the density of the liquid slag only slightly dependent on the structure, then one can use a simpler model to calculate its density in the liquid state. In this case molar volume (V) and molecular weight (M) of the mould powder are computed through Eq. 11:

$$\rho = \frac{M}{V} = \frac{\sum_i X_i.M_i}{\sum_i X_i.V_i} \qquad (11)$$

where X_i is the mole fraction, M_i is the molecular weight and V_i is the molar volume of component i.

The molar heat capacity is not affected by the structure, but rather by the composition. Thus, a good estimation of mould flux (Cp) can be obtained from the mole fraction (X_i) and the heat capacity (Cp_i) of each component (Eq.12):

$$Cp = \sum_i X_i.Cp_i \qquad (12)$$

If the mould powder is glassy, the value of Cp drops abruptly at glass transition temperature.

5. Experimental equipments and techniques to characterize mould powders

Experimental techniques are very important to characterize or previously evaluate the behaviour of a mould powder during the continuous casting process. The most important techniques are those that provide information about properties such as:

- Heat transfer
- Melting rate
- Viscosity / fluidity
- Critical temperatures

5.1. Heat transfer

Several operating problems and surface quality defects, which occur in the continuous casting process, are determined by the heat transfer through the flux layers. For this reason, it is important to perform measurements of thermal conductivity and compare the behaviours of the various types of fluxes used in the mould.

Many researchers have developed different experiments to measure thermal properties in melted fluxes trying to represent process conditions. Regardless of the measurement method used, the calculations are mainly based on the heat conduction laws (Eq. 13-15), allowing to determine the heat flow (q) [W/m^2], interfacial resistances (R) [m^2 K/W] and thermal conductivity of gap (k_{gap}) [W/m K]. In Eq. 13, k is the thermal conductivity of reference material and d the distance between the temperature measurement points.

$$q = k \cdot \frac{(\Delta T)}{d} \tag{13}$$

$$R = \frac{\Delta T}{q} \tag{14}$$

$$k_{gap} = \frac{q \, . \, d_{gap}}{\Delta T} \tag{15}$$

The main methods and devices reported in the literature are described below.

Schwerdtfeger et al. [26] simulated the gap between the steel and the copper mould, moving a cooled copper block to a surface of molten flux on a steel plate heated by electrical resistance. The temperature was registered by three thermocouples (two in the copper mould and one in the steel), which are used to calculate the effective thermal conductivity (k_{gap}) and the radiation and conduction components.

Mikrovas et al. [27] and Jenkins et al. [28] used the "finger test" based on the immersion of a copper cylinder in a molten flux bath. The cylinder is fitted with thermocouples placed strategically from which it is possible to calculate the heat flow and thermal conductivity of the system.

Yamauchi et al. [29] measured the thermal resistance of the powder through the device of Figure 8a which used an AlN plate as hot side and a steel block as refrigerated cold side with the ability to regulate the thickness of mould powder located between them.

The laser pulse method was employed by Mills et al. [30] to measure the thermal conductivity on solidified flux samples. The value is obtained from the estimation of the thermal diffusivity, density and specific heat capacity.

Similarly to [26, 29] in the device built by Holzhauser et al. [31] the sample is placed on a steel plate. The cold zone is provided by a cooled copper block with thermocouples located at strategic points to determine the thermal conductivity. The system is heated by means of electrical current (Figure 8b).

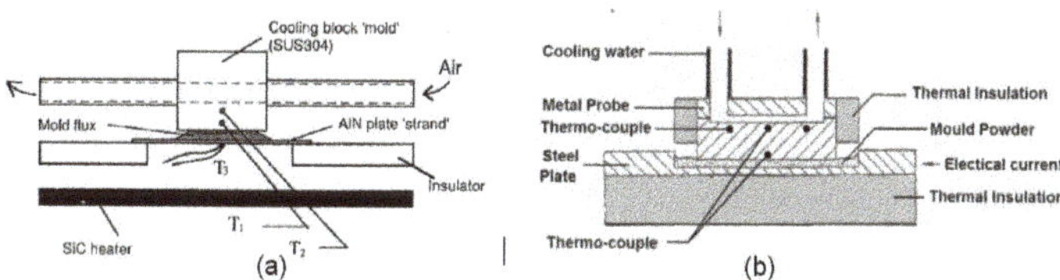

Figure 8. Experimental apparatus used to measure heat transfer through mould powder layers: (a) Yamauchi et al., and (b) Holzhauser et al.

Stone and Thomas [32] developed an equipment to simulate the mould conditions, based in a copper block and a steel plate to simulate the gap between mould and steel shell. The heat is applied on the steel plate by a torch. The molten powder is poured between the two plates. The different thermocouples placed in the equipment according to Figure 9 allow to calculate the thermal conductivity of molten flux using the equations 13 to 15.

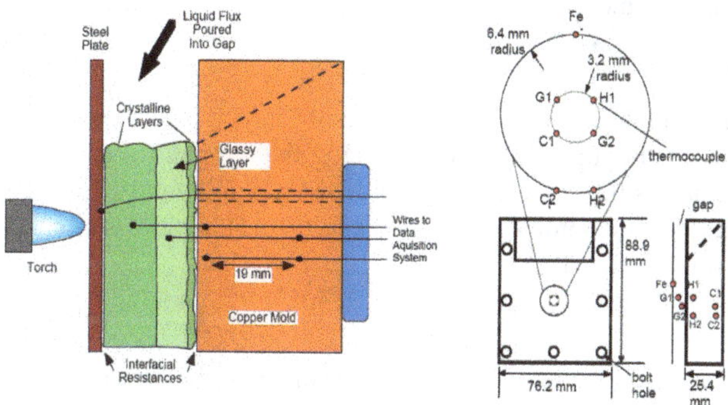

Figure 9. Heat transfer equipment developed by Stone and Thomas [32].

Brandaleze et al. [16] based on the Stone and Thomas design, made changes in both the positions of thermocouples and the cooling system, obtaining results in agreement with literature (Figure 10). Using this device Martin et al. [33] presented a comparison between a mould flux layer taken from a continuous casting machine, with another flux layer extracted from the heat transfer equipment. From the structural and microstructural analysis could be inferred that the thermal conductivity measurement is carried out under thermal conditions similar to those in the continuous casting process.

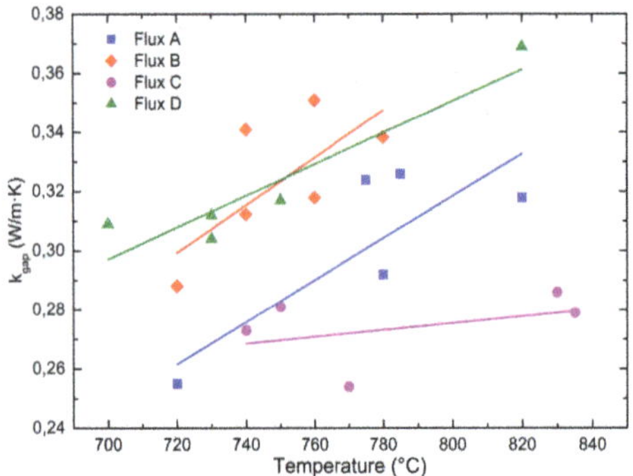

Figure 10. Heat transfer of four commercial mould powders.

5.2. Melting rate

The melting rate is an important property of the powder because it affects both the powder consumption and the depth liquid pool modifying the lubrication and heat transfer conditions. The main factor governing this property is the free carbon content. The C particles are not wetted by the molten flux and consequently separate the mineral particles delaying the agglomeration of the molten flux globules. For this reason, a higher content of free C promotes more time of agglomeration resulting in a lower melting rate.

There are basically two methods to measure this property:

i) Combustion capsules test

In this test 1.5 g of mould powder is placed in combustion capsules (porcelain) with one extreme closed and another open for easy viewing of the sample. Then, the capsule is heated inside a furnace and is observed through a horizontal window disposed for this purpose. The time taken to melt the sample is recorded and the melting rate is calculated. In this test the heat flow is unidirectional.

ii) Drip test

In this test the sample is placed in the conical base of a crucible and then molten mould flux drips out of the furnace. This molten flux is collected and weighted continuously by a balance placed at the bottom of the furnace.

5.3. Viscosity / Fluidity

As it was noted above the viscosity of mould flux has a decisive influence on the infiltration of liquid slag in the mould gap, which is probably the most important process in continuous casting because it affects the lubrication between the steel and the mould. The viscosity is also an important factor in the erosion of the refractory nozzle being a function of $1/\eta$.

High viscosity fluxes are frequently used to minimize slag entrapment. But in this case, the pressure developed by the molten slag in the mould-steel gap is high and can influence on the depth of the oscillation marks.

Several methods are used to measure the viscosity of mould powders [34, 35]:

i. Rotating cylinder method

ii. Oscillating method

iii. Inclined plane test

i) Rotating cylinder method

These viscometers consist of two concentric cylinders (Figure 11) the outer cylinder is usually a crucible and the inner cylinder (bob) is in movement. When a cylinder is rotated, it provides a velocity gradient and the torque developed is measured at different temperatures. There are two methods, (i) the rotating crucible method (RCR), in which the outer cylinder is rotated and (ii) the rotating cylinder method (RCyl) where the inner cylinder is rotated. Generally, all commercial instruments are of the latter type because of the simplicity of its construction. This method is the most widely used for this kind of materials.

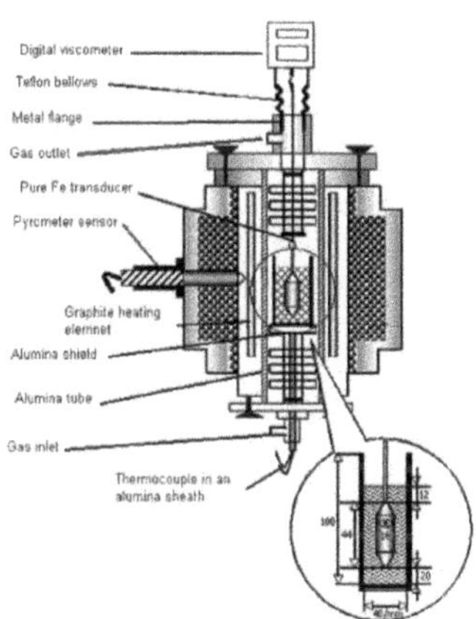

Figure 11. Rotating cylinder method used for measuring the viscosity of mould powders [36].

ii) Oscillating methods

The oscillating plate method is a relatively new method in which, subjected to a linear oscillating plate is submerged in the melt. As a result, there is a retarding force proportional to the viscosity of the fluid. When establishing a steady state, it records the amplitude of oscillation in air (φ_A) and in the melt (φ). This viscosity is derived from Eq. 16, where N is a constant, and$\rho$the density of the melt and G the constant load cell determined in calibration experiments.

$$\eta\rho = G\left[\frac{\varphi_A}{\varphi} - 1\right]^N \tag{16}$$

iii) Inclined plane test

This simple test has been used by some laboratories to estimate viscosities of molten fluxes. A mass of 10 g of powder is placed in a graphite crucible and then is melted at a specific temperature (T). The melted flux is maintained at that temperature for 15 min in order to achieve homogeneity. Then, the melt is fast cooled (quenched), pouring it onto an inclined plane. The length (L) of the ribbon is measured to have an estimation of the mould powder fluidity. In this case, the inverse of the length (1/L) is proportional to the viscosity of material at temperature T. Experimental trials in our laboratory indicated good reproducibility of results and very good relationship between ribbon lengths (L) and (1/η) for viscosities > 1 Poise (see Figure 12).

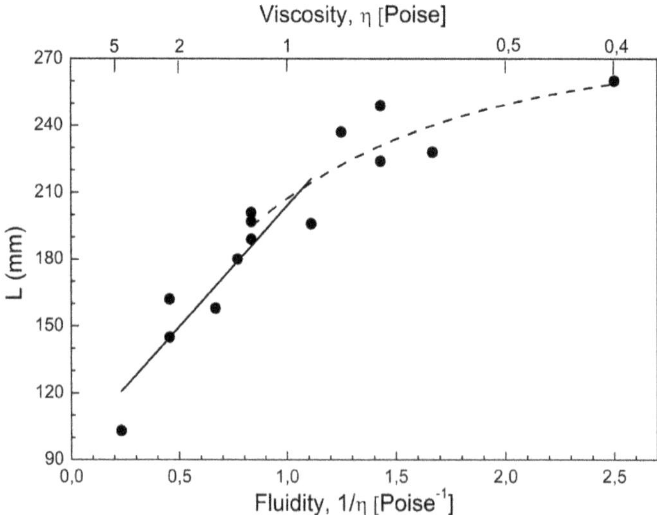

Figure 12. Slag ribbon length (L) as a function of fluidity (1/η).

5.4. Critical temperatures

The most widely used test to determine the melting range of a mould powder is the "high temperature microscopy test" (DIN 51730). The sample is pressed into a cylinder and placed in a furnace which continuously monitors the rate of heating and the changes in the sample shape. There are three critical temperatures determined by the cylinder morphology corresponding to the points of "softening", "hemisphere" and "fluidity" (see Figure 13). During the test of a commercial powder, the computer continuously analyzes variations in the shape of the sample and displays the resulting value of the critical temperatures.

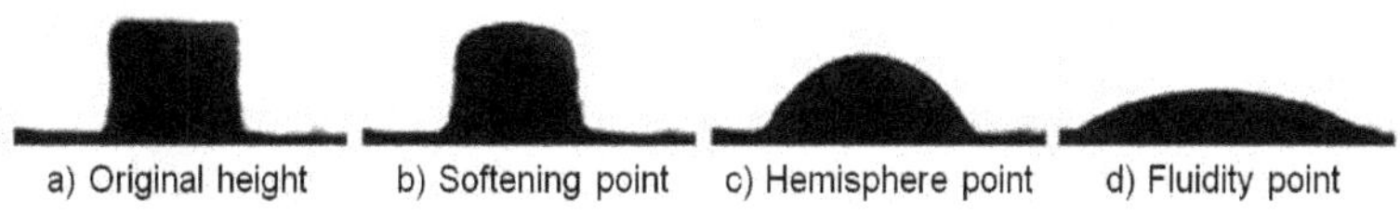

Figure 13. Image sequence showing critical temperatures of a commercial mould powder.

Another technique to determine the critical temperatures of the powders is by analyzing the ash fusibility. The test consists in prepare cones with the test material -mixed with a binder- and place them in a sample holder which is then inserted into the analyzer. Subsequently, the cones are heated to maintain a constant speed. Simultaneously, one sensor monitors the variation of the profile of the cones with temperature. At the end of the trial, the results are presented in form of four characteristic temperatures which are defined according to the morphology adopted by the cone: IT: initial temperature, ST: softening temperature, HT: hemispherical temperature, and FT: fluidity temperature.

6. Recent development in fluorine free mould fluxes

6.1. The fluoride evaporation problem associated with mould fluxes

As previously mentioned, many of the traditional mould fluxes used in continuous casting contain 4 to 10 wt % of CaF_2 in order to adjust their behaviuor according to the requirements of different steel grades casting process. At operation temperatures, harmful gas emissions (SiF_4, NaF) are produced and in many cases the gases in contact with water produce HF. These products can cause health problems to workers, affect the environment and they may cause damage to infrastructure of the plant (for example, to cooling system of the mould). Another aspect to consider is that losses of fluorides compounds also affect the chemical composition of the slag and may cause some changes in their behaviuor. For this reason, many researchers are searching substitute compounds for CaF_2 that can ensure the quality and behaviour of mould fluxes applied to slabs and long products casting [37-40].

Differential thermal analysis (DTA) and thermogravimetric (TG) data provide information about the kinetics of fluoride evaporation. By this technique, Persson et al. [38] report that the evaporation of these compounds occurs in the temperature range between 1400°C to 1600°C, in which the slags are one homogeneous liquid phase. CaF_2 is stable up to 900°C, so any emission occurs at temperatures above it. According to the estimates obtained in our laboratory using software FactSage, the temperature of the gas release during the decomposition of pure CaF_2 (at normal conditions) begins at 1153°C. The principal reactions that may occur are showed in Eq. 17 (by contact of fluorite and SiO_2) and in Eq. 18 (by the combination of CaF_2 and water vapor of the slag):

$$2\,CaF_{2\,(slag)} + SiO_{2\,(slag)} \rightarrow SiF_{4\,(gas)} + 2\,CaO_{\,(slag)} \tag{17}$$

$$CaF_{2\,(slag)} + H_2O_{\,(slag)} \rightarrow 2\,HF_{\,(gas)} + CaO_{\,(slag)} \tag{18}$$

It is known that the reaction mechanism involves the diffusion of cations and anions in the liquid slag, reactions of ions promotes the formation of SiF_4 (gas) with the consequent generation of bubbles. Finally the bubbles migrate to the liquid-gas interface and SiF_4 escapes to the atmosphere. The results of the investigation of Persson et al [38] show that the loss of fluoride depends on the temperature and composition of the slag, increasing at higher contents of SiO_2.

6.2. Compounds replacing fluorine: effects of Na_2O, B_2O_3 and Li_2O oxides

It is important to consider the role of CaF_2 in mould fluxes. One major objective of incorporating such a compound is to decrease the viscosity, the melting temperature and to control cuspidine precipitation during cooling. The latter effect is especially important in the processing of slabs where heat extraction control has a high incidence on the surface quality of the product.

In this chapter, the physical properties of mould fluxes containing fluorine in their composition have been developed. A contribution through a comparative study of fluxes with and without fluoride compounds to evaluate the effect of some oxides or compounds which can be used as substitutes of F is presented.

The main oxides considered as potential substitutes for CaF_2 are: Na_2O, Li_2O, B_2O_3 [37, 40, 41]. Several researchers studied the effects of these compounds on the viscosity and the initial temperature of crystallization T_{br}. The effect of CaF_2 on the increase of percentage of crystallinity is largely known. However, some researchers suggest that MgO and B_2O_3 can act in opposite manner [37]. For a better understanding of the effects that the new possible oxides additions can have on the behaviour of mould fluxes, a comparison of the obtained results on the behaviour of fluxes with and without CaF_2 in relation with viscosity, fluidity and crystallinity, is detailed.

In order to determine the effect on: viscosity, fluidity and melting behaviour of the mentioned oxides, different samples of fluxes were prepared in the laboratory for experimental tests: A (10% F), B (6% B_2O_3 and 4% Li_2O), C (10% B_2O_3) and D (6% B_2O_3), which simulate the behaviuor of one commercial mould flux identified as PC that contains 10% F. Powder PC is commonly applied in the slab casting. In Table 2, the chemical composition of the samples is presented.

Compound	A (10% F)	B ($6\% \; B_2O_3 + 4\% \; Li_2O$)	C ($10\% \; B_2O_3$)	D ($6\% \; B_2O_3$)	PC (10% F)
SiO_2	37.1	33.2	36.6	34.6	36.3
Al_2O_3	5.4	4.7	5.1	5	5.1
B_2O_3	-	5.8	9.8	5.8	-
CaO	30.6	28.6	31.2	29.6	30.9
Na_2O	12.6	18.6	12.3	19.6	12.7
K_2O	0.1	0.1	0.1	0.1	0.7
MgO	1.3	1.4	1.3	1.4	2.1
F	9.5	-	-	-	10.5
Li_2O	-	3.9	-	-	-
MnO	-	-	-	-	0.1
Fe_2O_3	3.4	3.7	3.6	3.9	1.6
IB	0.82	0.86	0.85	0.86	0.85

Table 2. Chemical composition (in wt %)of the samples with and without CaF_2.

i) Viscosity

The importance of ensuring good lubrication to avoid sticking problems has been previously mentioned. It is known that this leads to require adequate viscosity of the mould flux during operation. For this reason, we analyze the comparative results obtained using the Riboud model to estimate viscosity values and their correlation with temperature (Figure 14).

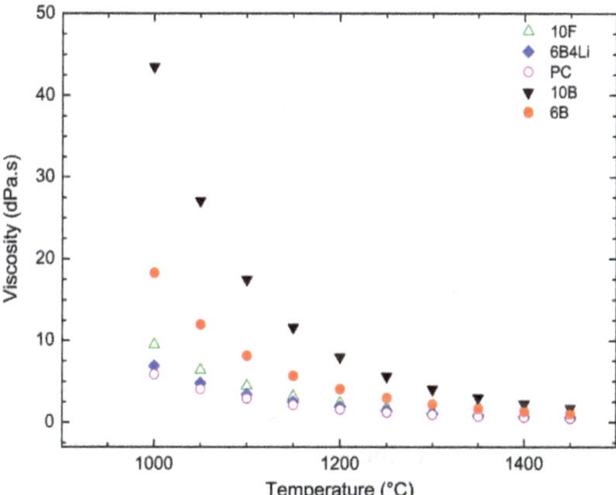

Figure 14. Correlation between viscosity and temperature of samples A, B, C, D and PC.

As it is observed the viscosities of the samples with (6% and 10% de B_2O_3) are the highest. Sample B that has 6% of the oxide is closest to the behaviour of commercial flux PC. It is noticeable that the addition of 4% Li_2O in sample B together with 6% B_2O_3, adjusts more precisely the viscosity behaviour with respect to PC flux. Higher contents than 6% of Li_2O cause a drastic decrease of viscosity and fluidity.

This suggests that the oxides considered in this study allow us to manipulate and adjust the viscosity of the mould fluxes to the required values for the processing of medium and low carbon steels. Furthermore, it is also possible to think in compensating a decrease of CaF_2 with a Na_2O increment in order to adjust the viscosity. In all samples, basicity values are around 0.85 such as PC.

The fluidity information obtained by the inclined plane test developed by Mills through the length of the layer (Lc) is consistent with the viscosity results reported in this chapter (Figure 15). The highest content of Na_2O in sample D, permits to justify the lowest fluidity obtained. Tandon et al. [42] studied the influence of high contents of Na_2O on the type of B-O bonds. The low oxide content promotes planar bonds and constitutes BO_3 but higher contents form stronger and tetrahedral bonds of BO_4.

As it is visualized, sample B (6% B_2O_3 and 4% Li_2O) is the one which presents a flow behaviour closer to the PC of reference. Samples with contents of 6% and 10% of B_2O_3 show a low fluidity because of their higher viscosity.

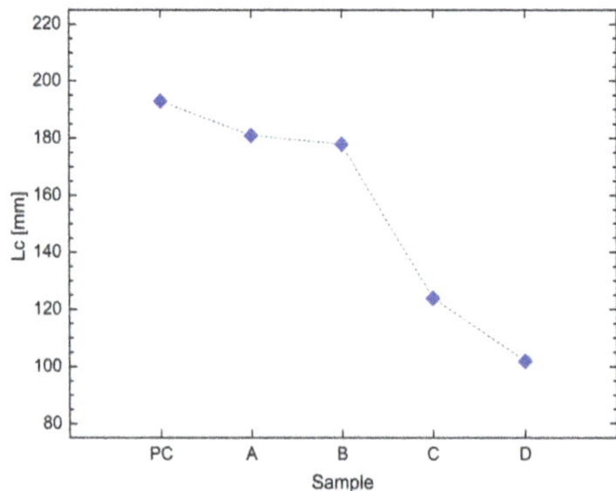

Figure 15. Fluidity behaviour of samples A, B, C, D and PC obtained by inclined plane test.

ii) Melting behaviour

The effect of the studied oxides on the melting behaviour of mould fluxes was also determined. In this case microscopy tests at high temperature (HSM) are carried out on all samples to determine the softening temperature (T_s), hemisphere temperature (T_h) and fluidity temperature (T_f). Figure 16 shows the results of the comparison between all the samples. Sample B (6% B_2O_3 and 4% Li_2O), is the one which presents the lowest critical temperatures.

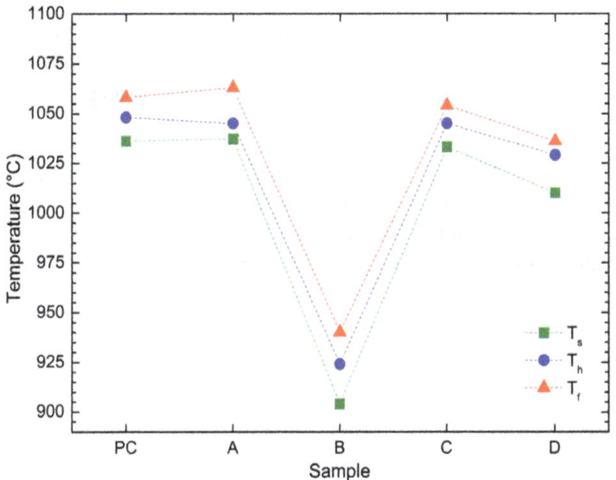

Figure 16. Melting behaviour of the samples PC, A, B, C and D.

Similar studies have been performed on mould fluxes applied in the processing of long products. In this application, mould fluxes are characterized by higher viscosities (2 to 3 Poise).

iii) Crystallization tendency

The heat extraction in the mould can be controlled by the crystalline proportion generated in the film of mould flux during the cooling stage. For this reason it is relevant to know the temperature at which the crystallization process begins (break temperature, T_{br}). Also, it is necessary to increase the knowledge of the crystallization mechanisms and tendency of mould fluxes at interest conditions.

The break temperature of the samples revealed that sample B (with 6% B_2O_3 and 4% Li_2O) presents a T_{br} = 1071°C and sample D (with 6% B_2O_3) a T_{br} = 1066°C. Both present a good agreement with PC sample in which T_{br} = 1064°C.

In the case of mould fluxes that are applied to long products casting, it is difficult to identify a clear change to verify the beginning of crystallization process because they are characterized by a high viscosity and a vitreous slag generation (or a supercooled liquid).

To evaluate the crystallization mechanisms of the samples, they were melted at 1300°C and then cooled drastically. These samples were identified as quenched (AQ). Then some of them were heat treated at different temperatures between 600°C and 870°C. All the samples were prepared for the microscopy study by light and electron scanning microscopy. Also, parts of the samples were ground to be analyzed by X-Ray Diffraction (XRD) and DTA.

The XRD results in PC at different temperatures show the evolution of the structure from a vitreous state to a crystalline one. The AQ sample is completely glassy. Nevertheless, samples treated at from 600°C produce a pronounced crystallization. In Figure 17a it is possible to observe the evolution of the crystal phases between 600°C and 850°C. By DTA, it was possible to identify the crystallization peaks of cuspidine (3 $CaO.2SiO_2.CaF_2$) present at all temperatures from 610°C, nepheline ($Al_4Ca0\ K_{0.8}Na_2Si_40_{16}$) at 729°C and villiaumite (NaF) at 854°C (Figure 17b).

Samples B and D, with (6% B_2O_3 and 4% Li_2O) and (6% B_2O_3) respectively, present different temperatures of crystallization. Sample B starts the crystallization at 610°C and sample D at 670°C. In Figure 18a, it is possible to observe the crystallization peaks of samples B and D and also the evolution of the crystallization peaks determined by DTA curves. It is found that the main phase in both cases is combeite ($Na_2Ca_2Si_3O_9$). In sample D the combeite crystallization peak is at 670°C and in sample B is at 610°C. The lower temperature in the crystallization peak of sample B could be due to the presence of Li_2O.

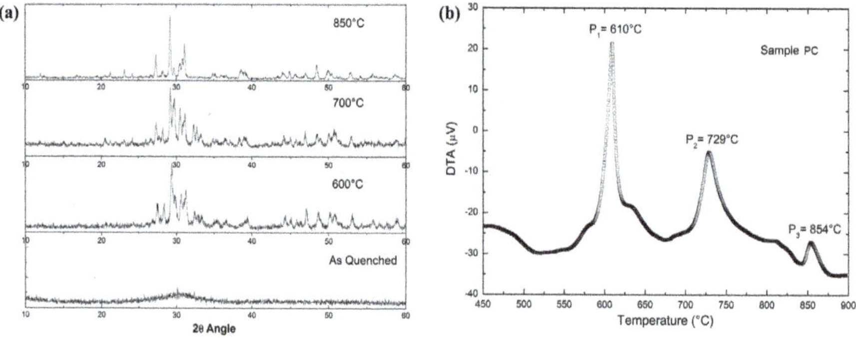

Figure 17. Crystallization evolution with temperature of sample PC: (a) XRD and (b) DTA.

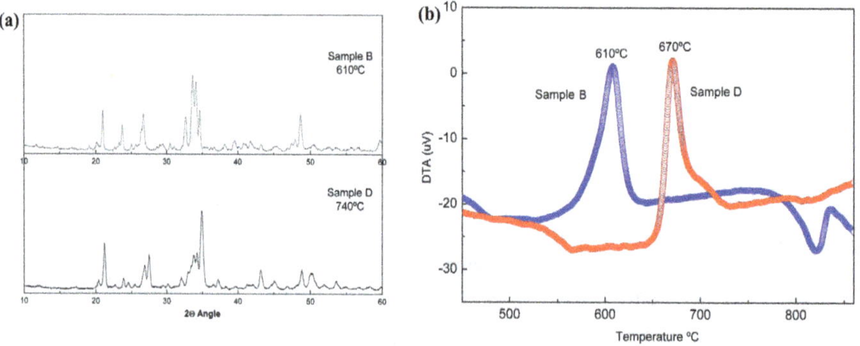

Figure 18. Crystallization evolution with temperature of samples B and D (a) XRD, (b) DTA.

Microscopy observations of all the samples permit to corroborate the information obtained by X ray diffraction and DTA curves. The crystallization mechanism begins at the surface of the samples where columnar crystals are developed. In samples PC and A crystals are constituted by cuspidine phase and in samples B and D by combeite phase. At higher temperatures (> 800°C), nuclei of irregular crystals appear in the inner part of the sample PC (Figure 19).

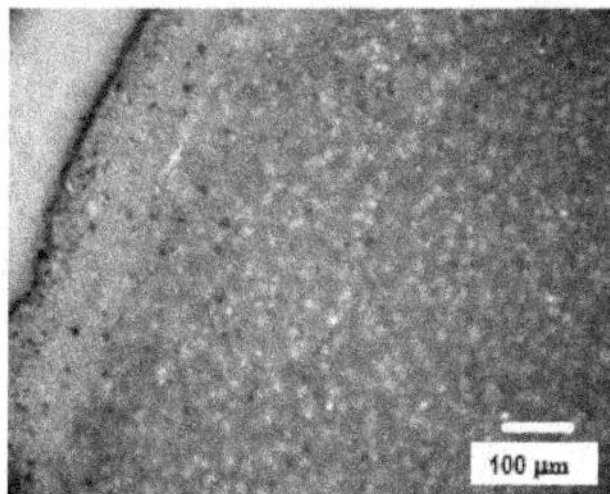

Figure 19. Morphology of sample PC at 850°C.

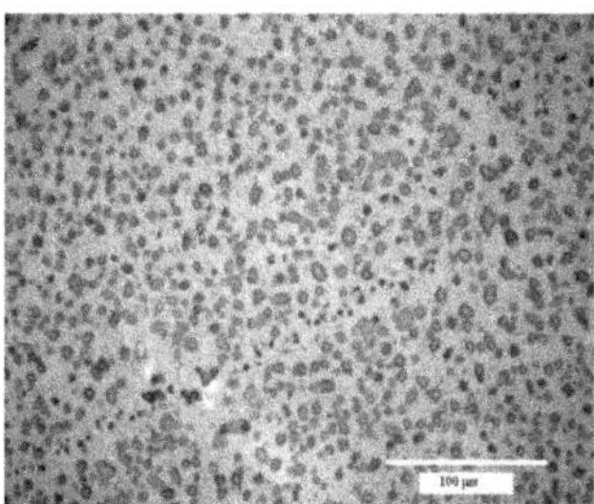

Figure 20. Liquid immiscibility phenomena observed at 610°C in sample B.

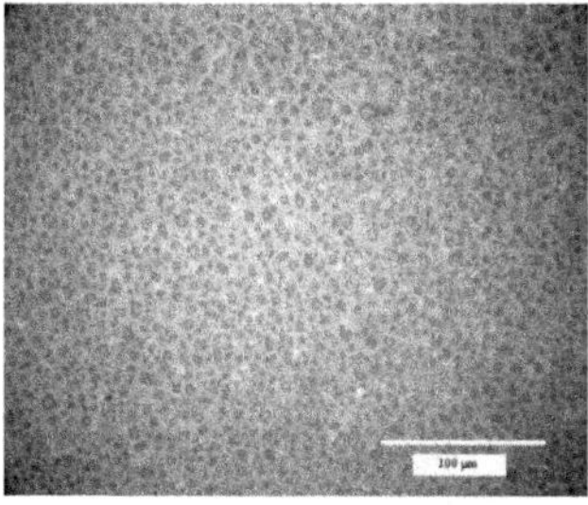

Figure 21. Liquid immiscibility phenomena observed at 680°C in sample D.

Samples B and D, present a liquid immiscibility phenomena (supercooled liquid effect), previous to the onset of the crystallization process (Figures 20, 21).

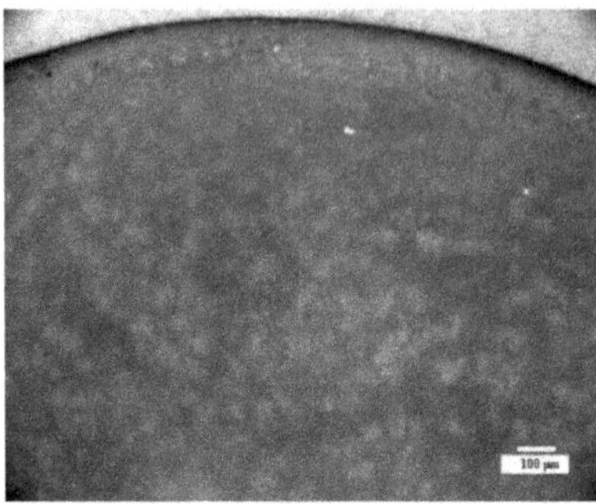

Figure 22. Morphology of sample B at 870°C.

In spite of liquid immiscibility and phase differences observed, the crystallization mechanism in the sample B at 870°C (Figure 22) is quite similar to sample PC. The presence of immiscible liquids phenomenon can be controlled by the degree of supercooling promoting a more homogeneous crystal nucleation.

Acknowledgements

The authors acknowledge the financial support of the Universidad Tecnológica Nacional (Argentina) and Ternium Siderar SAIC to promote the research in steel continuous casting process.

Author details

Elena Brandaleze[1], Gustavo Di Gresia[2], Leandro Santini[1], Alejandro Martín[1] and Edgardo Benavidez[1*]

*Address all correspondence to: ebenavidez@frsn.utn.edu.ar

1 Department of Metallurgy & DEYTEMA, Facultad Regional San Nicolás - Universidad Tecnológica Nacional, Argentina

2 Ternium Siderar SAIC, Argentina

References

[1] Mills, K., & Fox, A. (2002). Metals, Slags, Glasses: High Temperature Properties & Phenomena. *Mould Fluxes, Mills Symposium, The Institute of Materials*, 121-132.

[2] Thomas, B. (2001). Modeling of the Continuous Casting of Steel- Past, Present and Future. *Electric Furnace Conf. Proc. ISS*, 59, 3-30.

[3] Shin, H., Kim, S., Thomas, B., Lee, G., Park, J., & Sengupta, J. (2006). Measurement and Prediction of Lubrication, Powder Consumption, and Oscillation Mark Profiles in Ultra-low Carbon Steel Slabs. *ISIJ Int.*, 11, 1635-1644.

[4] Meng, Y., & Thomas, B. (2003). Heat Transfer and Solidification Model of Continuous Slab Casting: CON1D. *Met. Mat. Trans. B*, (34B), 5, 685-705.

[5] Wolf, M. (2003). Chapter 22. In: The AISE Steel Foundation, editors. Stainless Steels pp. 1-47.

[6] Mills, K. (1993). The Influence of Structure on the Physico-chemical Properties of Slags. *ISIJ Int.*, 33, 148-155.

[7] Waseda, Y., & Toguri, J. (1998). The Structure and Properties of Oxide Melts. World Scientific Publishing, Singapore.

[8] Mills, K., Chapman, L., Fox, A., & Sridhar, S. (2001). Round Robin Program for Slag Viscosity Estimation. *Scand. J. Metallurgy*, 30, 396-404.

[9] Riboud, P., Roux, Y., Lucas, L., & Gaye, H. (1981). Improvement of Continuous Casting Powders. *Fachberichte Huttenpraxis Metallweiterverarbeitung*, 19, 859-869.

[10] Iida, T. (2000). Accurate Prediction of the Viscosities of Various Industrial Slags from Chemical Composition. *J. High Temperature Soc.*, 25, 93-102.

[11] Brandaleze, E., & Benavidez, E. (2010). Effect of Different Oxides on Physical Properties of Complex Silicate Systems. 95° Physics National Meeting. Malargüe, Argentina.

[12] Moynihan, C. (1993). Correlation between the Width of the Glass Transition Region and the Temperature Dependence of the Viscosity of High-Tg Glasses. *J. Am. Ceram. Soc.*, 76, 1081-1087.

[13] Eriksson, R., Hayashi, M., & Seetharaman, S. (2003). Thermal Diffusivity Measurements of Liquid Silicate Melts. *Int. J. Thermophysics*, 24, 785-797.

[14] Hayashi, M., Ishii, H., Susa, M., Fukuyama, H., & Nagata, K. (2001). Effect of Ionicity of Non-bridging Oxygen Ions on Thermal Conductivity of Molten Alkali Silicates. *Phys. Chem. Glasses*, 42, 6-11.

[15] Hanao, M., & Kawamoto, M. (2008). Flux Film in the Mould of High Speed Continuous Casting. *ISIJ Int.*, 48, 180-185.

[16] Brandaleze, E., Martín, A., Benavidez, E., Santini, L., & Di Gresia, G. (2008). Development of an Equipment for the Measurement of Thermal Conductivity in Mould Fluxes. *Proceedings of 39th International Steelmaking Seminar*, ABM. In CD.

[17] Ozawa, S., Susa, M., Goto, T., Endo, R., & Mills, K. (2006). Lattice and Radiation Conductivities for Mould Fluxes from the Perspective of Degree of Crystallinity. *ISIJ Int.*, 46, 413-419.

[18] Nakada, H., Susa, M., Seko, Y., Hayashi, M., & Nagata, K. (2008). Mechanism of Heat Transfer Reduction by Crystallization of Mould Flux for Continuous Casting. *ISIJ Int.*, 48, 446-453.

[19] Mills, K. (1986). ACS Symposium Series 301 Mineral Mater and ash in coal. In: Vorres KS, editors. Estimation of Physicochemical Properties of Coal Slags. *Am. Chem. Soc.*, 195-214.

[20] Nakamoto, M., Tanaka, T., Holappa, L., & Hämäläinen, M. (2007). Surface Tension Evaluation of Molten Silicates Containing Surface-active Components (B_2O_3, CaF_2 or Na_2O). *ISIJ Int.*, 47, 211-216.

[21] Sridhar, S., Mills, K., Afrange, O., Lörz, H., & Carli, R. (2000). Break Temperatures of Mould Fluxes and Their Relevance to Continuous Casting. *Ironmaking and Steelmaking*, 27, 238-242.

[22] Brandaleze, E., Santini, L., Gorosurreta, C., & Martín, A. (2007). Influence of Carbonaceous Particles on the Melting Behaviour of Mould Fluxes at High Temperature. *Proceedings of 16th Steelmaking Conference IAS, Argentina*, 363-371.

[23] Wei, E., Yang, Y., Feng, C., Sommerville, I., & McLean, A. (2006). Effect of Carbon Properties on Melting Behavior of Mould Fluxes for Continuous Casting of Steel. *J. Iron Steel Res.*, 13, 22-26.

[24] Benavidez, E., Santini, L., & Brandaleze, E. (2011). Decomposition Kinetic of Carbonaceous Materials Used in a Mould Flux Design. *J. Therm. Anal Cal.*, 103, 485-493.

[25] Persson, M., Matsushita, T., Zhang, J., & Seetharaman, S. (2007). Estimation of Molar Volumes of some Binary Slags from Enthalpies of Mixing. *Steel Res. Int.*, 78, 102-108.

[26] Schwerdtfeger, K. (1983). Heat Transfer Through Layers of Casting Fluxes. *Ironmaking and Steelmaking*, 10, 24-30.

[27] Mikrovas, A., Agyropoulos, A., & Sommerville, I. (1991). Measurements of the Effective Thermal Conductivity of Liquid Slags and Mould Powders. *Ironmaking and Steelmaking*, 18(3), 169-181.

[28] Jenkins, M. (1995). Characterization and Modification of the Heat Transfer Performance of Mould Powders. *Steelmaking Conference Proceedings, ISS*, 78(3), 667-669.

[29] Yamauchi, A., & Sorimachi, K. (1993). Heat Transfer between Mould and Strand through Mould flux Film in Continuous Casting of Steel. *ISIJ Int.*, 33(1), 140-147.

[30] Mills, K. (1994). Thermal Properties of Slag Films Taken From Continuous Casting Mould. *Iron and Steelmaking*, 21(4), 279-286.

[31] Holzhauser, J. (1999). Laboratory Study of Heat Transfer through Thin Layers of Casting Steel. *Steel Research*, 70(10), 430-436.

[32] Stone, D., & Thomas, B. (1999). Measurement and Modeling of Heat Transfer Across Interfacial Mould Flux Layer. *Canadian Metallurgical Quarterly*, 38(5), 363-375.

[33] Martín, A., Brandaleze, E., Santini, L., & Benavidez, E. (2011). Study on a Mould Powder Layer Extracted During the Continuous Casting Process. *Proceedings of 18th Steelmaking Conference IAS*, 73-83.

[34] Mills, K. (1995). Viscosities of Molten Slags. In: Verein Deutscher Eisenhüttenleute editors. Slag Atlas, Second Edition. Mills: Verlag Stahleisen GmbH, pp. 349-352.

[35] Brooks, R., Dinsdale, A., & Quested, P. (2005). The Measurement of Viscosity of Alloys- a Review of Methods, Data and Models. *Meas. Sci. Techol.*, 16, 354-362.

[36] Persson, M., Görnerup, M., & Seetharaman, S. (2007). Viscosity Measurements of Some Mould Fluxes Slags. *ISIJ Int.*, 47(10), 1533-1540.

[37] Fox, A., Mills, K., Lever, D., Bezerra, C., Valadares, C., & Unamuno, I. (2005). Development of Fluoride-free Fluxes for Billet Casting. *ISIJ Int.*, 45(7), 1051-1058.

[38] Persson, M., Seetharaman, S., & Seetharaman, S. (2007). Kinetic Studies of Fluoride Evaporation from Slags. *ISIJ Int.*, 47(12), 1711-1717.

[39] Li, G., Wang, H., Dai, Q., Zhao, Y., & Li, J. (2007). Physical Properties and Regulating Mechanism of Fluoride-free and Harmless B_2O_3 Containing Mould Flux. *Journal of Iron and Steel Research International*, 14(1), 25-28.

[40] Brandaleze, E., Benavídez, E., Peirani, V., Santini, L., & Gorosurreta, C. (2010). Impact of Free Fluor Fluxes on Nozzle Wear Mechanisms. *Advances Science and Technology*, 70, 205-210.

[41] Kim, G., & Sohn, I. (2012). Influence of Li_2O on the Viscous Behaviour of CaO-Al_2O_3-12 Mass% Na_2O-12 Mass % CaF_2 Based Slags. *ISIJ Int.*, 52(1), 68-73.

[42] Tandon, S., Agrawal, R., & Kapoor, M. (1994). Viscosity of Molten Na_2O-B_2O_3 Slags. *J. Am. Ceram. Soc.*, 77(4), 1032-1036.

Chapter 6

Permanent Molding of Cast Irons – Present Status and Scope

M. S. Ramaprasad and Malur N. Srinivasan

Additional information is available at the end of the chapter

1. Introduction

We have only one earth and we must protect it. It is no more an option but an imperative that we adopt proactive measures to protect the earth and move towards Greener world. The official UN website lists 10 sectors for a greener planet. One of the sectors, is, Industries.

Industries drive economic growth, but they also produce pollutants and can exhaust natural resources. They also generate a lot of waste. If we do not curb the same, the planet may soon become chocked with rubbish.

Despite all the developments, foundry industry is way far from green. The situation is worse in the case of sand foundries. Sand foundries, in addition to producing hazardous air pollutants in the form of dust and fumes, also generate a lot of used sand as waste. Sand disposal is a serious problem and expensive. Our planet is threatened to become a dump yard for used foundry sand unless some feasible solutions are developed.

Sand foundries consume more energy too, thus resulting in higher fuel consumption, in turn leading to higher CO2 emissions. A Strong Energy Portfolio is needed for a strong economy of any nation.

Permanent molding (using a *reusable metal mold*, instead of a *dispensable sand mold*) offers a greener technology. While this technology is used to fairly good extent, for low temperature non-ferrous alloys, its application for the production of high temperature ferrous castings is rather limited. The technology of PM of ferrous metals was born almost a hundred years ago, but has not made much progress. Although some progress is seen in the last two decades, it is nowhere near desirable. The slow growth is vastly attributed to poor mold life. There is an urgent need to develop better mold materials.

This paper reviews in detail, the past developments in permanent mold technology for cast iron (including some research work done by the present authors). The present status of the technology is briefly discussed. Some plans for future work are suggested.

1.1. Permanent Molding Process (PM)

In Permanent Molding Process the molten metal is poured repeatedly into a reusable, refractory coated, metal mold, to produce a large number of shaped castings. This is unlike all the variants of the conventional Sand Casting process (SC), which use a dispensable mold. The repeated usage of the mold is the main advantage of the PM process.

It is very essential to make the following clarifications at the outset.

• The word Permanent does not mean that the molds last forever. In fact, the useful / service life of the mold depends largely on the pouring temperature, the material of the mold and the complexity of the component being cast [1]. The other factors are: casting weight, the thermal cycle, mold preheating, mold coating, gating design, cleaning, storage & handling, and, whether the operation is manual or automated. The end use of the casting also has a bearing (If the structural function of a casting is the only criteria, and not its appearance, a mold can be used longer before discarding) [2].

• Although, by and large, the permanent molds are metallic, graphite molds, used at times, also come under the category of Permanent Molds [2].

• The cores employed may be either metallic or made of sand. When sand cores are used, it is called a Semi-Permanent Molding (SPM) process.

• Permanent Molds are used in a number of variants of casting processes like Gravity Die Casting (GDC), Low Pressure Die Casting (LPDC), High Pressure Die Casting (HPDC), Centrifugal Casting (CFC), Squeeze Casting (SC) and Continuous Casting (CC).

• Throughout this paper, the terminology Permanent Molding is used to mean Gravity Die Casting only.

• Some foundrymen call it Chill Casting Process (CCP) since the metal mold cools the casting rapidly.

1.2. Advantages of PM

In addition to the main advantage over the sand casting process as mentioned above, the PM process offers several other distinct advantages like:

• Higher productivity (7-10 tons / man / month as against 3.5 tons / man / month in the case of sand casting process) [3],

• Better repeatability, dimensional stability, geometric fidelity and near - net shaped castings.

• Denser castings (finer grain structure), and superior surface finish that reduces the post-casting cleaning operations. Better surface finish also renders improved static bending and fatigue properties.

• Closer dimensional tolerances and hence lower machining costs,

• Elimination of sand (less polluting) and hence no costly sand handling equipment (& its maintenance),

• Reduced floor space and the ease of mechanization for mass production,

• Better process control due to the flexibility in design for heating and cooling of any particular location in the mold;

• Possibilities of incorporating certain design features for achieving a higher casting yield.

• The process is more energy efficient than sand casting process since the heat remains within the process loop.

1.3. Disadvantages of PM

There are several disadvantages in employing PM as compared to SC. The serious limitations are with regards to:

• The limitation on types of alloys that can be handled,

• Size, Shape and Section thickness of the castings,

• The batch size that can be economically handled. Since the tooling costs are relatively high, the process can be prohibitively expensive for low production quantities [2].

1.4. Few Other Issues Concerning PM

The flowability (fluidity) and fillability of metal in metal molds is poorer compared to sand casting process. Permeability of the mold is *zero* which calls for extremely carefully designed Air Venting System.

Due to the faster heat extraction, the rigidity of the metal mold (and metal cores), as also due to the thermal expansion / contraction problems associated with the metal molds (and metal cores), the stresses developed in the castings during the solidification is much higher than in the sand castings. This calls for a very careful mold and core design as well as proper casting extraction method.

Air Gap formation is one unique feature applicable to metal molds and this has a significant effect on the mode and hence the heat transfer rate through the mold. Since the structure of the solidifying casting partly depends upon the freezing rate, a thorough understanding of the behavior of air gap formation is very vital for satisfactory design of the mold and operating parameters. The pattern of air gap formation also affects the location of the shrinkage within the casting [4].

Unlike in the case of sand casting process, where the metal after preparation and treatment can be poured into several molds in one go, in the case of PM process the metal is often held for a while (sometimes for hours) for repeated pouring into a set of dies. Holding the metal for long has its own associated quality issues (temperature drops and fading effect of certain melt treatments).

1.5. Where Does PM Stand Today?

Although Permanent Mold casting ranks second to sand casting in terms of popularity, the tonnage produced by the process is only a small percentage of that made by sand casting [2].

1.6. March Towards Green Foundries

Recent years has witnessed some serious attempts made towards green foundry operations [5-10].

Today's Global Green Initiative has prompted manufactures, including foundrymen, worldwide, to seriously look into Environmentally Benign Manufacturing (EBM) [5]. Foundry industry is one amongst a very few others that consume a lot of energy and also produce considerable amount of dusts & fumes, and wastes. The sector has an uphill task in going greener.

The speech presented by Gigante, as the American Foundry Society Hoyt Memorial Lecture for 2010 touches upon the issue of The Green Assault in foundries [6].

The 2002 Annual Report on Metal Casting Industry of the Future published by the US Department of Energy [7] says that as per the priorities outlined in the Metal casting Technology Roadmap of USA, 2/3rd of research funding goes toward improvements in manufacturing processes, where greatest opportunities for energy saving exist. Additional research funding is going to improvements in material performance (thereby reducing scrap and increasing yield), as well as to address environmental needs such as recycling of foundry spent sand. According to this report, Metal Casting is one of the most energy intensive industries in the United States and it is very critical to the to the U.S. economy as 90% of all manufactured goods contain one or more cast metal components and that the metal castings are integral in U.S. transportation, energy, aerospace, manufacturing, and national defence. Situations are likely to be similar in most other countries.

Technikon LLC, a privately held contract research organization in California operates the Casting Emission Reduction Program (CERP), a cooperative initiative between the Department of Defence (U.S. Army) and the U.S. Council for Automotive Research (USCAR). During 2004 - 2007, Technikon has published a number of reports [8-10] based on detailed studies carried out on connected topics like:

the sources of various Hazardous Air Pollutions or HAPs – both organic and inorganic (metallic), in different foundry operations[8], Monitoring Systems for HAPs [9], Energy Reduction in Foundry operations[10], the development of economically feasible permanent Mold system for high temperature alloys like iron, steel, Nickel, and Titanium[1]. The conclusions of these studies give a very good indication of the task ahead of foundry industry to become Green.

A study of the above reports give a hint that foundry industry will now be under a constant scanner and they will face never - ever - seen pressure due to stricter & newer environmental acts that are emerging globally. Foundries will be compelled to reduce emissions of fumes and dust so as to comply with these stricter norms. Further, their operations must be improved or changed to become more and more energy efficient to reduce the fuel consumption. It appears that all the future developments in the field of foundry will be dictated more by this Green Initiative than any other factor.

1.7. Foundry Scenario From the Above Perspective

On a worldwide average, sand castings account for almost 80% of the castings produced. Despite advancements in the foundry technology, sand casting operation is far from Green in the following respects and hence is a serious hindrance to *The March Towards A Green Planet.*

• Sand casting foundries emit a lot of dust and fumes causing environmental pollution and health hazard to operators. This is in addition to the problem of heat normally involved in any foundry (Inadequacy of labor force to work in such environment has already affected the foundry sector).

• Sand costs and sand transportation costs are constantly going up [1]. Sand mining may face restrictions in future.

• Sand reclamation systems are energy intensive and expensive to operate & maintain.

• Sand disposal is a serious problem and is expensive. Our planet is threatened to become a dump yard for used foundry sand unless some feasible solutions are developed.

• HAPs' monitoring systems are also expensive to operate and maintain [1].

• Foundries in general, and sand casting foundries in particular, may be eventually forced to move to remote areas (where infrastructure may be inadequate). Sand transportation cost may also go up as a consequence.

• As mentioned earlier, sand casting operation is less energy efficient compared to PM process.

• As per the statistics available, mold & core making, and shot blasting operations consume almost 27% of the total energy cost in a foundry. This will be far less in the case of PM process. Even if PM process uses sand cores, the organic emissions would be relative only to the amount of core [8].

These above mentioned issues are prompting foundrymen worldwide to seriously consider possibility / feasiblity of converting some sand castings to equivalent PM castings. Holmgren and Naystrom [11] strongly advocate that for a Green Foundry, one must not only use the Best Available Technique (BAT), but also evaluate and create better and better techniques (through Practice - Oriented R & D) for a good environment. One obvious approach is of course the increased utilization of Permanent Molds, which almost eliminates a sand waste stream [1]. In fact, for some castings, minor changes can permit conversion to PM castings thereby giving the above - mentioned benefits with regards to reducing HAPs, in addition to considerable cost savings [2]. The present authors firmly believe that in the very near future, such environmental issues will bring about *Compelled-process-Changeovers*. This will bring additional opportunity to PM process. This applies not only for non - ferrous castings but to ferrous castings as well (mainly, cast irons).

This brings us to our main topic of *Permanent Molding of Cast Irons.*

1.8. Permanent Molding of Cast Irons

The application of PM for ferrous alloys has been rather limited. The published literature on the subject is also very little. The subject is addressed only here and there in some publications, only occasionally, covering some very general aspects. It appears that a thorough understanding of the subject is somewhat lacking and that this subject has not been given its due attention. Most foundrymen raise their eyebrows in disbelief at the mention of cast iron production by PM process!!! This clearly shows that the technology has not been popularized to the extent it deserves and there is a serious lack of awareness.

However, it is well in place to mention here that there are a few publications [12,13] that give an indication that PM Cast Iron castings are produced in reasonable quantities in several countries of Former Soviet Union (almost 15 %), Eastern Europe, Germany and Japan, in a small way in USA and Canada, and a few Asian countries. Lerner [13] mentions that although the technology of PM of cast iron originated on the U.S. soil, the process has been more widely embraced overseas. According to him, in Europe, 6-8 % of all iron castings are made by PM, and, that the growing use of the process is also seen in China and India. However, beyond such general information and a minimal statistics quoted here and there, no detailed information is available on this technology, both in terms of research and practice.

Considering the great potential that this technology has, particularly in the context of going Green as discussed above, there is an urgent need to work on improvements in the process. The very first step is to bring the awareness on this technology amongst the broader spectrum of foundry community. The authors of this paper are constantly working in this direction with reasonable success.

In what follows, the authors present a brief review of the work done world over, in the past – in the chronological order. They share their own findings based upon their research and practice.

2. Work Done So Far on The PM of Cast Iron

• *15thCentury* - Cannon balls of iron were made in two part metal molds at the end of 15 th century and a patent covering this process was taken out in Germany in 1898 [14].

• *1920s* - Holly Corp cast the Ford Model " T " carburetor of gray iron in PM [12,15,16]. They sold the company to Eaton Corporation, Michigan, in 1930. From then onwards, the process is called by the name Eaton Process. (During that period, Forest City Foundry was the only other making substantial use of the process of cast iron PM [12, 17].

• *Aug.1925,* Walter Anderson of USA got a patent for developing an improved permanent mold for cast iron [18]. The invention related to the design of the in-gates and air vents to enable proper filling of the metal.

• *1932* - The Ferrous Permanent Mold (FPM) Process was patented by Eaton Corporation [13].

• *1959* - The very first significant publication on Eaton Process [19] provided valuable practical information on the process – the iron poured, mold material, die operating parameters, the coatings, heat treatment, the structure & properties of the end product and finally mold failure modes. The paper also provided some valuable information on the type of castings made by the process using a twelve - station turntable Eaton Permanent Mold Machine, with varying speeds of rotation.

• *1965* - A publication from Foseco [14], provided very useful practical information on the process parameters. In addition, the paper gave some information on mold design (gating, venting and feeding). The author also discussed the influence of the mold weight / casting weight ratio (WR), mold temperature & pouring temperatures, mold coating, mold cooling parameters and the casting removal, on the mold life. Casting defects common to the process were also highlighted. It was clearly spelt out that in addition to the design and chemical composition of the mold, WR is also very important.

• *1966 - 1973.* Although not directly a part of the present topic, it may not out of place to make a mention of few developments in other variants of PM casting of ferrous materials. Some of the experiences gained through this can possibly be made use of in the Gravity Die Casting Process also.

1. Progress made by Lamp Metals& Components Dept., General Electric Co., Cleveland on the Pressure Die Casting (PDC) of Ferrous materials (gray iron, malleable iron, ductile iron, and various steels) using molds made of unalloyed - pressed & sintered molybdenum [20-24].

2. Southern Research Institute, Birmingham, Alabama, USA successfully employed graphite permanent molds for gray and ductile iron castings [25]. The paper claims that the cost benefit and quality of end product of this process, as compared to sand casting process, is very attractive. This is in addition to lesser emission, better safety and lesser health hazards.

3. The successful development of pressure die casting of ferrous materials in Federal Die Casting Co., Chicago and its expansion unit in Ireland. Tungsten and molybdenum were used for the molds to overcome the temperature problems [26].

4. A publication from Poland [27] indicated the usage of Shaw Process for producing the permanent molds (molds for pouring both ferrous and non-ferrous alloys). Traditional methods of making the permanent molds by means of machining semi finished cast products with considerable allowances for machining are time consuming, expensive, requires specialists and special equipment. Reduction/elimination of machining of mold working surface brings about some savings in mold material, labor cost and investment cost. Considering the cost of molding materials used in Shaw Process, the ceramic slurry is used only for that part of the mold that is a direct reproduction of its working surface, which in turn corresponds to the outer surface of the final casting. This is a very useful information for implementation.

• *1967* - A book by Fisher [28] devoted a chapter on the technology of PM of cast iron. The chapter addressed the issues like cast iron composition, mold material, mold coating, die & pouring temperatures.

• *1968* - Yet another important publication from Eaton Corp [15] provided various practical aspects of the process.

• *1968 - 1970* - Skrocki and Wallace at Case Western Reserve University did research on various aspects of PM of cast iron. They studied the effect of mold coatings, mold & pouring temperatures, velocity and the casting section thickness on the filling ability of the metal [29]. They also made a significant contribution to the understanding of solidification behavior of cast iron at high cooling rates that are encountered in PM process [30-31]. This understanding, in turn provided very valuable information on the resulting structure and properties of the end product, under different operating conditions.

• *1970* - Chapter on Permanent Molding in Metals Hand Book, Vol.5, 8th edition [32] gave a brief description of the process of PM of cast iron. A publication, " First Annual Summary of recent literature on PM Casting of Cast Iron " by Schoendorf [33] which appeared in the same year gave very valuable information on the subject.

• *1970 - 1973* - Ramesh [34] undertook a long-range study on various aspects of casting hyper eutectic cast iron in metal molds.

• *1972* - A publication from Zuithoff et al [35] indicated that there was a steady increase in production of cast iron permanent mold castings in the East European countries. The report also mentioned that in the then U.S.S.R increasing quantities of nodular cast iron were produced using PM process. The paper further provided some interesting statistics indicating that in the Western Europe and U.S.A. also there was a rising trend in the production of cast iron castings by PM process (in England, 2.5 % of total output in 1957 and 4 % in 1967; and in U.S.A. 1.5 % in 1957 to 5% in 1967).This clearly indicates that the growth in volume of permanent molded cast iron castings between 1920 and1967(nearly 5 decades), has been very insignificant, more so considering the world average. This point is noteworthy and deserves detailed probing into the reasons behind.

• *1973* - The authors of the present paper published a detailed analysis of the past literature on the subject [36].

The analysis showed that the process of cast iron PM was still not fully exploited commercially, the progress appeared quite slow, and that there was still a vast lack of knowledge on the thermal and metallurgical aspects of permanent molded cast irons. The reasons for slow progress were attributed to the following.

a) The pouring temperatures involved are higher there by putting a higher demand on the metal for the mold.

b) Cast iron as an alloy, though very easy to cast, it is very difficult to understand in terms of the behavior. The structure and properties of cast iron not only depend upon the Chemical Composition, Melt Treatment and Heat Treatment but also vastly on the cooling rates during solidification. Cast iron is a section sensitive alloy. The matrix structure and the graphite morphology could vary from one extreme to the other. Further, It is possible for the same casting to have several combinations of graphite forms and matrix, at different locations, which means that the properties such as strength, ductility, machinabilty, wear resistance, damping capacity, and others could be subject to variation over rather wide limits. Since these properties are a consequence of the structure, which in turn is related to solidification (cooling rates), it was felt essential to generate knowledge on these aspects of PM of cast iron.

Considering this gap in knowledge, the present authors, then at the Indian Institute of Science, initiated a 3 year long research project. The parameters studied included the size and shape factor of the casting, composition of the metal, the mold & pouring temperature, mold wall thickness, the coating material & thickness, and the melt treatment. The effect of these parameters on the solidification, structure of graphite & matrix and strength & hardness were studied in great depth.

The magnitudes of the several process variables for the above research project were so chosen after a careful analysis of the earlier literature cited above, as to conform as closely as possible, with those employed by the previous investigators, as well as in industrial practice.

The main data drawn from the earlier literature are summarized in Table 1. All the relevant details regarding the various experimental conditions employed in this research project are set out in Table 2 and 3.

Out of the above study, large amount of valuable data was generated on the effect of these parameters on the air gap formation time, solidification time, solidification rate, the mold temperature distribution, the heat extraction rate, the resulting microstructure, tensile and hardness properties. The microstructures were studied not only with optical microscope but also with Scanning Electron Microscope (SEM). The SEM studies revealed a lot more information. In addition to understanding the matrix and the graphite structure as separate entities, it was possible to understand the pattern of the interface between the matrix and the graphite and how smooth or otherwise the graphite – matrix interface is. The type of this interface appeared to have a strong influence on the strength properties. With slower solidification, although the graphite is coarser, the strength was higher presumably due to smoother interface that is likely to reduce the stress concentration.

The findings of the above research have already been reported in several publications by the authors [37-42].

Since most of the data and the analysis of the above research have already been published, all those are not covered at length in this paper. Only a few important findings are presented in brief. Very large amount of data has been generated on the thermal behavior of the molds. It must be appreciated that this research was conducted in 1974-75, almost 37 year back. With the present day advancement in the various computer simulation techniques, one can generate these data fairly accurately. Hence, for these thermal aspects, only some typical graphical representations and a summary are given. However, many SEM microstructures (not exhibited in the earlier publications) are presented for the benefit of the readers, since the microstructure part cannot be so easily / accurately predicted by the use of a software.

1	Material of cast iron poured	Hypereutectic cast irons. (Carbon Equivalent, C.E in the range of 4.20 to 4.60) are invariably used for permanent molding [3,14-17,19,28-35]
2	Mold Material	Cast Iron [3,14-16,19,28-32,35]. In fact most recommend a cast iron of composition same as the alloy cast [15,16,19,28,32].One recommends special alloy cast iron and Ductile Iron [14] for achieving better life.
3	Mold Coating	Most investigators recommend a primary coating consisting of a mixture of China Clay, sodium silicate and water, with a secondary coating of Acetylene Soot [14-16,19,28-34].
4	Mold Temperature	Most recommend a temperature range of 150-250°C [14-16,19,35]. However some recommend slightly higher temperature of upto 350°C [3,32.]
5	Pouring Temperature	Most recommend 1250-1350°C [14,32], while a few recommend upto 1400°C [3,17]
6	Mold wall Thickness	The normally employed mold wall thickness is 12.50 to 31.00 mm and the widely used Volume Ratio (Volume Of the Mold / Volume of the Casting) is about 5.00 [19].
7	Inoculation of the metal	Invariably all the melts are inoculated before pouring into the mold.
8	Heat Treatment of Castings	Normally castings are given annealing treatment (heat uniformly and rapidly to 860°C, hold sufficiently long to secure equilibrium between Austenite, Cementite and Graphite (normally about 75 min. for castings not exceeding 25 mm wall thickness), cool slowly to ensure breakdown of Cementite to Ferrite and Graphite – say at the rate of 3° per min., between 860°C and 600°C) [14,15,32]. Annealing results in uniformity in hardness and grain structure that gives many machining advantages like machining with greater feeds and speeds and longer tool life. Normally, it is difficult to retain a sharp corner or a smooth thread during machining of annealed gray cast iron due to the pullout of coarse graphite flakes. Such problems are not faced in PM cast iron castings owing to very finely dispersed under cooled graphite structure.

Table 1. Process Variables – Data from past literature.

1	Alloys Poured	% C - 3:45, % Mn - 0.6, % P - 0.27, % S - 0.09 and % Si - (a) 2.42 *, (b) 3.00, (c) 3.62 *
2	Mold Material	%C-3.5, % Si - 3.2, % Mn - 0.55, % P - 0.36, % S - 0.042.
3	Mold Coatings	a) Primary coat: China clay : Sodium Silicate : Water (4:1:20 by weight)-0.2 mm thick.
		b) Secondary coat : Acetylene soot-0.1mm thick.
4	Test Castings**	a) Cylinders: 150mm heights. Cylinder dia (D_c, mm) -- 37.5, 62.5, 87.5 and 112.5 **.
		b) Plates: 150mm width x 125mm height. Plate thickness (t_p, mm) -12.5, 18.75, 25.00 and 31.25.
5	Test Molds**	Mold Wall thickness(MWT),mm of plate & cylindrical molds-12.5, 18.75, 25.00 and 31.25.
6	Mold Temperature, (M.T, °C):	150, 200, 250 (300 and 350 in a few cases only)
7	Pouring Temperature, (P.T, °C):	1250, 1300 and 1350

Table 2. Process variables employed in the Research Project.

Batch No.	1	2	3	4	5	6	7
% Si	3.00	3.00	3.00	3.00	3.00	2.42	3.62
M.T. °C	250	200	150	150	150	150	150
P.T. °C	1350	1350	1350	1300	1250	1250	1250

Notes: * % Si of 2.42 and 3.62 were used only for limited combinations as shown in Table 2.

** Combination of Cylinder of dia.112.5 mm and test mold wall thickness of 12.5mm was not poured since the Volume ratio VR (volume of mold / volume casting) is too low.

Table 3. Combination of % Si, M.T, P.T. for different experiments.

A) Findings on Solidification, Structure and Properties of the Castings

a) Solidification time

The plots of the solidification time of test castings (T, sec.) against the corresponding volume to surface area ratio (V / SA) indicated that there exists a relationship of the form $T = K (V/SA)^n$ (where K is a constant) as in [42] when the casting size alone is varied. The value of n is constant for a given casting shape, being 1.8 for plates and 1.6 for cylinders, irrespective of the mold wall thickness, mold temperature, pouring temperature and the silicon level. The value of K, however, increased with increase in initial mold and pouring temperatures and with decrease in mold wall thickness. Variations in the silicon level did not change the value of K. it is very well known that similar equation holds good in the case sand castings the value of n being 2, irrespective of shape.

b) Microstructure of castings

The relationships between the type of graphite and the solidification time, & the type of matrix and the solidification time are shown in Fig. 1 [42]. If solidification time is reckoned as a measure of the cooling rate of the casting, then it is evident from this figure that the type of graphite changes from under cooled type to flake type as the cooling rate is progressively decreased from a high value (Figs.2-3 and 4-6 and Table. 4).

In addition, the matrix changes from predominantly ferritic to a mixture of ferrite and pearlite, and again to predominantly ferritic. At very high cooling rates however, some pearlite is associated with ferrite (Fig. 1).

The observation of undercooled graphite at the surface in all castings but for those cooled very slowly, and the presence of flake graphite in gradually increased quantities towards the centre in larger castings in the present series of investigation, is in well in keeping with the trend noted above.

The matrix also changes in a predictable manner from the surface to the centre on the basis of the above consideration. Thus the microstructures of these gray cast iron castings can be predicted with confidence on the basis of heat conduction considerations. It is interesting to note that the experimental results of Skrocki and Wallace [30] are in accordance with this in respect of castings poured into molds preheated to different temperatures.

There appear to be ramifications in a given type of graphite when the structure is observed by scanning electron microscopy. However changes within a given type of graphite (undercooled or flake) also occur in a predictable manner on the basis of heat conduction considerations. Thus, as the cooling rate is progressively decreased from a high value, heavily branched undercooled graphite (Fig.7-10, 17-18) changes to rounded undercooled graphite (Fig.13-14, 33). Further reduction in cooling rate results in the appearance of flake graphite with a moderate degree of branching (Fig.15-16, 19-24,27-28,38) and at very low cooling rates coarse flake graphite (Fig.25-26, 29-30, 34-36) and some with surface protuberances (Fig. 37) is observed in the microstructure.

The SEM structures showed that in fact the graphite formed shows variety of interesting patterns like branching, curling, twisting, bending, folding, coarse graphite, smooth graphite, graphite with surface protuberances, etc., under various operating conditions. This is possibly a subject in itself, with a vast scope for further investigation. To give an idea to the readers on this aspect, several SEM pictures are presented. Those who are practicing PM of cast iron may be able to relate some of these features to their own observations, and throw some light.

The matrix changes observed in the castings led to the postulation that diffusion distance, rate of diffusion of carbon, and surface area offered for the diffusion of carbon are all important considerations in determining the type of matrix present in a permanent mold gray cast iron casting.

c) Eutectic Cell count:

Plots of eutectic cell count values at the centre of the casting vs. solidification time show appreciable scatter especially at low solidification times [38]. It is nevertheless evident that the eutectic cell count decreases with decrease in cooling rate of the casting.

d) Tensile Strength And Hardness:

Fig. 39 shows that the tensile strength gradually decreases with increase in solidification time until about 180 seconds and the decrease thereafter is much less marked. As seen in Fig. 1 castings with solidification times longer than 180s have a predominantly ferritic matrix associated with flake graphite at their centre. It is therefore evident that with this type of structure the tensile strength is not appreciably reduced despite the coarsening of the graphite as well as the matrix. One factor which could be of importance in leading to such behavior may be the smoothening of the leading edge of graphite which could be responsible for reduced notch sensitivity. Figure 40 shows the effect of variation of %Si on the tensile strength.

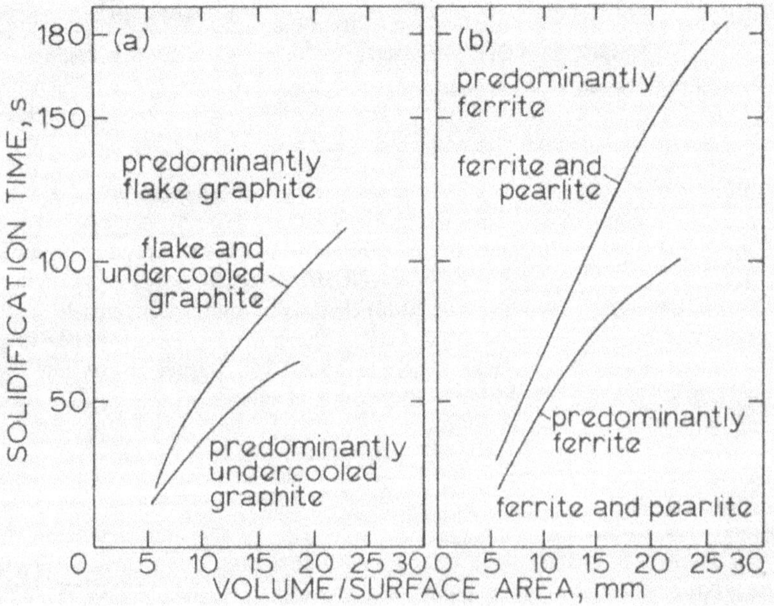

Figure 1. Variation in the graphite and matrix structure in gray cast iron [42].

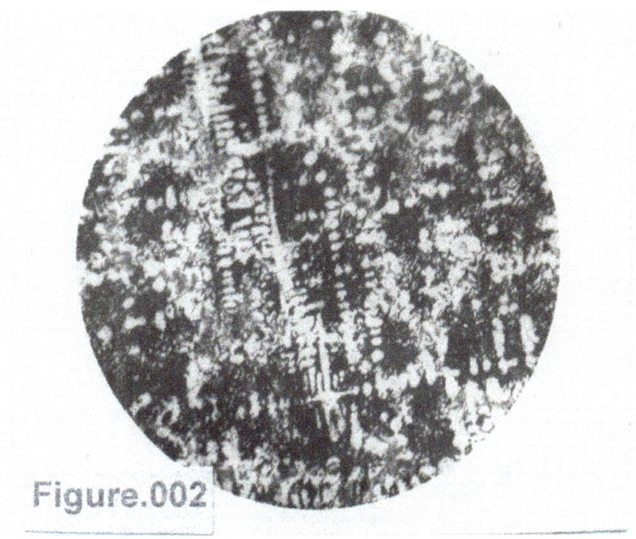

Figure 2.

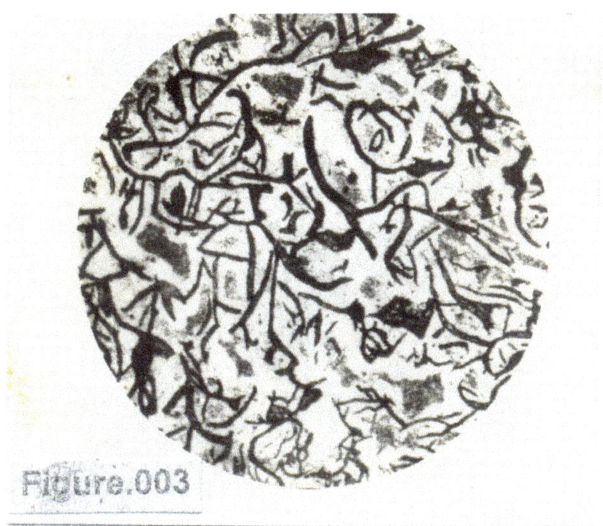

Figure 3.

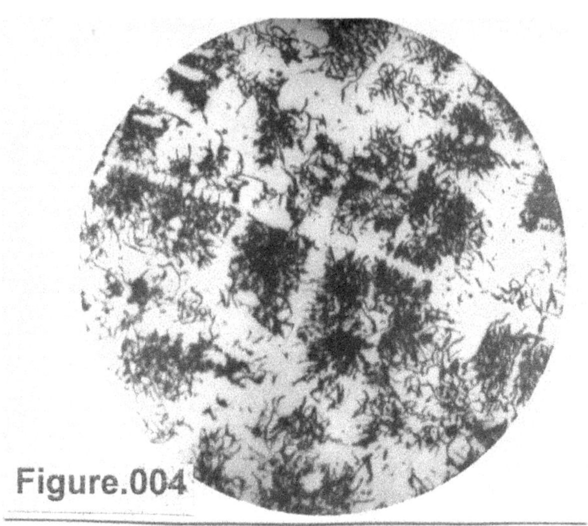

Figure 4.

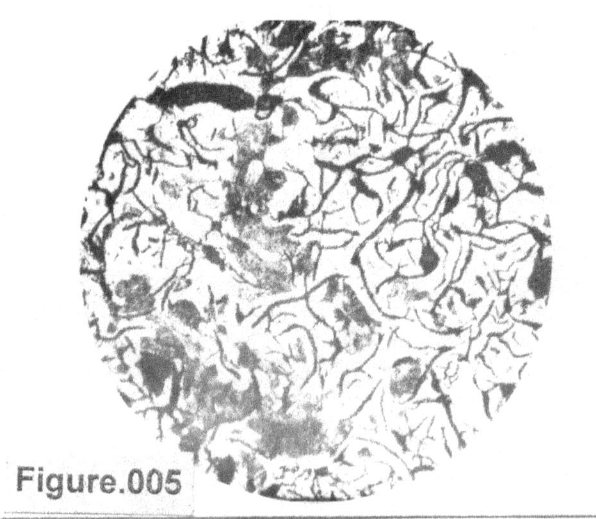

Figure 5.

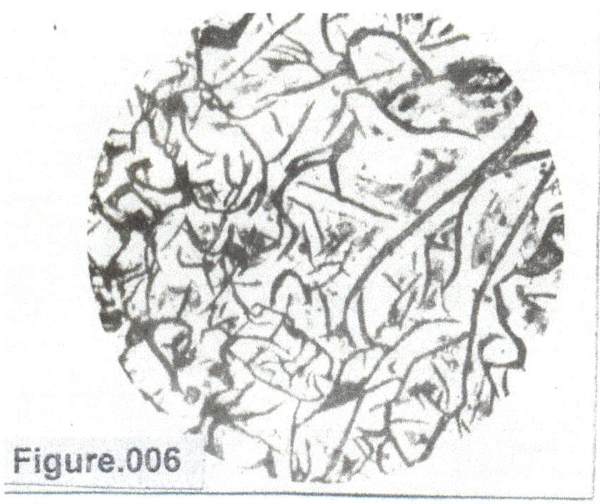

Figure 6.

Further it can be seen from Fig. 1 that castings with solidification times less than 180 sec. may have a variety of graphite - matrix combinations. Since the tensile strength falls continuously with increase in solidification time in this range (Fig. 39) it is to be surmised that factors tending to increase the notch sensitivity such as the coarseness of graphite of a given type, increased pearlite spacing, and coarseness of ferrite override the beneficial effect of the smoothening of the leading edge of a given type of graphite as the solidification time is increased in this range. Fig. 40 shows the effect of % Si on tensile strength. Lower the % Si, higher is the strength, in the range studied.

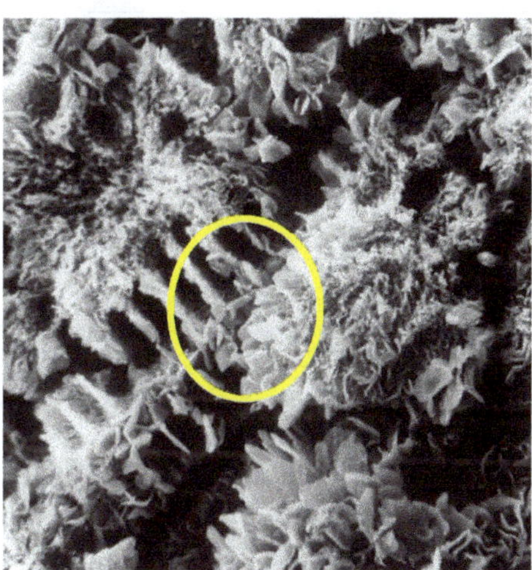

Figure 7.

Figure.	Shape	Size	MWT	M.T °C	P.T °C	% Si	Location	Magnification
2	Plate	12.50	31.25	1350	150	3.00	Surface	100
3	Cylinder	87.50	31.25	1350	250	3.00	Centre	100
4	Cylinder	87.50	12.50	1350	300	3.00	Surface	100
5							Intermediate	100
6							Centre	100
7	Plate	12.50	31.25	150	1250	3.62	Surface	420
8								2100
9	Plate	12.50	31.25	150	1350	3.00	Surface	420
10								2100
11	Plate	12.50	31.25	150	1250	3.62	Surface	420
12								2100
13	Cylinder	62.50	31.25	250	1350	3.00	Surface	420
14								2100
15	Cylinder	87.50	12.50	250	1350	3.00	Intermediate	420
16								2100
17	Cylinder	87.50	12.50	150	1250	3.00	Surface	420
18								2100
19	Cylinder	87.50	12.50	150	1250	3.62	Centre	420
20								2100
21	Plate	12.50	31.25	150	1250	3.62	Centre	420
22								2100
23	Cylinder	87.50	12.50	350	1350	3.00	Surface	420
24								2100
25	Plate	31.25	12.50	150	1250	3.62	Centre	420
26								2100
27	Cylinder	62.50	18.75	250	1350	3.00	Centre	420
28								2100
29	Cylinder	62.50	31.25	250	1350	3.00	Centre	420
30								2100
31	Plate	12.5	25.00	150	1350	3.00	Surface	2100
32	Plate	12.50	31.25	150	1250	3.62	Centre	420
33	Cylinder	112.50	31.25	250	1350	3.00	Surface	2100
34	Cylinder	112.50	31.25	250	1350	3.00	Centre	2100
35	Cylinder	87.50	12.50	150	1250	3.00	Centre	2100
36	Cylinder	87.50	12.50	250	1350	3.00	Centre	2100
37	Cylinder	87.50	12.50	350	1350	3.00	Centre	2100
38	Cylinder	87.50	12.50	150	1350	2.42	Centre	2100

Table 4. Values of Casting Parameters applicable to microstructures (both Optical and SEM).

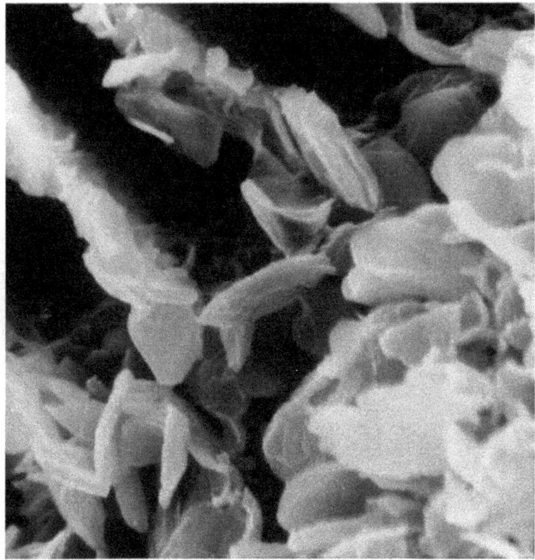

Figure 8.

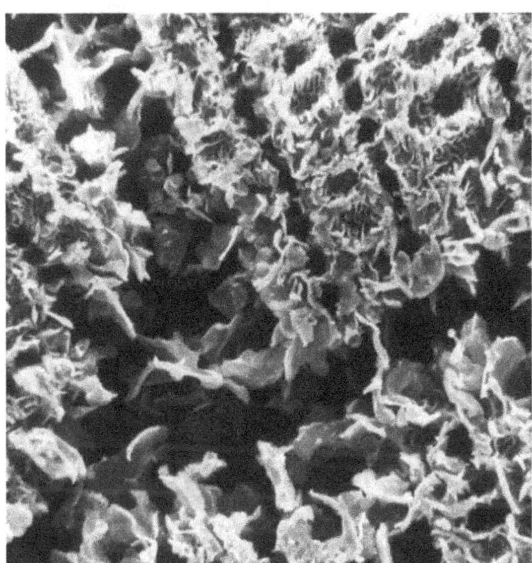

Figure 9.

Figure 10.

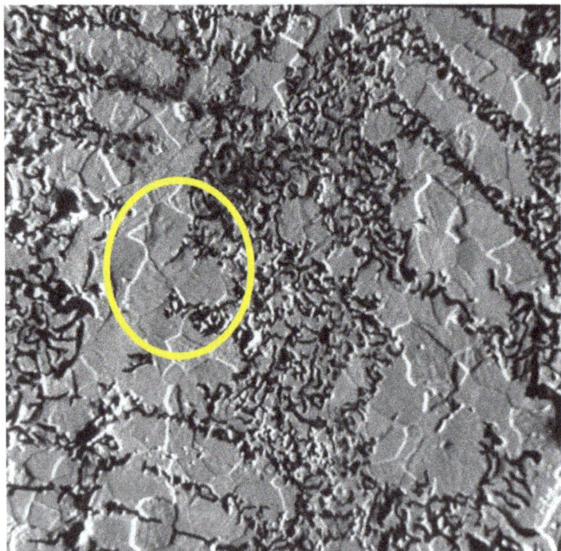

Figure 11.

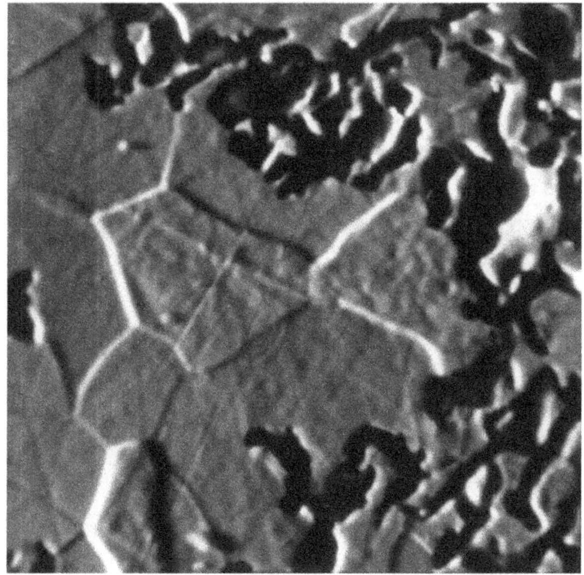

Figure 12.

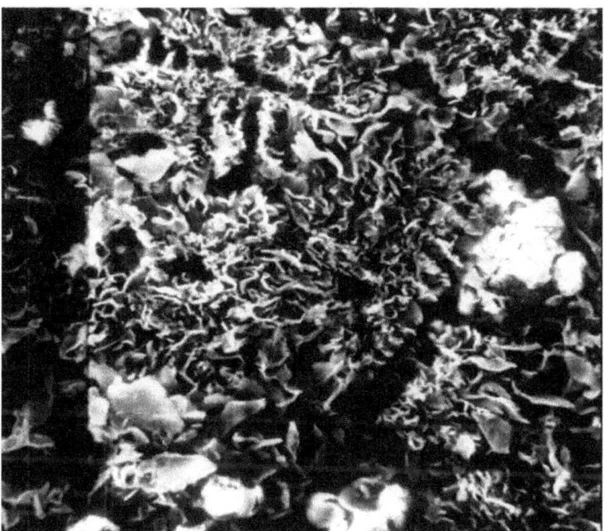

Figure 13.

Figure 14.

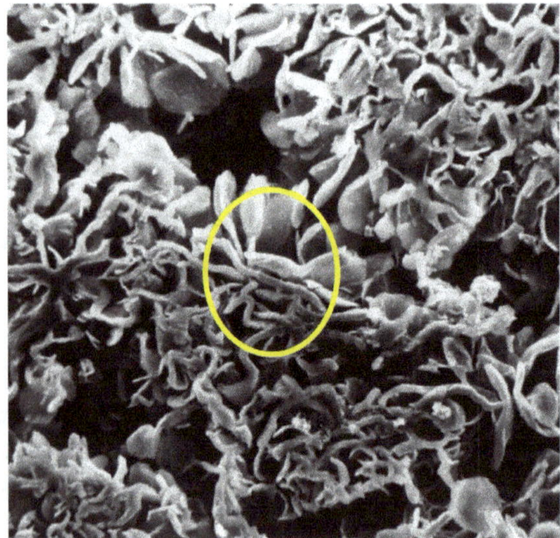

Figure 15.

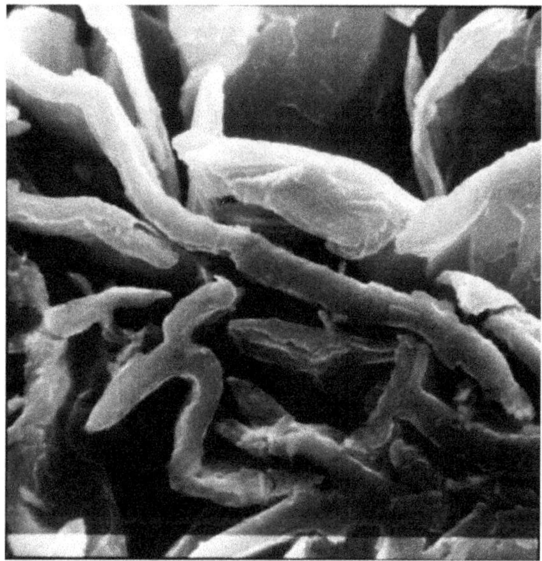

Figure 16.

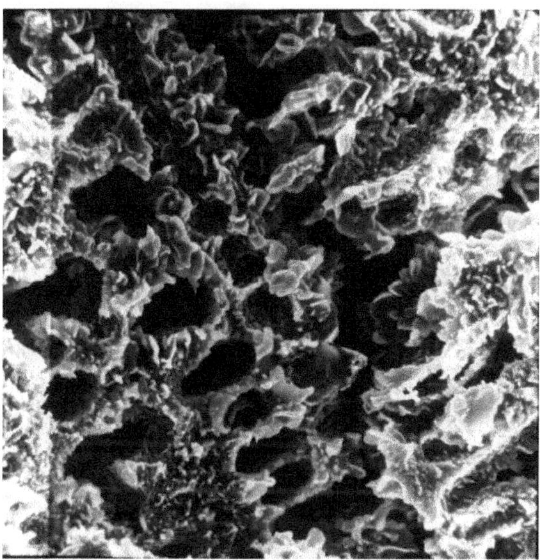

Figure 17.

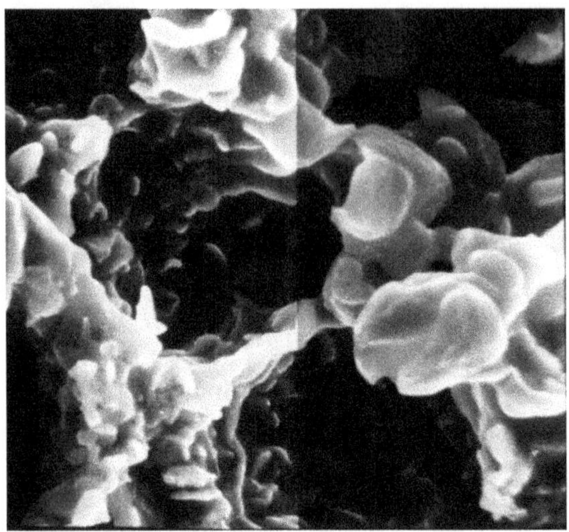

Figure 18.

Figure 19.

Figure 20.

Figure 21.

Figure 22.

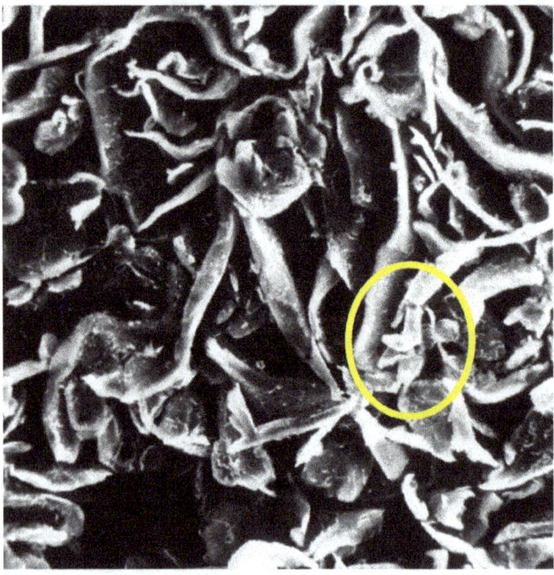

Figure 23.

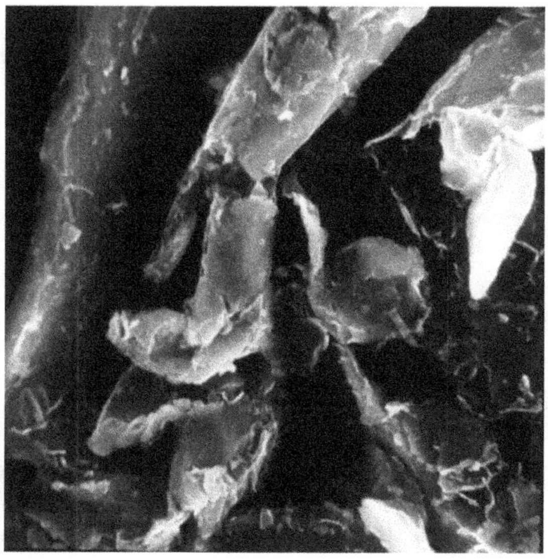

Figure 24.

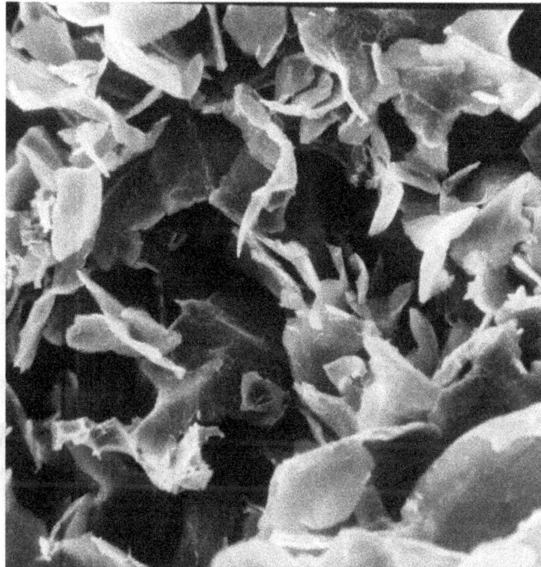

Figure 25.

Figure 26.

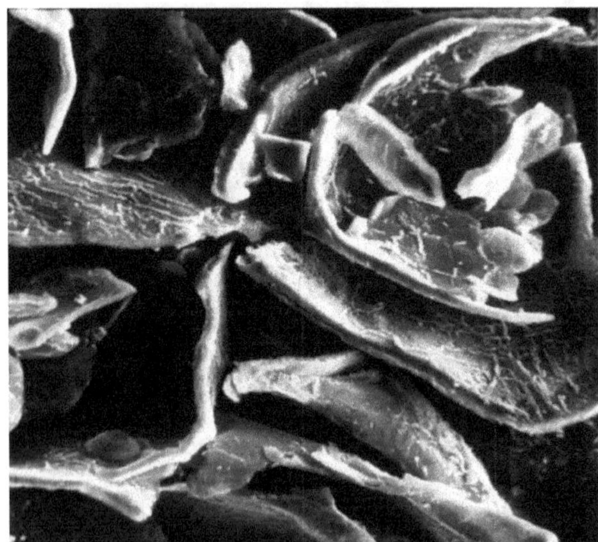

Figure 27.

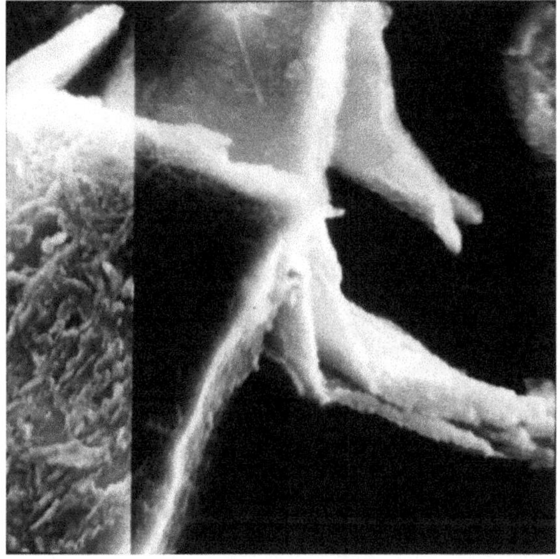

Figure 28.

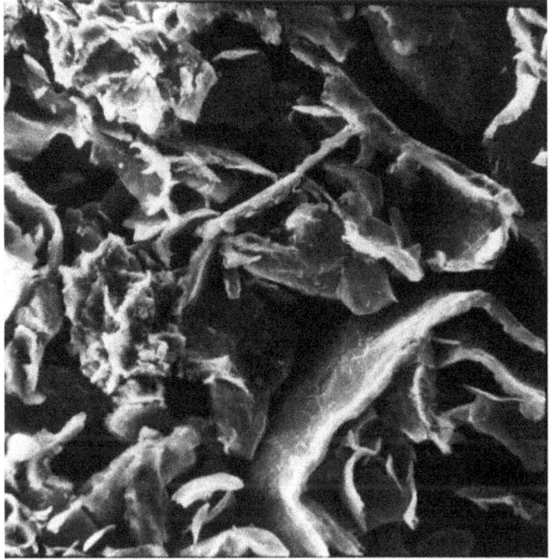

Figure 29.

Figure 30.

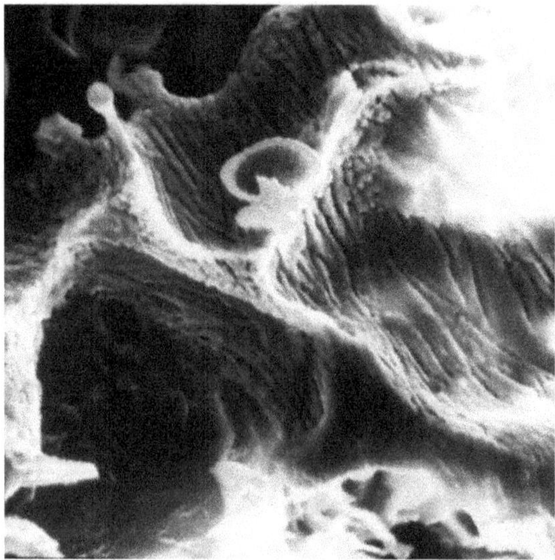

Figure 31.

Figure 32.

Figure 33.

Figure 34.

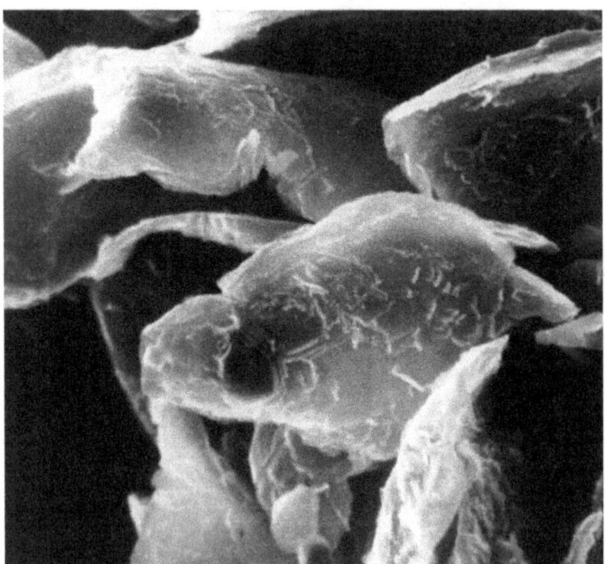

Figure 35.

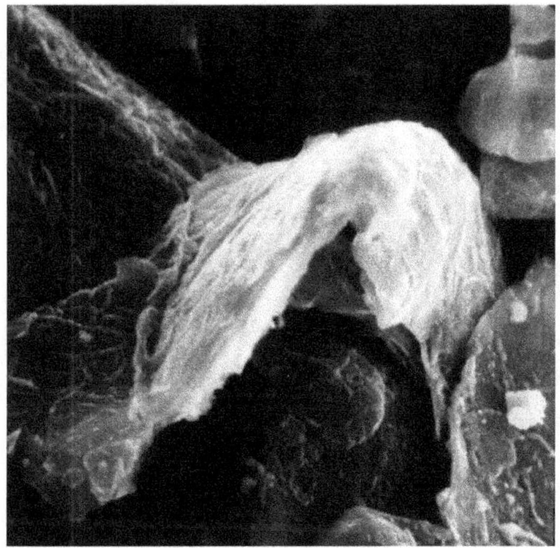

Figure 36.

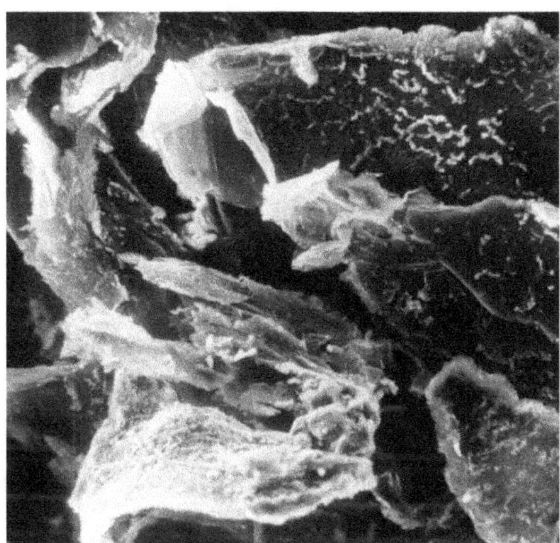

Figure 37.

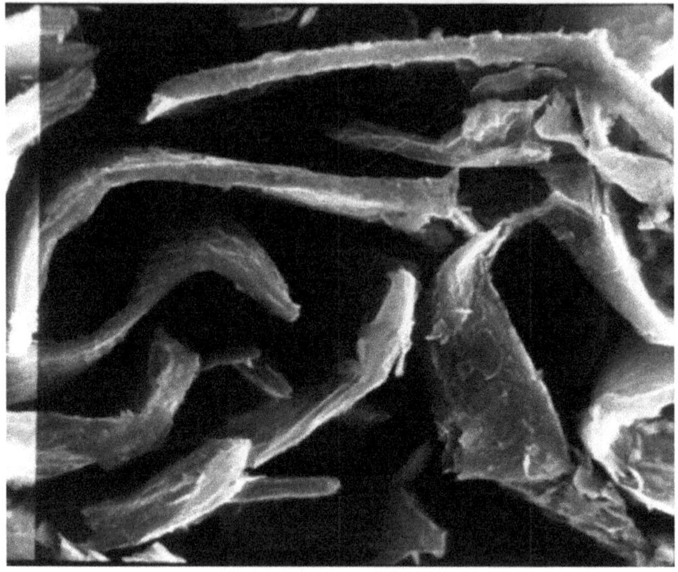

Figure 38.

The Hardness values bear very similar relationship with solidification time (Figs. 41 and 42).

B) On The Thermal Behaviour Of Molds (Figures 43 to 46)

Poor life of the molds has been the major reason for the slow progress of PM of ferrous and other high temperature alloys. The life of the mold is basically governed by the thermal cycle. Hence, a thorough understanding of the thermal behavior of the molds as affected by the operating parameters is very vital for the process designer. The thermal behavior also governs the extent and location of the defects in a given casting.

Studies on the thermal behavior aspects of metal molds during cast iron solidification indicate that the Volume Ratio (VR) is an important parameter in determining the thermal behavior of the metal molds. All the thermal behavior aspects considered (the interface temperature prior to air gap formation and during the final stages of solidification, air-gap formation time and the heat absorbed by the mold at the end of solidification), decrease gradually with an increase in the volume ratio but this decrease is not significant beyond a particular volume ratio. At a given volume ratio, an increase in either the mold or the pouring temperature causes an increase in the magnitudes of the above, thermal behavior aspects whereas the thermal behavior aspects are not significantly affected by the silicon content of the iron poured, in the range studied.

• *1977* - In his AFS 1977 Hoyt Memorial Lecture, Rassenfoss talked about Mold Materials for Ferrous castings [43]. His observations were very relevant to present subject. He highlighted the huge costs involved in sand molding and sand reclamation. He also touched upon the problem of used sand disposal – that the dumpsites are getting farther and farther from the foundries and adding to the transportation costs. Problem of used sand dumpsites causing ill effects on the ground water and the streams nearby was serious, he observed.

Considerable effort and cost are involved in preparing the sand for Dump Worthiness. The advantages of PM process for ferrous castings on energy usage and environment were highlighted. He touched upon the various features of the Eaton Process. It was mentioned that although considerable efforts have been made to avoid chill formation in as-cast PM cast iron castings, no dependable practice has been obtained and for that reason all the castings need to be given an annealing treatment prior to machining / shipment. He quoted that Eaton and Kubota ltd. employ a high carbon equivalent (CE) for the permanent mold.

The use of Molybdenum dies for PDC of steel casting, and the usefulness of Graphite molds for ferrous castings were covered in his lecture. It was mentioned that graphite has a very low coefficient of thermal expansion and that it does not crack either on heating or cooling, and it does not heat check under even the most severe heating cycles. The problem with graphite is fragility and hence needs careful handling. According to him, in the US, about 16% of all iron castings are made in metal molds, and about 12% of all steel castings are made in graphite molds. He concludes by saying that with the economic and ecological advantages of PM, efforts will continue to adapt it to a greater amount for ferrous casting production in the future.

• *1982* - A publication by Cast -Tec Ltd [12], Ontario revealed that Europe was far ahead of USA in the production of iron castings in metal molds. The author reported that a study of literature accumulated at AFS and BCIRA could lead to assumption that USSR produced more than 15 % of their ferrous castings in metal mold. The same was found to be true with Eastern European countries. The author also quoted that West Germany and Japan also cast considerable amount of cast iron and ductile iron castings in metal mold. The author claims that the process gained popularity and became practically viable after being dormant for nearly 60 years!!

• *1984* - A publication from Russel Cast-Tec [44] indicated that the company was able to offer PM castings in several grades of gray iron and Spheroidal Graphite (SG) irons. They claimed casting yield of over 90%. SG Iron PM castings gave much higher nodule count than the corresponding sand casting. Their experience also showed that austempered SG irons could be produced in pearlitic grades by the PM process without the need for Nickel or Molybdenum alloy additions.

Another paper [45] jointly published by Cast-Tec Ontario, and Russel Cast-Tec., UK, claims a substantial cost reduction in PM process, compared with high - speed sand molding both in the casting and the product finishing. Reduced maintenance cost and rejection level have also been reported. To a question posed - "The advantages claimed of PM sound like a foundryman's dream. Why isn't it in general use?" - their answer was that in the past, many foundries were discouraged by high mold cost and poor mold life, and that has been the major hurdle. Their success, they claim, came from improvements in this area – one is the use of improved coatings and the other is the cleanliness of the iron poured that offers better fluidity that allows the mold to be filled easily at a lower temperature than the normal. This is a very significant point to be noted by those seeking similar improvements.

• *1988* - A book on cast iron by Roy Elliott [46] devoted a full chapter for PM of cast iron. Factors affecting the final microstructure of the casting and the mold life are discussed. Mention is made of the study conducted by Henych and Gysel, [47] on the thermal performance of the various die materials, and the merits of a high performance mold made from a copper-base alloy is highlighted.

• *1989* - Another publication by M/s Cast –Tec [48] revealed that the technology developed by them (Cast-Tec's Permanent Mold Technology - a patented system) became a viable reality in 1978 and that a large number of components of gray and ductile iron castings – that included compressors, engine crankshafts, connecting rods, steering knuckles and components, hydraulic components, pump housing, pulleys, brake rotors, refrigerator cranks, electric motor end frames, hub type castings, brake carriers, golf club heads, pipe fittings, etc. were being produced on a regular basis since then.

• *1990* - The authors of the present paper introduced the technology of PM of cast iron in a newly established automotive component manufacturing company (Allparts Castings Limited), in Kenya. The components manufactured included brake rotors & brake drums (for various Japanese, European and American models of cars, light commercial vehicles and trucks), and a variety of engineering components. The weight range covered was from 5 kg to nearly 100 kg. A typical brake drum made by them using PM process is shown in Fig. 47.

Test reports (from Germany) confirmed the superior properties and field reports confirmed superior performance of these PM products compared to sand cast equivalents. In the case of both brake rotors and drums, the users confirmed better braking efficiency.

A project on PM of cast iron, as a part of the Masters Degree Program, University of Nairobi, was carried out at the plant [49], and considerable data were generated under production conditions. The findings were presented in a workshop under UNIDO Innovation Technology Management Program in Nairobi, in 1977 [50].

During 1990 to 2002, PM cast iron tonnage poured at Allparts castings Limited was in excess of 15000 tons.

Some of the practices followed at Allparts Castings Limited, and Experience gained.

• All the components made were of hypereutectic cast iron, inoculated prior to pouring.

• Molds were made of desulfurized, hypereutectic cast iron with % S less than 0.05,

• Most mold were made of 2 parts, either top-bottom or left-right type.

• All the castings were top poured (through the riser). To improve die life, the portion where the metal stream first strikes the bottom mold, was made of a separate replaceable metal insert, and in some cases, made of a dispensable pad of Shell sand.

• The mold coating used was a water base silica flour spray, sprayed for each pour. Where situation demanded, a thick shell resin sand coating was employed to reduce the cooling rate.

• Mold temperature of 200-250°C, and Pouring temperature of 1300-1350°C was employed in most cases. However, in some special cases, to achieve a slower cooling rate, a higher mold temperature was employed through continuous external heating with gas.

• Draft angle provided in the casting was minimum 1° for easy extraction. Easy extraction meant less of stresses in the castings.

• The castings were removed from the mold in red - hot condition and cooled under a layer of sand. This was to get an annealing effect, without resorting to costly and time - consuming heat treatment process.

• Multi-Part metals cores were used in most cases. In some special cases, sand cores were used (hollow cores wherever strength of the core permitted, to reduce sand usage). In cases where the core was in contact with the working surface (like in the case of Brake Drums), the working surface of the metal core was provided with a large number of 1mm deep pockets, in thermosetting resin sand was filled (this resin sand layer was replaced for each pour). In such cases, each mold had two sets of metal cores.

• The mold failure was invariably due to thermal fatigue cracks (Fig 48). Where the mold crack area corresponded to machined surface of the casting, the mold was not discarded at the initial appearance of cracks, but continued in production until the cracks become too severe and unmanageable, or the die broke into pieces. Moreover, minor, hairline cracks get covered by the mold coating.

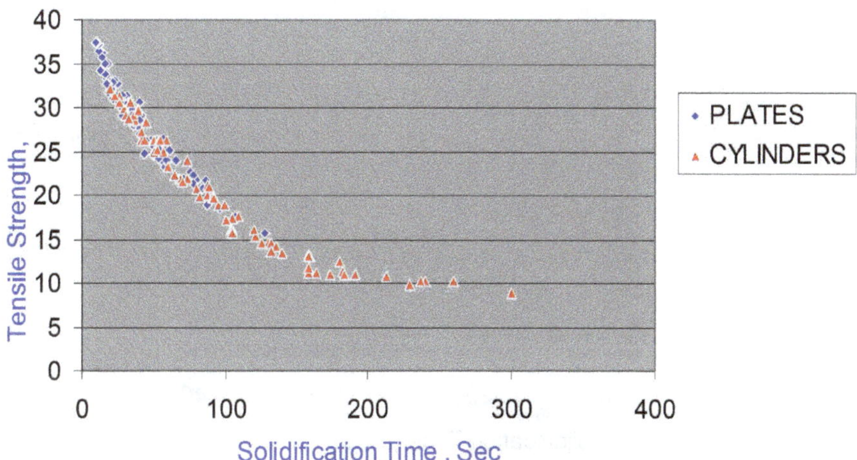

Figure 39. Tensile Strength (Kg / Sq. cm) Versus Solidification Time, for both Plates and Cylinders. % Si = 3.00.

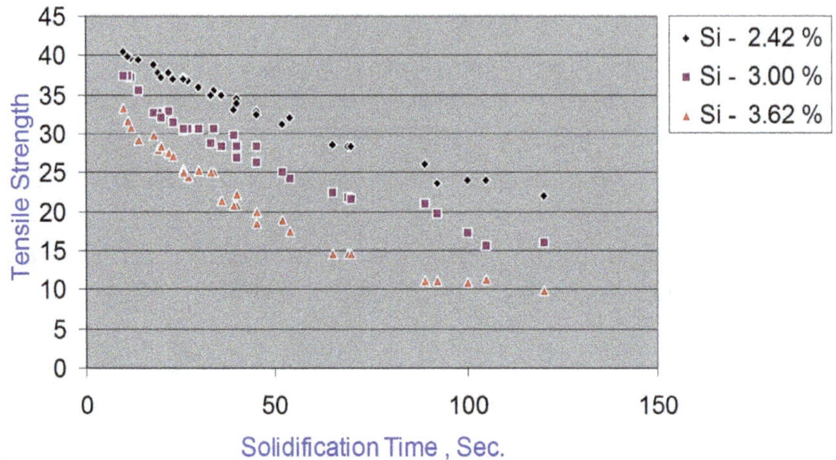

Figure 40. Tensile Strength (Kg / Sq. mm) Versus Solidification Time for both Cylinders and Plates, for Different Si %. M.T. = 150°C, P.T. = 1250°C.

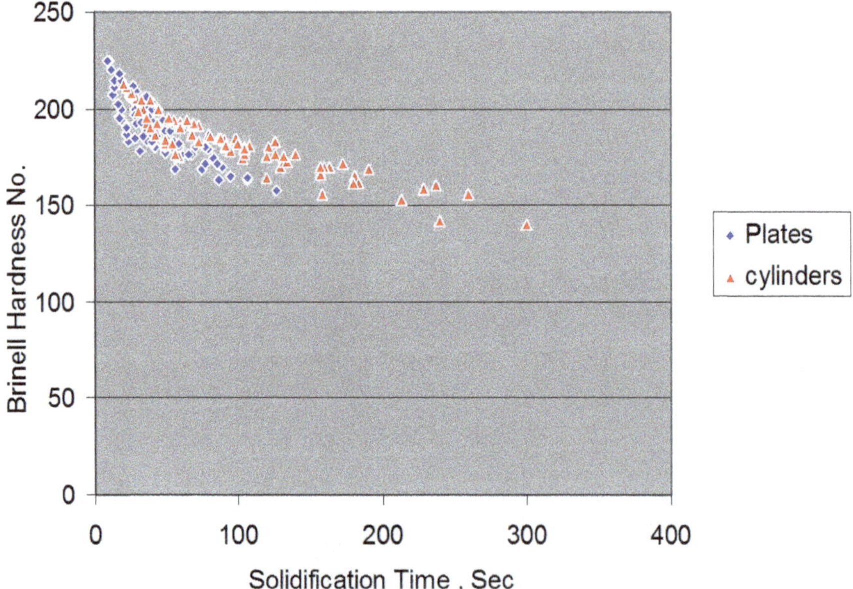

Figure 41. Brinell Hardness Versus Solidification Time for both Plates and Cylinders, % Si = 3.00.

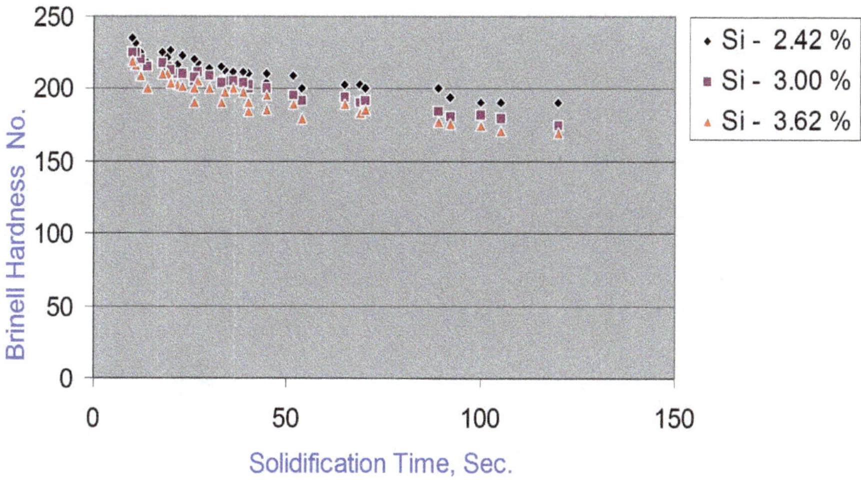

Figure 42. Brinell Hardness Versus Solidification time for both Cylinders and Plates, for different Si%, MT=150°C, PT=1250°C.

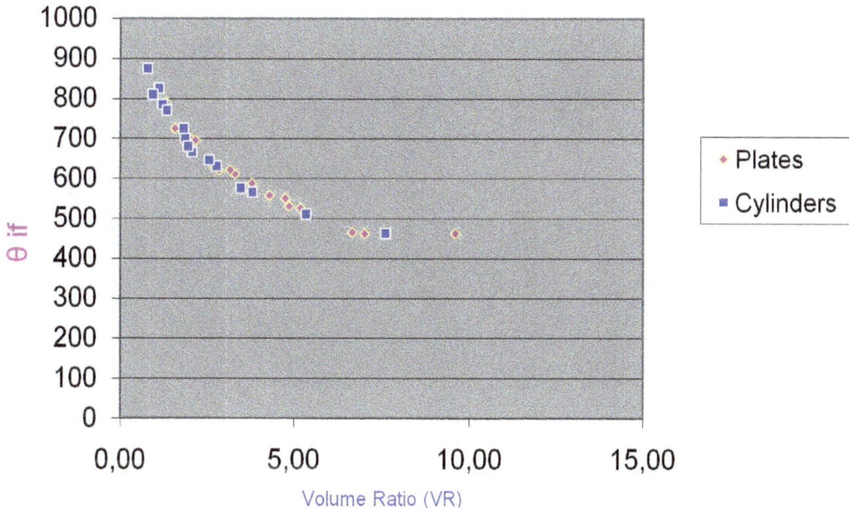

Figure 43. Interface Temperature (θif °C) During The Last Stages Of Solidification Versus Volume Ratio (VR). % Si=3.00,M.T. = 150°C, P.T. = 1250°C.

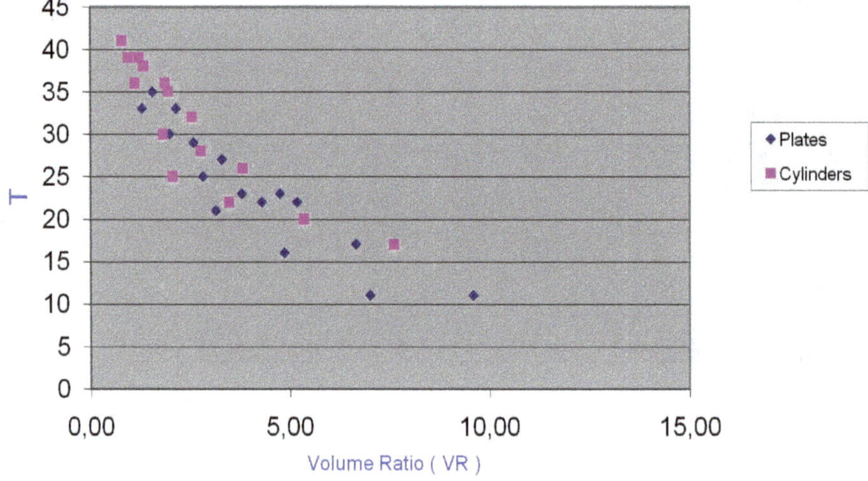

Figure 44. AirGap Formation Time (T, Sec) Versus Volume Ratio (VR). % Si=3.00, M.T.= 150°C, P.T.= 1250°C.

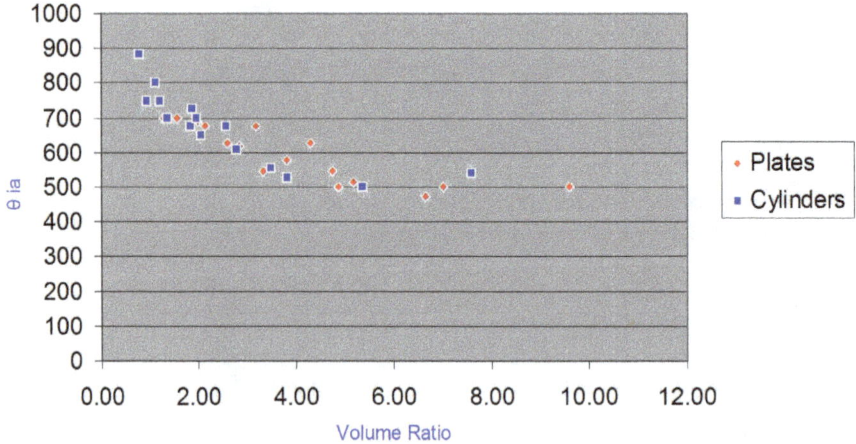

Figure 45. Interface Temperature, °C,Prior To Air Gap Formation (θia°C) Versus Volume Ratio (VR). % Si= 3.00, M.T.= 150°C, P.T.= 1250°C.

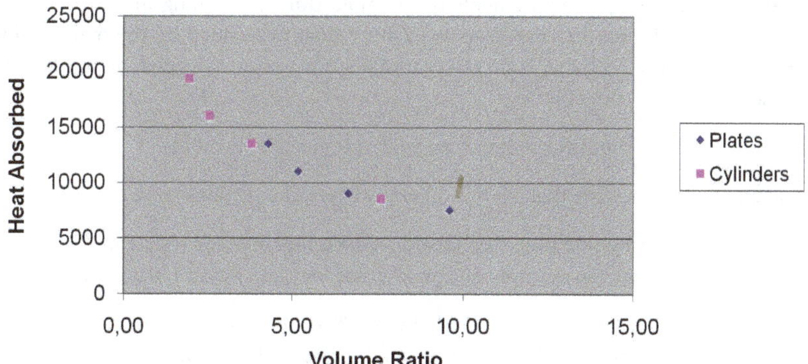

Figure 46. Heat Absorbed by the mould at the end of solidification (K.Cal / Sq.M) Versus Volume Ratio.

Figure 47. A Brake Drum made by PM process.

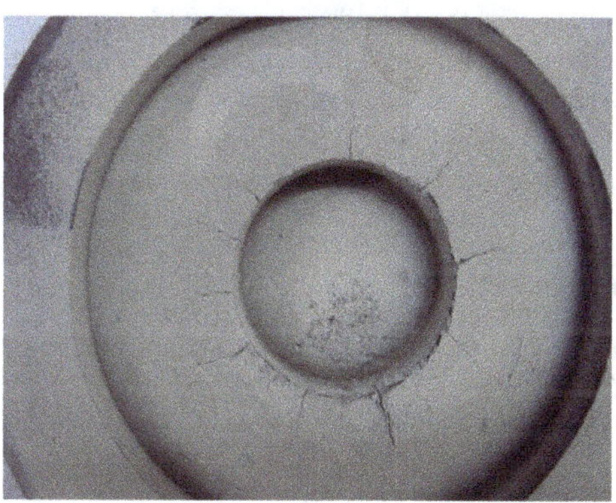

Figure 48. Typical Thermal Fatigue Cracks in a PM.

• Considering the strength requirements of the mold during handling in hot condition, in most cases the mold wall thickness was kept more than demanded by thermal considerations. Hence most dies with cracks on the working surface were salvaged, by re-machining. Multiple salvaging was possible.

• The totally damaged mold were simply re-melted to make new molds.

• The casting yield was more than 95 %, and many cases, the castings were riser-less.

• The parts produced by this process showed a much higher wear resistance compared to equivalent sand cast part. An example of a brake rotor for Land Rover 110 is shown in Fig. 49.

• Where the specification demanded a little higher % of pearlite, addition of small % of Sb and Cu were tried as per the hints given in the literature [13,46] and the results were extremely encouraging.

• In some very thick castings, even under the fast cooling conditions, it was not possible to achieve predominantly Type D graphite on the working surface as specified. Here again, a hint given in one publication [13] came to the rescue – addition of 0.1 % Ti settled the matter to the fullest satisfaction.

• Generally Brake Rotors and Brake Drums made from sand castings are machined all over to achieve a good dynamic balancing. It was found that in PM castings, with machining on only working surface and the fitting surface, and leaving the rest as-cast, a good balancing was still possible. Even on the machined surfaces, the machining allowance in most cases was 1mm only.

• Quality of both the castings and the machined components was extremely good - in most cases, the overall rejection was under 2%. Machinability was very good – higher speeds & feeds, good surface finish, retention of sharp corners and edges, smoother thread formation, reduced tool consumption, and so on. Normally Brake Rotors and Drums are removed from the vehicle many times during its life, for re-skimming the working surface. In the case of those with threaded bolt - holes, the threads get damaged easily during removing and fixing. The feedback from customers showed that such thread wearing in PM cast components was virtually absent, where as it was quite common in sand cast equivalent. The thread formation in PM castings is very smooth due to fine Type D graphite, where as in sand castings with coarse Type A graphite smooth threads are not possible due to graphite pullout [15].

• The PM components were at least 30 % cheaper than the equivalent sand cast components, as applicable to Kenyan conditions.

•*1996 - A very valuable publication* (possibly the most informative of all the publications, touching upon both the thermal and metallurgical aspects), on Ferrous Permanent Mold (FPM) process, by Lerner [13] highlighted the various developments in the recent years. Advantages of the process in terms of Cost, Quality, Energy Reduction and Environmental Issues have been addressed. The author mentions that in addition to superior casting finish and dimensional tolerance, the process has various other distinct advantages like:

a) Gas and shrinkage porosity-free structures for leakage-free castings needed in hydraulic and gas components' applications. Pressure tests routinely performed on these castings showed little or no rejection.

b) Reduced production time, reduced finishing costs, elimination of sand and sand handling, and improved dimensional accuracy and stability.

c) Castings have a history of exceptional machinability, very low rejection on machining, ability to hold close tolerances.

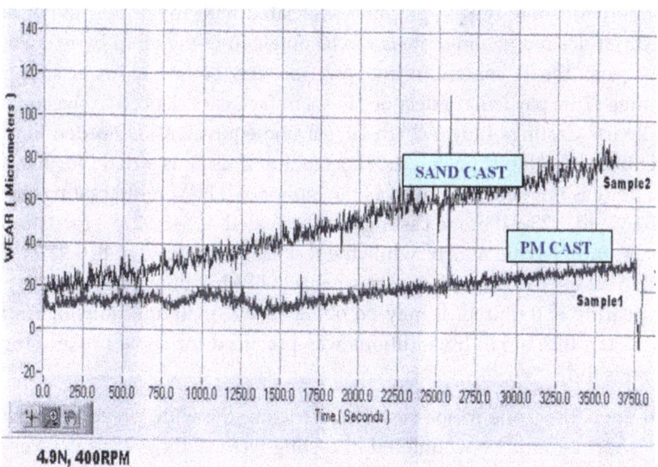

Figure 49. Typical example of relative wear pattern of a brake rotor of Land Rover 110 - cast in sand mold and permanent mold.

The author reports that PM gray iron castings can give 30000 psi tensile strength with 147-201 BHN hardness in a fully ferritic matrix containing predominantly type D graphite. Basically the castings are strong yet machinable. For SG iron PM castings, the amount of Mg that has to be added is less than for sand castings. This results in lower residual content, which in turn results in controlled shrinkage, improved nodularity thus enhancing mechanical properties and better overall casting quality.

Some statistics provided by the author on the production volumes of PM castings world over is very useful indicator of the progress made in recent years. The figures are as follows:

Europe - 15 foundries with estimated annual production of 35000 tons, Eastern Block (former Soviet Union, Czech Republic, Poland, Hungary, Bulgaria) – 650000 tons, a new German owned foundry in Brazil – 12000 tons of gray iron and 6000 tons of ductile iron, Japan – at least 6 foundries, 18000 tons, two Japanese built foundries – one in Malaysia and the other in China with a combined production of 6000-8000 tons, two foundries in India with low volumes, a few foundries in Canada and U.S.A (including Perm Cast in Kentucky – the original Eaton Corp., Honda of America, Anna, Ohio).It is reported that Honda of America began producing ductile iron steering knuckles on an automatic FPM line (Quick Cast Knuckle – QCK) in the fall of 1995 and casting production via this process is of the order of 22 tons per day. The author has provided list of components made by these several above foundries in addition to a very detailed list of FPM castings made by former USSR.

The author has also touched upon some metallurgical aspects PM cast irons. In addition to the value of C, Si, Mn, P and S specified for PM gray cast iron, he has touched upon the addition of small quantity of Ti (0.02 to 0.10 %). He states that Ti is essential for providing the under-cooling required to meet ASTM Specification A 823-84, that calls for predominantly type D graphite with some type A graphite associated with the center line or around sand cores. However, if desired cooling rate is can be obtained in the mold by using a more effective cooling system, the Ti content in the base iron may be on the lower side of the above mentioned range (This particular effect of Ti was in fact, experienced in the commercial production at Allparts Castings Ltd). A high CE (carbon equivalent) is needed to regulate chill depth and reduce sink / lap type defects. Inoculation, if used, is strictly for the chill control, as type A graphite is not desired, observes the author. All FPM mold castings are heat treated as per ASTM std. 823-84. Some castings are annealed at 843-927°C for 1 hr and furnace cooled to obtain fully ferritic matrix, while the rest are normalized at 816-927°C for 1 hr and air quenched. The microstructure of a normalized FPM usually has 10-30% pearlite. If a higher % of pearlite is required, it may be obtained by small additions of Antimony (Sb). Taking a hint from this, small Sb additions was practiced for some brake rotor castings at Allparts Castings Ltd.

According to the author, one major obstacle restricting the widespread adoption of FPM is the relatively short mold life encountered in casting ferrous alloys (this is a very significant point to note for future research work). This problem is reduced by the use of Lined Permanent Molds (LPM) where the working surface of the mold is lined with a thin layer of slurry or sand mixture depending upon the alloy poured. This practice not only increases the mold life, but also reduces / eliminates carbides in the structure (again, taking a hint from this paper, such methods were employed for some components at Allparts Castings Ltd., with a great degree of success). However, if high wear resistant chilled iron microstructure is desirable, like in automotive camshaft applications, the portions corresponding to the eccentrics are not lined and the molten metal comes directly in contact with mold. The author says that LPM process is quite popular in former Soviet nations.

The author mentions that the thermal effects of the liquid metal flow in the mold are the major factors in determining the mold life as well as the casting quality. This fully justifies the earlier study conducted by the present authors on the thermal behavior of metal molds.

Learner adds that by and large, a gray iron with type A graphite is recognized as a good material for the mold. Research to improve mold life showed that the highest resistance to thermal shock was exhibited by Cr-Mo containing gray iron. The same was the experience at Allparts Castings Limited as mentioned earlier on. Type A graphite raises the thermal conductivity of the mold, while Cr and Mo increase the metallic matrix heat and thermal fatigue resistance.

• *2004* - Technikon LLC, that operates the casting Emission Reduction Program (CERP) published the findings of their research on the durability of metal molds used for high temperature alloys like Iron, Steel, Nickel and Titanium [1]. They observed that the principal drawback to the application of PM to castings of these high temperature alloys is a short mold life. The shortened life is caused by the thermal shock when the molten metal is poured, as well as wear produced in the removal of the previously used mold coatings. This durability problem is the main reason behind the slow progress of PM of ferrous castings. The publication covers methodology used to candidate alloys for evaluation as a high temperature permanent mold insert material. The results of the manufacturing of the test die blocks / coupons by a process known as laser consolidated powder deposition, for each of the candidate alloys is discussed in the report. The findings clearly indicate that the service life (cycles) of the permanent mold drops as the pouring temperature is increased. A case involving iron metal mold in which castings were made of different alloys are presented (Table 5). Eight different mold materials have been compared in respect of conductivity, hardness, melting point, phase / volume change, eutectic reaction, cost, machinability and repairability. Further studies are planned.

Alloy System	Melting Temp (°F)	Casting Runs / Service life
Titanium	3270	250
Iron	2802	500
Nickel	2651	700
Copper	1981	4000
Aluminium	1220	100000
Magnesium	1202	110000
Zinc	787	500000

Table 5. Melting Temperature of Alloys Poured Versus Estimated Service Life (cycles) for Iron Molds.

• *2008*: The PM process for cast iron was established at Abilities India Pistons& Rings, Ghaziabad in the out skirts of New Delhi. The castings made are presently limited to some of their own in-house requirements of fixtures for machining. The company is now working on the prospects of developing various PM cast iron components for domestic and export market. The fist step taken towards this is educating the customers on the on the merits of the process.

Some of the microstructures (both Optical and SEM) observed in the various production castings are given in Figures 50 to 57.

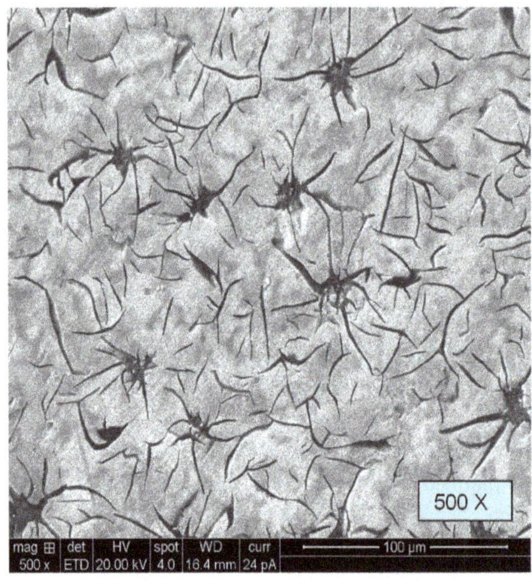

Figure 50. Flake graphite adjacent to the core in a hollow cylindrical casting.

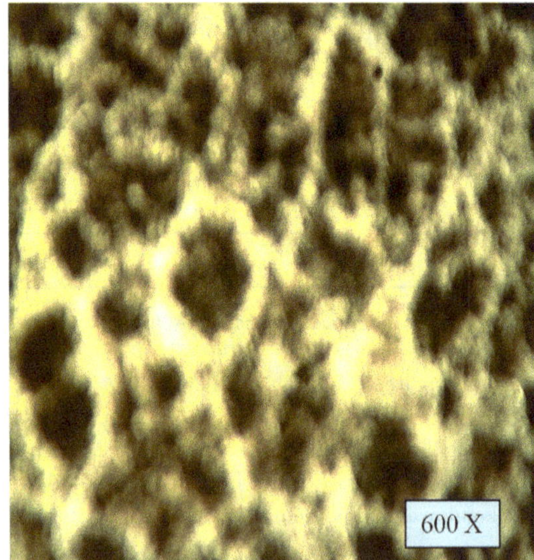

Figure 51. Eutectic Cells.

3. Way Forward Towards Enhancing the Production of PM Cast Iron

It becomes the sacred duty of all researchers and practitioners of foundry, to work together in this direction, create awareness and share their experiences to make the Permanent Molding of Cast Irons a totally viable process for mass production. Foundry industry has to work harder, to be recognized as a sustainability leader by other industries and the public.

An International Expert Committee consisting of leading foundry personalities may be formed, to work out modalities to bring awareness on the subject, collect detailed statistics through world foundry associations, and to suggest practice based research programs, with some time bound plans of action. The development of better mold materials and ways to improve the mold life need to be tackled on priority.

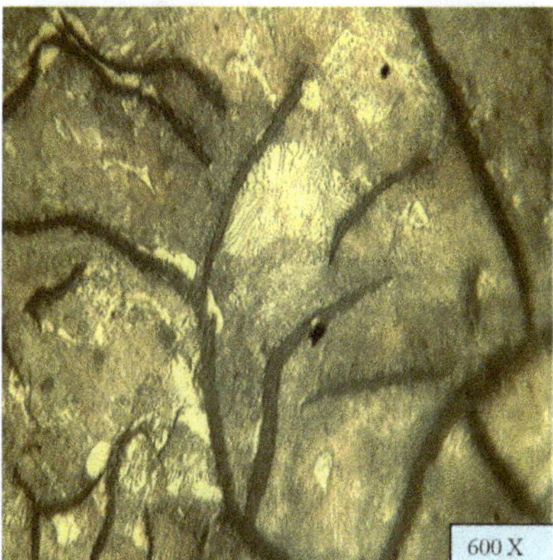

Figure 52. Flake graphite in a pearlitic matrix adjacent to the core in a hollow cylindrical casting.

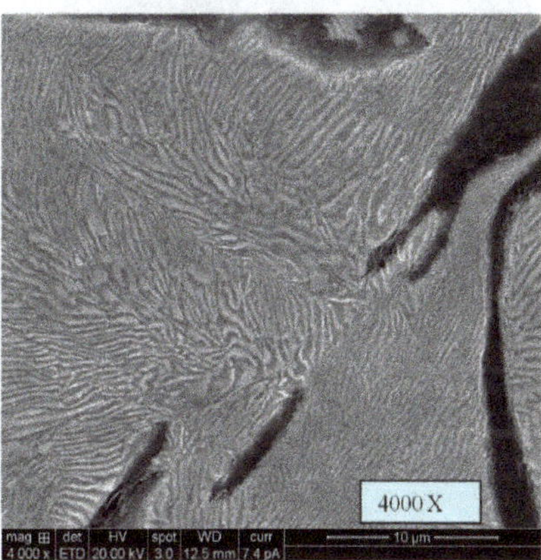

Figure 53. Pearlitic matrix adjacent the core in a hollow cylindrical casting.

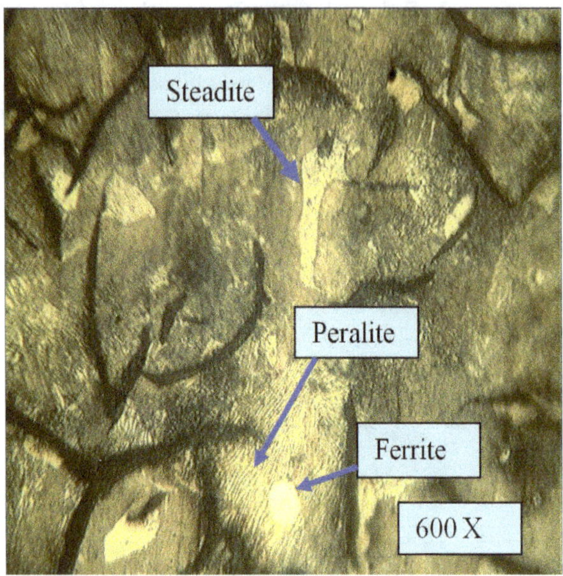

Figure 54. Steadite, Pearlite and Ferrite.

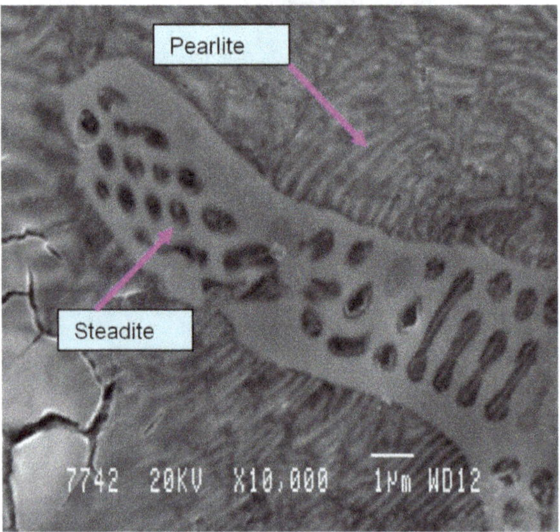

Figure 55. Steadite and Pearlite.

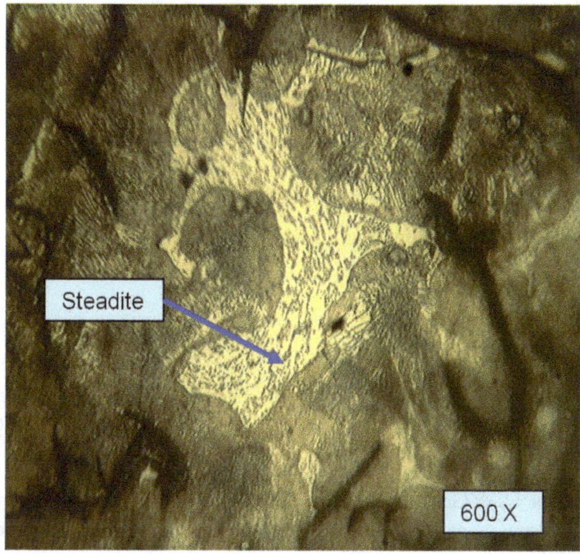

Figure 56. Steadite.

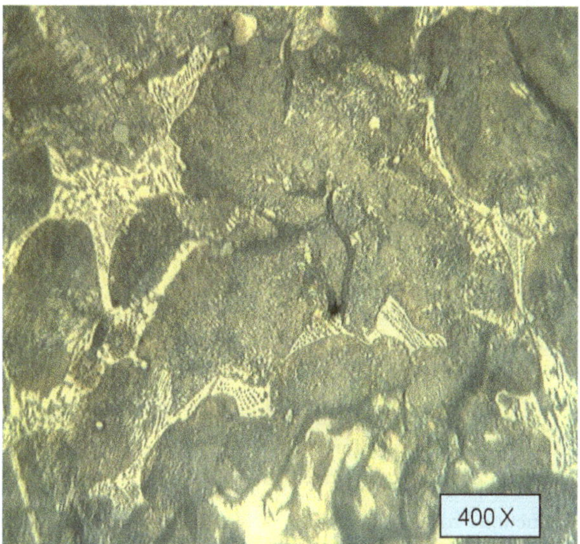

Figure 57. Steadite network.

Author details

M. S. Ramaprasad[1*] and Malur N. Srinivasan[2]

*Address all correspondence to: vaidehi_ramaprasad_sk@yahoo.co.in

1 Foundry Consultant, Bangalore, India

2 Department of Mechanical Engineering, Lamar University, Beaumont, USA

References

[1] Technikon.LLC and Material Applications Branch of US Army Research laboratory (2004). Permanent Mold Technology. Document # 1410 - 160Under CERP (Casting Emission Reduction Program), Oct.

[2] West, Charles. E., & Grubach, Thomas. E. (1998). (Aluminum Company of America), Chapter on Permanent Mold casting, ASM Metals Hand Book, Volume 15 - Castings- Fourth Printing, March.

[3] Teillet, Rafael. M. (1962). Permanent mold casting. *Mod. Casting*, 41, May, 103.

[4] Henzel, J. G. Jr., & Kaverian, J. (1966). Gap Formation In Permanent Mold Castings. p. 373, Trans AFS.

[5] WTEC (2001). Panel Report on “Environmentally Benign Manufacturing” International Technology Research Institute (TRI- Loyola College,Baltimore, MD 21210,USA), April.

[6] Gigante, Gary. (2010). How Can We become a Practically Green Foundry Industry ? “- AFS Hoyt Memorial Lecture for 2010.

[7] U.S. Department of Energy (2002). Metal Casting Industry of the Future: 2002 Annual Report.

[8] “CERP Organic HAP Emission Measurements For Iron Foundries and Their Use in Development of an AFS HAP Guidance Document“, Technikon # 1412-317 NA, Under CERP (Casting Emission Reduction Program), Jan 2006 & Aug 2007.

[9] “Second Verification of SIVL (Systems Integration and Validation Laboratory): Triboelectric Particulate Monitors- Monitor A“, Technikon # 1411- 234- A, Under CERP (Casting Emission Reduction Program), July 2005.

[10] “Energy Reduction Projects“, Technikon # 1411- 815, Under CERP (Casting Emission Reduction Program), September 2005.

[11] Holmgren, Mats., & Naystrom, Peter. (2008). "The Green Foundry" presented at 68th World Foundry Congress, Chennai, India, Feb.

[12] Clark, Antony P., & Cast-Tec Ltd. Ontario. (1982). "A new Development in ferrous Permanent Mold Casting", Modern casting, June.

[13] Lerner, Yury S. (1996). "Another Approach to Iron Castings", p. 48, Modern Casting, November.

[14] Francis, J. L. (1965). "Gravity Die Casting of Cast Iron", Foundry trade Journal, Vol. 118, p. 443, Apr.

[15] Frye, George. (1968). Eaton Gray Iron Castings by Permanent Mold Process. Modern Casting, Vol. 54, p. 52, Oct.

[16] A quick picture of the Eaton Permanent Mold Process for producing gray iron castings. A booklet published by Eaton Corporation, Michigan, USA.

[17] Clark, D. (1968). "Permanent Mold Aluminium and gray iron Castings at Forest City Foundries" Modern Castings, p. 65, Oct.

[18] Patent granted to Walter S. Anderson for development of permanent mold for cast iron. (1925) Published in Metal Founding, Aug.

[19] McClelland, H. U. (1959). "Grey Iron Permanent Molding", Modern Casting, Vol. 35, p. 68, Apr.

[20] Miske, Jack C. (1966). "Ferrous Die Casting" A report on work progress, p. 10, FOUNDRY, July.

[21] Miske, Jack C. (1966). "Ferrous Die Casting" A report on work progress, p. 190, FOUNDRY, Oct.

[22] Barto, R. L., Hurd, D. T., & Stoltenberg, J. P. (1967). "The pressure Die casting of Iron and Steel" p. 181, Trans. AFS, Vol. 75.

[23] Hurd, D. T. (1967). "Ferrous Die Casting" A Continuing Report, p. 127, Foundry, Nov.

[24] Die Casting of Nodular Iron (1968). p. 137, Foundry, Aug.

[25] Bates, C. E. (1972). "Profit Potential in Permanent Mold Iron Castings" Foundry, Vol. 100, p. 49, Nov.

[26] Ferrous Die casting In Ireland (1973). Foundry, p.91, Nov.

[27] Lusniak Lech, Ludmila. (1973). "Casting Permanent Molds in Poland" p. 93, Foundry, October.

[28] Fisher, T. P. (1967). "The Technology of Gravity Die casting", A Text Book, published by George Newnes Ltd., First Edition.

[29] Skrocki, R. R., & Wallace, J. F. (1968). "Permanent Molding of Iron and Steel Castings" p. 581, Trans AFS, Vol. 76.

[30] Skrocki, R. R., & Wallace, J. F. (1969). "Control of Structure and Properties of Irons cast in Permanent Molds" Part I, p. 296, Trans AFS, Vol. 77.

[31] Skrocki, R. R., & Wallace, J. F. (1970). "Control of Structure and Properties of Irons cast in Permanent Molds" Part II, p. 239, Trans AFS, Vol. 78.

[32] Permanent Mold Casting (1970). p. 265, Metals Hand Book, Vol. 5, 8th Edition, published by ASM.

[33] Schoendorf, P. (1970). "First Annual Summary of recent literature on Permanent Mold Casting of Cast Iron" Giesserei, Vol. 57, p. 715, Oct.

[34] Ramesh, K. (1973). "Studies on Hyper Eutectic Cast Iron cast in Metallic Molds" M.Sc Thesis, Indian Institute of Science.

[35] Zuithof, A. J., et al. (1972). "The section sensitivity of cast iron permanent mold castings" Cast Metals Research Journal, p. 83, Issue 2, Vol. 8, June.

[36] Ramaprasad, M. S., & Srinivasan, M. N. (1975). Permanent Molding of Cast Iron. *Indian Foundry Journal*, 21(6), 1-7.

[37] Ramaprasad, M. S. (1976). PhD thesis, Indian Institute of Science.

[38] Ramaprasad, M. S., & Srinivasan, M. N. (1977). Eutectic Cell Structure in Permanent Molded Cast Iron. *Castings (Australia)*, 23(9/10), 27-35.

[39] Ramaprasad, M. S., & Srinivasan, M. N. (1977). Permanent Molding of Cast Iron-Thermal Behaviour. *The British Foundryman*, 70(12), 359-364.

[40] Ramaprasad, M. S., & Srinivasan, M. N. (1977). Studies on Structure and Strength of Permanent Molded Cast Iron. *Proceedings of the Department of Atomic Energy Symposium on Structure- Property Correlations and Instrumental Techniques in Materials Research, Rourkela*, 73-87.

[41] Ramaprasad, M. S., & Srinivasan, M. N. (1977). "Graphite Morphology in Permanent Molded Cast Iron" Paper sent for presentation at the Seminar on Structural Modification in Cast Iron, TMS-AIME Fall Meeting, Chicago.

[42] Ramaprasad, M. S., Narendranath, C. S., & Srinivasan, Malur N. (1983). Some Aspects of Grey and Spheroidal Graphite Iron Cast in Metallic Molds. *Proceedings of the ISI Conference on "Solidification in the Foundry and Cast House, University of Warwick, U.K., The Metals Society*, 336-344.

[43] Rassenfoss, J. A. (1977). "Mold Materials for Ferrous castings" AFS Hoyt Memorial Lecture for 1977.

[44] Gravity Diecasting of Iron- The Russel Cast-Tec Process (1984). p. 77, Foundry Trade Journal, Feb.

[45] Clark, A. P., & Godsell, B. C. (1984). Answers to Questions about Ferrous Permanent Molding. p. 25, Modern Casting, Feb.

[46] Elliott, Roy. (1988). Chapter on permanent Molding, Test Book "Cast Iron" Butterworth & Co (Publishers) Ltd.

[47] Henych, I., & Gysel, W. (1982). "Development of the high performance die as a basis for a mechanized permanent mold casting process for medium weight cast iron parts" paper presented at 49th International Foundry Congress, Chicago.

[48] Cast Tec Ltd., Ontario (1989). "Producing As-cast ferrous Permanent Mold castings" Modern casting, May, (special GIFA issue).

[49] Onyuna, M. O. (1994). Masters Degree Thesis, Dept. of Mechanical of Mechanical Engineering, University of Nairobi.

[50] Ramaprasd, M. S., Patel, Bhupendra C., Patel, Devendra B., Patel, Paresh B., Patel, Pradip B., & Patel, Dilip M. (1977). "The Development of Cast Iron Permanent Molding Technology at Allparts castings Limited, and Experiences Gained" Presentation made as a participant of "UNIDO Innovation Technology Management Program", Nairobi.

Chapter 7

Control Technology of Solidification and Cooling in the Process of Continuous Casting of Steel

Qing Liu, Xiaofeng Zhang, Bin Wang and Bao Wang

Additional information is available at the end of the chapter

1. Introduction

Solidification and cooling control, which is a key technology in the continuous casting process, has a quick development in recent years, and meet the modern requirements of the continuous casting process on the whole. However, the control models and cooling technology need constant development and improvement due to the trend toward delicacy and full automation in continuous casting. This chapter discusses the hot ductility, the thermophysical properties, the solidification and cooling control models and nozzles layouts for secondary cooling, besides these, the planning for the process of steelmaking-rolling, which are closely related with solidification and cooling in continuous casting process.

2. Research on the thermal physical parameters of steels

This section summarizes formulae for calculating thermal physical parameters of steel slabs, including the liquidus temperature, solidus temperature, thermal conductivity, and so on. The database of thermal physical parameters including thermoplastic was specially established and embedded in the control model of the solidification and cooling, which is convenient to query data and update operation for technical staffs. Moreover, based on the thermoplastic parameter database, the target surface control temperature of slab is determined for the production of various grades of steels. And the database is helpful for users to acquire more accurate results of the heat transfer model.

2.1. Research on thermoplastic of steels

Thermoplastic is a key researching content of high-temperature mechanical property of steels. The hot ductility curve of steel should be known in order to make slab avoid "fragile pocket area" during straightening process. Generallyin order to get that useful date, the slab samples will be tested at high temperature by Gleeble tensile testing when the test condition is similar to actual continuous casting process.

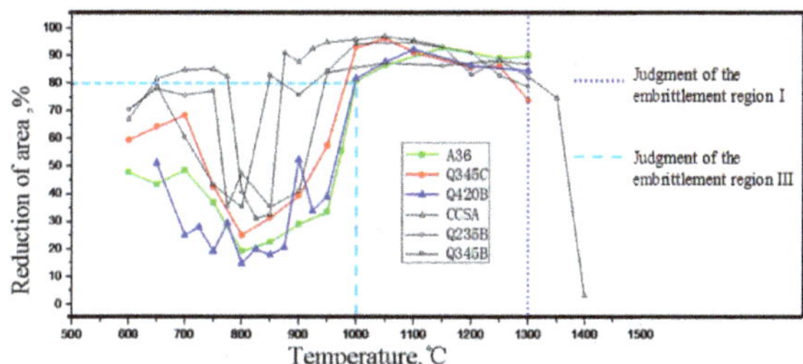

Figure 1. Reduction of area with temperature for some steel grades

According to the experimental results shown in Fig.1, for Nb steel such as A36, it is known that in the embrittlement region◎, temperature range is between melting temperature and 1330 ℃ from the hot ductility curve. Considering the high crack sensitivity of Nb steels, the temperature range of A36 in the embrittlement region ◎ is 600 ℃~ 1000 ℃ when taking the R.A. = 80% as the brittle judgment,in order to ensure the slab has great plasticity. Thus, this brittle judgment can effectively prevent or reduce crack source generation by controlling the slab surface temperature.

It is generally known that the surface temperature fluctuations of slab are impossible to avoid completely during solidification and cooling process. When the temperature fluctuation is large, cracks of some steels such as Nb steel with highly crack sensitivity are easily brought compared with common steels in the process of continuous casting. Therefore, it is proposed especially that the area reduction is more than 80% (the traditional opinion is 60%) for controlling slab surface temperature in each segment exit. Then it should decrease specific water flowrate, cooling intensity and casting speed, in order to effectively prevent crack of Nb steel in the process of continuous casting. Otherwise, it can properly increase withdrawal speed and specific water flowrate for slab casting of steels without Nb to improve the productivity. Generally speaking, the cooling for slabs should avoid the embrittlement region ◎ temperature range as far as possible during straightening process.

As so far, a lot of scholars have tested and researched on hot ductility of many kinds of steels. We can acquire these useful thermoplastic parameters from the literature when needed. Even so, most secondary cooling control systems are difficult to adapt to so many kinds of steels produced by each caster in actual production, due to the difference cooling characteristics of steel grades, especially for new steel production. In author's opinion, the database of hot ductility should be set up by sorting and summarizing this useful dataFig. 2. At the same time, the database is embedded in the secondary cooling control system in order to acquire the corresponding reference and guidance for different kinds of steels and set suitable target surface temperatures by means of querying data from the database.

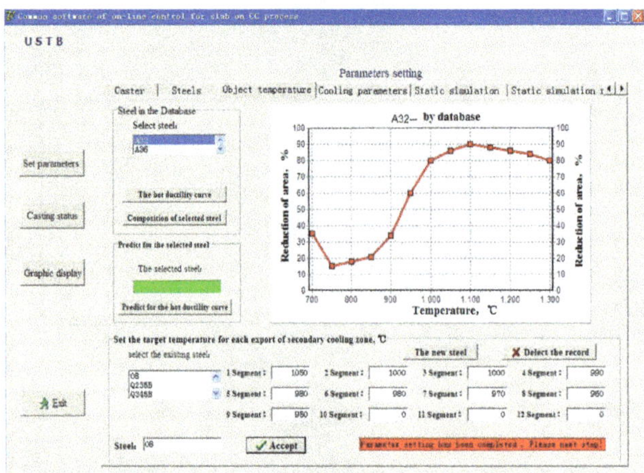

Figure 2. The software interface of the database for hot ductility of steels

The hot ductility of steel is mainly influenced by the chemical composition or technical conditions. Thus, the mathematical model has been established for predicting the reduction of area with chemical composition. The multiple linear regression analysis method has been applied to this model, which was conducted from 24 groups tested data in the similar experiment condition. Moreover, the model considers 12 elements as the independent variables and the reduction of area as the dependent variable.

Gleeble test condition should be similar to deformation and cooling straightening of the industrial operating condition in continuous casting process as far as possible. Mintz's research suggests that the strain rate is $10^{-3} \sim 10^{-4}$ /s during straightening process. Therefore, this study adopted that strain rate as the rule to select hot ductility of steels from literature..Meanwhile,the cooling rate is 3 ℃ / min.

Besides, because the molybdenum has little impact on thermoplastic of steel and the data of nitrogen content is less than 0.005% basically. Thus, these two elements are ignoredand 12 elements such as C, Si, Mn, P, S, Al, Nb, Ti, V, Ni, Cr, and Cu have been used in regression computation.

Regression methods include the forward method, the backward method and the stepwise regression. The stepwise regression method is adopted extensively, as it can obtain better regression subsets of arguments and a high level of statistical significance. Howeverin this paper, the backward method is selected in order to make the regression reflects the influence of the elements as accurate as possible.

This regression analysis applies SPSS 13.0 software selecting backward method. And the model has been established with comprehensive consideration of three aspects, such as the number of elements, the statistic, the actual impact of the elements on hot ductility and so on. Formula is as (1):

$$\varphi_T = A + \sum (B_i \times [i]) \tag{1}$$

In formula (1):

φ_T —The reduction of area at temperature T;

A—Real constant;

[i]—The mass percentage of the element i;

Bi—Multiplication coefficient of the element i.

T℃	A		B_C	B_{Si}	B_{Mn}	B_P	B_S
700	114.36		-97.23	-20.46	-13.61	99.33	-734.17
750	148.67		-252.98	8.70	-49.85	478.56	-929.03
800	69.00		-143.38	—	-11.38	—	—
850	11.92		—	—	26.90	477.17	—
900	55.21		—	—	22.69	383.60	-1410.3
950	82.51		—	—	—	—	—
1000	96.00		-114.47	—	—	597.10	—
1050	89.75		-71.84	—	7.86	356.09	—
1100	90.50		-58.72	—	5.80	222.55	—
1150	77.01		33.91	9.63	6.44	—	—
1200	75.26		47.94	15.72	—	—	—
1250	85.22		—	18.30	—	393.09	-775.12
T℃	B_{Al}	B_{Nb}	B_{Ti}	B_V	B_{Ni}	B_{Cr}	B_{Cu}
700	-625.63	-483.49	336.86	—	—	-13.75	—
750	-717.20	—	1609.30	-877.75	140.19	-35.70	-244.36
800	—	-168.32	1134.63	-382.10	57.49	—	-131.32
850	—	-898.73	1712.58	-299.43	—	—	—
900	-222.5	-1070.4	953.0	-576.9	80.36	-38.67	—

T℃	A		B_C	B_{Si}	B_{Mn}	B_P	B_S
950	-251.49	-835.70	1317.37	-360.55	183.88	-41.38	—
1000	—	-447.60	732.06	-161.26	403.80	-88.33	-282.22
1050	—	-441.53	290.78	-91.64	274.55	-68.82	-162.25
1100	114.88	-362.14	—	—	200.44	-44.85	-136.52
1150	77.20	-418.90	405.60	—	76.43	-29.22	—
1200	—	-49.51	—	59.79	-42.96	—	51.38
1250	-143.42	-73.51	—	—	75.94	-35.42	-86.62

Table 1. A, Bi values of formula (2)

The accuracy of regression model needs significant tests. Several important significant test statistics indexes of the regression model are as follows

F: F inspection value; the bigger the F value, the better the significance level is.

R^2Multiple correlation coefficientsreflect regression effect quality: the greater the R^2, the better the regression result is. Generally, R^2 equaling to 0.7 or so can give a positive attitude.

Ra^2: Multiple correlation coefficients after adjustment. Formula is as (2):

$$R_a^2 = 1 - \frac{n-1}{n-p-1}\left(1 - R^2\right) \tag{2}$$

Sig: Significant level value; the smaller value, the better result is.

Specific details are shown in Table 2. The significant level value, Sig at different temperatures is all less than 0.1 except for 900 ℃, and it means that the accurate probability of the predicted values is more than 90%. Multiple correlation coefficients, R^2 is more than 0.5 which indicates the better significant of the model.

T℃	Used date	Number of elements	R^2	R^2_a	standard deviations	F	Sig
700	24	9	0.746	0.582	9.9	4.557	0.006
750	24	11	0.860	0.732	12.5	6.700	0.001
800	24	7	0.505	0.289	11.6	2.333	0.076
850	15	5	0.773	0.647	9.1	6.129	0.010
900	24	9	0.526	0.221	18.8	1.725	0.174
950	24	6	0.521	0.352	15.5	3.086	0.031
1000	21	8	0.656	0.426	6.5	2.856	0.050

T℃	Used date	Number of elements	R^2	R^2_a	standard deviations	F	Sig
1050	21	9	0.739	0.526	4.4	3.163	0.028
1100	21	8	0.660	0.433	4.2	2.913	0.047
1150	21	8	0.698	0.497	3.4	3.467	0.026
1200	21	6	0.688	0.554	2.8	5.140	0.006
1250	21	8	0.724	0.540	4.1	3.936	0.017

Table 2. Statistics in significant test of regression model

In order to prove the accuracy of the hot ductility prediction model, the tested data selected from literatures, which is outside the regression analysis samples data, has been compared with the prediction model for pipeline steels and weathering steels.

The chemical composition of two steel grades is shown as Table 3. Test conditions for the strain rate is 1.0×10^{-3} / sand the cooling rate is 3 ℃ / min.

type of steel	C	Si	Mn	P	S	Al	Nb	Ti	V	Ni	Cr	Cu
weathering steel	0.094	0.295	0.4	0.076	0.005	0.033	—	—	—	0.22	0.53	0.29
Pipeline steel	0.054	0.224	1.6	0.008	0.002	0.037	0.054	0.013	0.042	0.17	—	0.18

Table 3. The chemical composition of pipeline steel and weathering steel

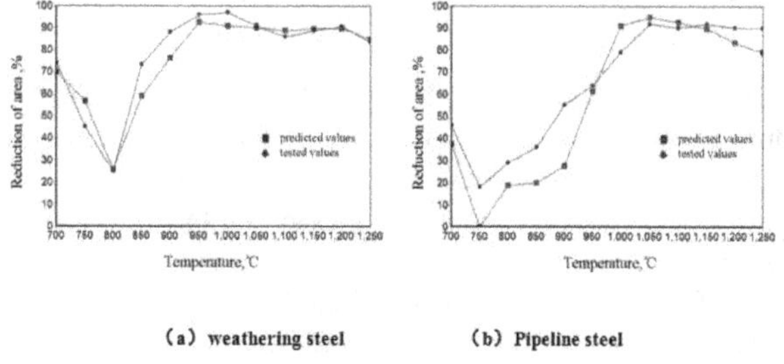

Figure 3. Comparison of hot ductility between predicted values and tested values

The curve of predicted values is very close to tested values and they have the same tendency by comparison from the Fig.3. It should be aware that the predicted values will be difficult in exact conformity with the tested values due to test conditions and test errors. Therefore, it shows that prediction model of thermoplastic established in this paper has a better practicability. Even so, the model has some limitations because of less regression sample data of only 24 groups. But with more studies on hot ductility, the model will evolve further.

2.2. Formulae for thermal physical parameters

The thermophysical property parameters of steel such as density, conductivity coefficient, specific heat capacity, latent heat, liquidus temperature, and solidus temperature are essential for calculating the heat transfer model. Although these parameters can only be acquired accurately by tests, the thermophysical properties of a new steel grade can also be approximately calculated from the chemical composition with the requirements of more steel grades to cast.

2.2.1. Liquidus temperature

The liquidus temperature of steel plays a very important role in metallurgical production and related scientific research. The lowest superheat may be achieved during the process of continuous casting if an accurate liquidus temperature of steel is obtained. This is described as it is useful to acquire a fine grain structure and higher quality of slab for steel plants. The accurate liquidus temperature of steel is also required for scientific investigation of solidification processes of molten steel by numerical simulation. Research shows that the main reason why the liquidus temperature of steel is lower than the melting point of pure iron is the presence of impurities and alloying elements. Generally speaking, there are two ways to obtain the liquidus temperature of steel for the research: firstly, as a standard method for determining transformation temperature of materials, a differential thermal analysis (DTA) measurements is conducted, and a number of studies have used DTA for the determination of liquidus temperature; secondly, the more common method, is to select the appropriate model according to the different kinds of steel. On the basis of the analysis of Fe-i binary phase diagram, a new calculation model for liquidus temperature of steel is established in this study.

The different effects of 11 elements (C, Si, Mn, P, S, Ca, Nb, Ni, Cu, Mo,Cr) on the melting point of pure iron were investigated and 11 groups of discrete data (A_C, A_{Si}... A_{Cr})that isthe value of liquidus temperature was decreased or increased together with the content of element i increase (or decrease) by 0.1% mass fraction in Fe-i binary phase diagramwere obtained. Then, each group data was fitted to obtain the mathematical formula (ΔT_{lc}, ΔT_{lsi}, ...ΔT_{lMo}). Finally, the model of steel liquidus temperature can be establishedintroducing the mathematical formulae of each element into the Eq.(3).The calculation model for steel liquidus temperature developed in this study is as follows

$$T_l = T_0 - \sum\left(\frac{\partial T_l}{\partial C_i} \times [\%C_i]\right) \tag{3}$$

Where

T_l —Liquidus temperature of steel, ℃;

T_0—Melting point of pure iron, ℃, the general value range is 1534~1539℃, and T_0 is 1538℃ in this study;

$\partial T_l/\partial C_i$—The changing rate of liquid isotherm to the content of element i on Fe-i binary phase diagram;

$[\%C_i]$—The percentage content of element i.

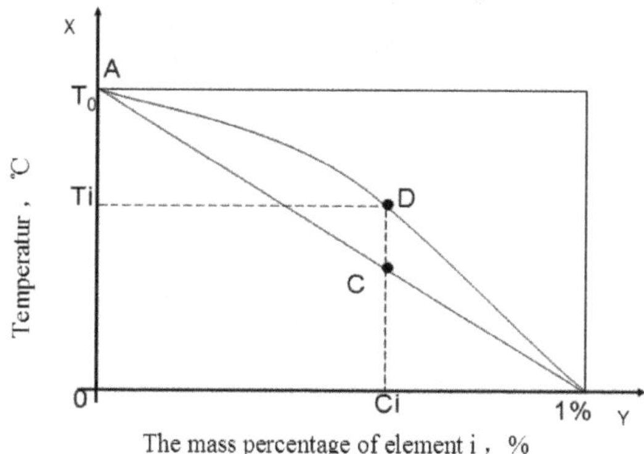

Figure 4. The influence of element i on the liquidus temperature

In Fig. 4, the X axis represents the mass percentage of element i and the Y axis represents temperature. The curve ADB is the change in the actual liquidus temperature with the content of element I; however, most research on liquidus temperature assumed that the influence of each element on reduction value of the melting point is kept linear relation (shown as the straight-line segment AB). Therefore, the calculation is easy to result in deviation. For instance, when the content of the element i is C, the liquidus temperature is the value corresponding to C (where point C corresponds to the liquidus temperature according to traditional models), however the actual liquidus temperature is T_i (corresponding to point D). Therefore, the deviation is the line segment CD. As a result, the traditional calculation model for liquidus temperature of steel is likely to have a large error when steel has many elements.

Owing to drawbacks of the general models for liquidus temperature calculation, a new model is needed. After differentiated the Fe-i binary phase diagram, new temperature coefficients of each element in the molten steel is obtained, a new calculation model for liquidus temperature is established. The margin of error with the use of this universal model is likely to be less than that with traditional models. All the alloying elements of steel or cast iron influence the liquidus temperature; however, the element which has the greatest effect is carbon. Considering an example of the phase diagram of Fe-C and amplifying the part of interest will help explain this.

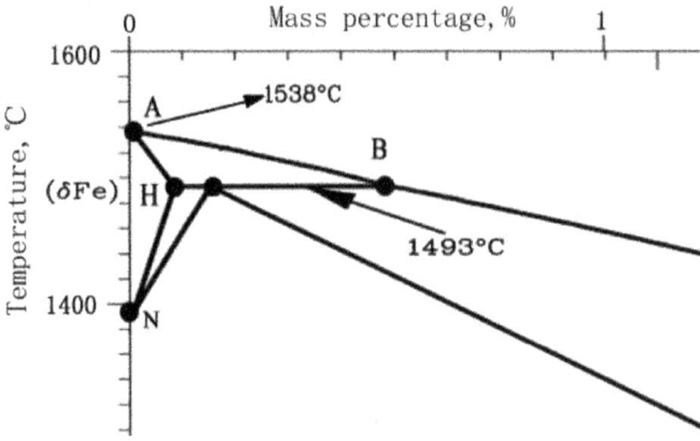

Figure 5. Partial Fe-C binary equilibrium phase diagram enlarged

Processing of the curve AB in Fig.5 by Photoshop software shows the influence of carbon content on the liquidus temperature (Table 4).

CContent, %	0.01	0.02	0.03	0.04	0.05	0.06	0.07	0.08	0.09
ΔT_l,℃	2.00	2.50	3.00	3.40	4.40	4.90	5.70	6.10	6.70

Table 4. Impact of carbon content on the liquidus temperature

The data in Table 4 are fitted with the least square methodand the calculation formula for the influence of carbon content on steel liquidus is established and expressed as:

$$\Delta T_k = 32.15[\%C]^2 + 62.645[\%C] - 0.8814 \tag{4}$$

$$T_l = 1538 - \begin{pmatrix} 31.15[\%C]^2 + 62.645[\%C] + 0.609[\%Si]^2 + 2.0678[\%Si] - 0.0674[\%Mn]^2 \\ +5.3464[\%Mn] + 20[\%P]^2 + 9[\%P] - 1.7724[\%S]^2 + 24.775[\%S] + 1.1159[\%Nb]^2 \\ +5.3326[\%Nb] - 0.0758[\%Ca]^2 + 3.1313[\%Ca] + 0.0379[\%Ni]^2 + 5.2917[\%Ni] \\ +0.6818[\%Cu]^2 + 2.5955[\%Cu] + 0.0214[\%Mo]^2 + 3.2214[\%Mo] + 0.0359[\%Cr]^2 \\ +1.1402[\%Cr] + 10.797 \end{pmatrix} \tag{5}$$

In the same way, for Si, Mn, P, S and other elements, their binary phase diagrams are processed, and different formular for each elements influence on the steel liquidus are obtained. Finally, a new model for calculating steel liquidus temperature is set up by synthesizing, which is verified with some testing liquidus temperature of steel, shown as Eq.(5).

It has been proved that errors between liquidus formula (5) with others are all less than 4 ℃.

2.2.2. *Thermal conductivity coefficient*

Thermal conductivity coefficient of steel solid-phase is relevant to temperature and elements. For carbon steels and stainless steels, the expression is shown as Eq. (6). Moreover, due to the great influence of liquid convection in liquid core, the equivalent thermal conductivity coefficient is used for liquid-phase.

$$\begin{aligned}\lambda^S &= 20.76 - 0.009T - 3.2627[C] + \begin{pmatrix} 0.0124 - 2.204\times10^{-4}T \\ +1.078\times10^{-7}T^2 + 7.822\times10^{-4}[Cr] - 1.741\times10^{-7}T\cdot[Cr] \end{pmatrix}[Cr] \\ &+ \begin{pmatrix} -0.5860 + 8.354\times10^{-4}T - 1.368\times10^{-7}T^2 + 0.01067[Ni] \\ -1.504\times10^{-5}T\cdot[Ni] \end{pmatrix}[Ni] - 0.7598[Si] - 0.1432[Mn] \\ &-0.2222[Mo]\end{aligned} \tag{6}$$

$$\lambda^L = m\lambda^S \tag{7}$$

$$\lambda^{SL} = \lambda^S f_S(T) + \lambda^L(1 - f_S(T)) \tag{8}$$

Where,

$\lambda^L \lambda^S \lambda^{SL}$— the conductivity coefficient of liquid phase, solid phase and mush zoon respectivelyW/(m ℃)

T —Temperature℃

[i] —The mass percentage of the element i%

$f_S(T)$—Solid fraction

m —Equivalent coefficient.

2.2.3. *Density*

The density with high temperature of carbon steels is relevant to the carbon content and temperature. Its solid, liquid density can be used formula (9), (10) to calculate.

$$\rho_S = \frac{100\left(8245.2 - 0.51(T + 273)\right)}{\left(100 - [C](1 + 0.008)[C]\right)^3} \tag{9}$$

$$\rho_l = 7100 - 73[C] - \left(0.8 - 0.09[C]\right)(1550 - T) \tag{10}$$

But for stainless steels, it is strongly related to Cr, Ni, Mo, Mn, Si and other major elements, the expression is shown as formula (11), (12)

$$\rho_s = \begin{pmatrix} 79.6\%[Fe] + 78.3\%[Cr] + 85.4\%[Ni] + 76.9\%[Mn] + \\ 60.2\%[Mo] + 47.1\%[Si] \end{pmatrix} - 0.5(T - 25) \tag{11}$$

$$\rho_l = \begin{pmatrix} 69.4\%[Fe] + 66.3\%[Cr] + 71.4\%[Ni] + 57.2\%[Mn] + \\ 51.5\%[Mo] + 49.3\%[Si] \end{pmatrix} - 0.86(T - 1550) \tag{12}$$

$$\rho_{sl} = \rho_s f_S(T) + \rho_l\left(1 - f_S(T)\right) \tag{13}$$

Where

$\varrho_s\varrho_l\varrho_{sl}$—The density of solid phase, liquid phase and mush zoon respectivelykg/m^3

T—Temperature℃

[i] —The mass percentage of the element i%

f_S(T) —Solid fraction.

Moreover, formulae for specific heat and latent heat have been described in many research literatures. They will not be mentioned in this chapter.

Because thermal physical parameters have an important influence on the accuracy for the heat transfer calculation model, the database of thermal physical parameters, such as liquidus temperature, conductivity coefficient, density and so on, has been established by summarizing, which can provide accuracy "basic parameters" for the "targeted" solidification and heat transfer numerical model.

3. Control models for secondary cooling in continuous casting process

Secondary cooling control, which is a key technology in the continuous casting process, not only determines the productivity of a caster, but also significantly influences the quality of the slab. Currently, nearly all secondary cooling control systems are based on a heat transfer model of solidification during continuous casting, which makes the control process more quantitative. At present, there are several popular control models for secondary cooling in continuous casting, such as the parameter control model, effective-speed control model, online thermal model, and models that are combinations of these. These control models have their respective advantages and meet the modern requirements of the continuous casting process on the whole. Even sothe mathematical heat transfer model of solidification is an important base for secondary cooling control, so authors will briefly introduce it before expounding the control models of continuous casting.

3.1. Heat transfer model

The mathematical heat transfer model of solidification during continuous casting is composed of heat conduction equations, initial conditions, and boundary conditions. The heat conduction along a strand is usually neglected. The unsteady two-dimensional equation of heat transfer is shown as below:

$$\rho c_p \frac{\partial T}{\partial \tau} = \frac{\partial}{\partial x}\left(\lambda \frac{\partial T}{\partial x}\right) + \frac{\partial}{\partial y}\left(\lambda \frac{\partial T}{\partial y}\right) + q_v \tag{14}$$

Where, q_v is internal heat source, which is latent heat (J kg^{-1}) here and can be equivalent to the equivalent specific heat capacity or effective thermal enthalpy to simplify the conduction equation. The heat transfer model is the basis for the quantitative method of controlling the secondary cooling water, and many models of secondary cooling control have been developed. Some popular control models are reviewed in this chapter, as follows.

3.2. Control models for secondary cooling

3.2.1. Parameter control method

The parameter control method requires determination of the target surface temperature curves for different steel grades; calculation of the control parameters A_i, B_i, and C_i for every secondary cooling zone such that the strand surface temperature coincides with the target surface temperature; and building a mathematical model as in equation (15).

$$Q_i = A_i V^2 + B_i V + C_i \tag{15}$$

With the wide use of continuous temperature measurements of molten steel in the tundish and growing research on the influence of the temperature of secondary cooling water on slab cooling, the superheat and the temperature of the secondary cooling water have been considered to be the important factors for controlling the surface temperature of the slab. The secondary cooling water flow rate needs to be adjusted according to these two factors, so equation (15) can be modified and improved, whereby equation (16) is presented as follows:

$$Q_i = A_i V^2 + B_i V + C_i + D_i \Delta T + F_i \tag{16}$$

Where, D_i is the adjusting parameter for the water flow rate based on superheat, and F_i is the adjusting parameter of water flow rate based on the temperature of the secondary cooling water, which changes with the season.

The water flow rate in the parameter control method changes with the variation of casting speed continuously, and are controlled according to the theory of the solidification of the slab and the practical conditions. The control pattern can be run in an automatic, manual, or semi-automatic way. Indeed, the parameter control method requires little investment but has strong applicability, and so is still widely applied in steel plants. However, the control method shows an apparent disadvantage: the parameter control method cannot keep the stability of the slab surface temperature in the unsteady casting state, such as in the case of a change in the submerged entry nozzle and the hot exchange of the tundishes. Therefore, an improved control method called the "effective speed" method has been developed based on the parameter control method.

3.2.2. *Effective speed control method*

The effective speed control method is derived from the residence time control method of slabs. As shown in Fig.6 (the calculating model of residence time of a slab), the slab is divided into a number of small slices, each slice is pulled forward at the casting speed of the slab, and new slices are generated at regular intervals. The residence time can be regarded as approximately the same for a slice. Once the slice is pulled out of the end of the secondary cooling zone, it is no longer tracked, and is deleted from the computer memory. The data for each slice are updated at regular intervals; this includes the distance from the meniscus to the position of the slice and its running time in the caster, which is called the "residence time."

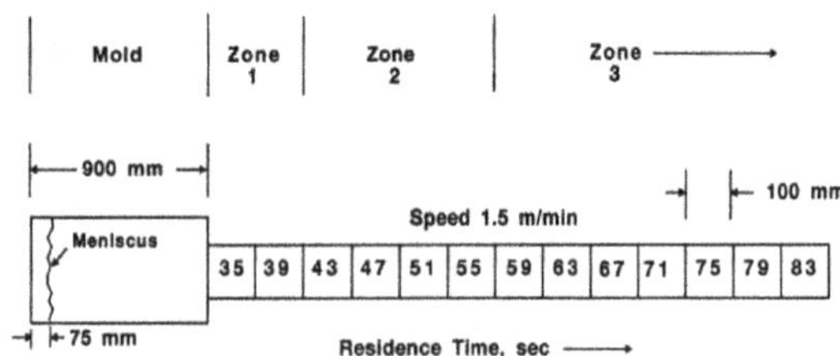

Figure 6. The calculation model for the residence time of a slab

The relationship between the residence time and the water flow rate can be converted into the relationship between the average speed and water flow rate. The average speed of one cooling zone can be calculated from equation (17):

$$V_{ai} = \frac{S_i}{\frac{1}{n_i} \cdot \sum_{j=1}^{n_i} t_{rij}} \tag{17}$$

It has been shown that a modified effective speed based on the average speed can be used to reduce surface temperature fluctuations and improve the safety of continuous casting operations. Effective speed can be calculated by equation (18):

$$V_{ei} = \varepsilon_i V_{ai} + (1 - \varepsilon_i) V \tag{18}$$

Where, ε_i is the weighting coefficient, which is between 0 and 1, and depends on the distance from the center of the zone to the meniscus: i.e., the longer the distance, the higher is the value.

The equation of the effective speed control method is constructed by replacing the real-time speed in equation (15) with the effective speed, shown as below:

$$Q_i = A_i V_{ei}^{\ 2} + B_i V_{ei} + C_i \tag{19}$$

Fig.7 and Fig.8 show the fluctuation of the slab surface temperature calculated by the computer simulation using the parameters control model and the effective speed control model, both with the same speed conditions. Compared with the parameters control model, the effective speed control model can keep the surface temperature smoother in the unsteady casting state of speed fluctuations, such as the change of the submerged entry nozzle.

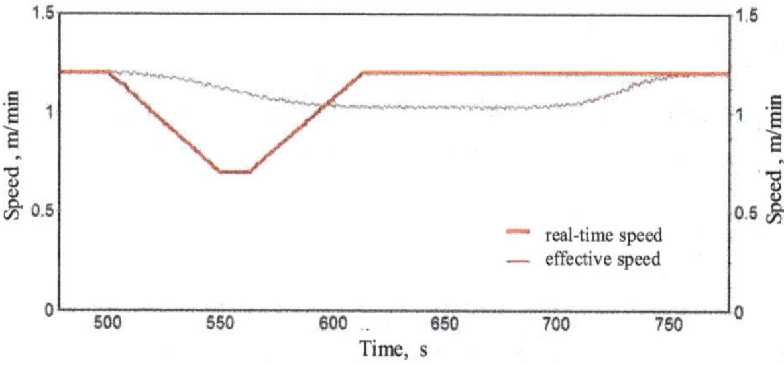

Figure 7. Fluctuation of the effective speed of the third zone due to a fluctuation in the casting speed

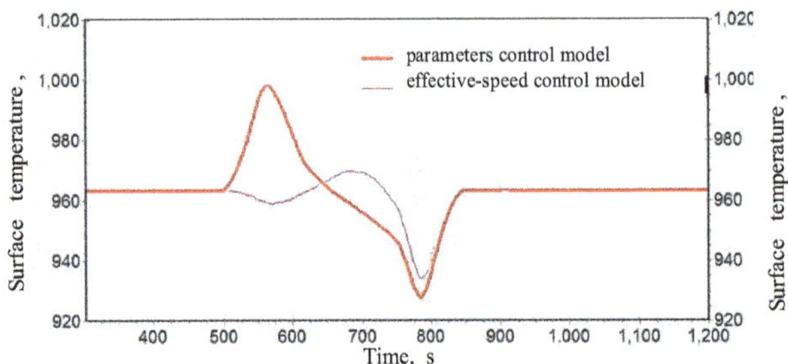

Figure 8. Comparison of the surface temperature fluctuations at the end of the third cooling zone (4.2 m from the mold meniscus) for the two control methods(Section is 220 mm×1600 mm, casting temperature is 1818 K, peritectic steel)

The parameter control method and the effective speed control method are both based on an off-line thermal model. With advances in computational technology and the reduction of computational costs, online calculation of temperature profiles is no longer a problem. The online thermal model control method is based on an online simulation model of heat transfer and controls the water flow rate of secondary cooling zones through real-time calculation of the slab temperature profile.

3.2.3. *Online thermal model control method*

The online thermal model control method can be described as follows. The online simulation model of heat transfer calculates the real-time temperature profile of the slab at certain intervals, and the water flow rate of the secondary cooling zones is controlled by the deviational value of the target temperature and calculated temperature.

The water flow rate control relies only on the feedback of the surface temperature calculated by the online thermal model and has a hysteresis quality. The stability of the control system is poor, because the control system has a strong dependency on the accuracy of the calculated surface temperature. Therefore, the online thermal model control method needs to be combined with other feed-forward control methods, for example, a combination of the online thermal model with the effective speed control method, shown as equation (20). In this control method, the surface temperature is controlled through setting the basic water flow rate with effective-speed and fine-tuning it with the deviational value between the target temperature and the calculated temperature to further reduce this deviation.

$$Q_i = f_1(V_i) + f_2(\Delta T_{fi}) \tag{20}$$

Where, $f_1(V_{ei})$ is the water flow rate calculated using the effective speed model, and $f_2(\Delta T_{fi})$ is the water flow rate calculated based on the deviation value between the target temperature and the calculated temperature using the online thermal model.

This control method is characterized by good stability and accuracy and no delays, thus it enables the surface temperature to be controlled around the target values. Fig.9 shows the fluctuations of the measured surface temperature at a position 5.0 m from the meniscus with fluctuations in the casting speed in a continuous casting process. It can be seen in the figure that the surface temperature of the slab is controlled around 920°C despite strong fluctuations of the casting speed.

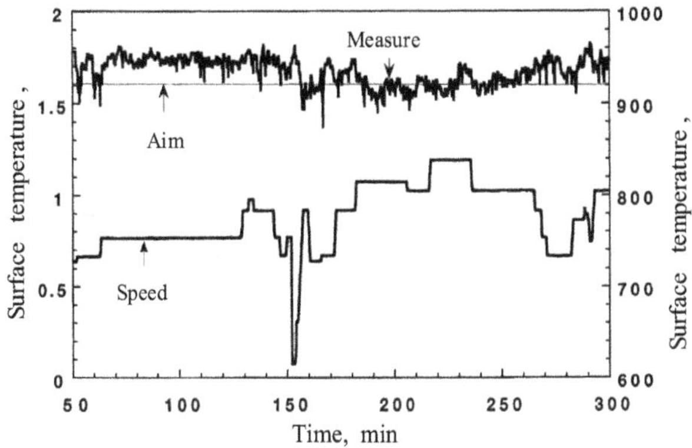

Figure 9. Measured fluctuations of the surface temperature at a position 5.0 m from the meniscus

The online thermal model can calculate the real-time surface temperature of a slab, but due to the inevitable deviation between the calculated temperature and the actual temperature, the actual surface temperature can only be obtained by measurement. Therefore, while a system of water flow rate control that relies only on the feedback of the measured surface temperature is not commonly adopted, a thermometer combined with an online thermal model can be applied as one of the main tools of secondary cooling control. In this case, the feedback value is not directly used to control the water flow rate, but to dynamically adjust the parameter A in equation (9), which reflects the relationship between the heat transfer coefficient and water flow rate, and to eliminate the temperature error – the difference between the calculated temperature and the measured temperature. The thermometer does not need to be working continuously, rather, the online thermal model can be corrected with temperature measurements at certain intervals; thus the expenditure of thermometers is improved, and the accuracy of the online model and precision of the secondary cooling control is ensured.

3.2.4. Synthetical model dynamic control method based on online temperature measurement

In order to build a new secondary cooling control model that integrates the advantages of the control methods mentioned above, the concept of effective superheat is put forward, and the synthetical model dynamic control method based on online temperature measurement is established in this study.

Effective superheat is obtained by modifying the average superheat. In the parameter control method, the water flow rate compensation according to superheat is based on the real-time superheat, and this control model can meet the control requirements of the surface temperature only in the case of small fluctuations in the casting temperature. However, if the fluctuations are large, the surface temperature of the slab will not be controlled. In order to achieve accurate water flow rate compensation according to superheat, the initial superheat in the meniscus of the slab should be obtained, and thus the average superheat needs to be applied. In the residence time model of the slab (shown in Fig.6), the computer not only calculates the residence time of each slice, but also stores the data of the initial superheat of each slice when it is generated. The average superheat $\triangle T_a$ of one zone is the average value of the initial superheats of all the slices in this cooling zone. The average superheat represents the initial superheat of the slab in a cooling zone, but there are shortcomings in applying this method. Because of the upper and lower convection from the liquid core, the temperature of the liquid steel in the mold influences the temperature profile of a slab with a liquid pool. Furthermore, the shorter the distance of a cooling zone from the mold, the stronger is this effect. In addition, the water flow rate of the cooling zone closer to the mold cannot be adjusted in time when using the average superheat, thereby a breakout may happen if the casting temperature suddenly rises in the continuous casting process. Therefore, with regards to the effective speed, the average superheat should be corrected, and the effective superheat $\triangle T_e$, derived, as shown in equation (21):

$$\Delta T_{ei} = \lambda_i \Delta T_{ai} + (1 - \lambda_i) \Delta T \tag{21}$$

Where,ΔT_{ei} is the effective superheat of zone i ℃;ΔT_{ai}, average superheat;ΔT, real time superheat; andλ_i the weighting coefficient, which ranges from 0 to 1. The weighting coefficient depends on the distance from the cooling zone to the meniscus: the further the distance, the greater is the value. The value is 1 at the cooling zone of the solidification endpoint.

Fig.10 and Fig.11 show the surface temperature simulated fluctuations of a slab with conditions of no water flow rate compensation according to superheat, water flow rate compensation according to real time superheat and water flow rate compensation according to effective superheat under the same casting temperature conditions. It is evident that when the superheat rises sharply when the casting speed is stable, the surface temperature is not well controlled with no water flow rate compensation according to superheat. When the casting temperature rises, the surface temperature increases. Moreover, with water flow rate compensation according to real time superheat, the surface temperature undergoes large fluctuations although it can return to the temperatures close to those before the casting temperature rise. In the mode of water distribution based on the water flow rate compensation according to effective superheat, not only does the surface temperature almost return to what it was before the increase of the casting temperature, the temperature fluctuations are also much smaller, showing better control of the surface temperature.

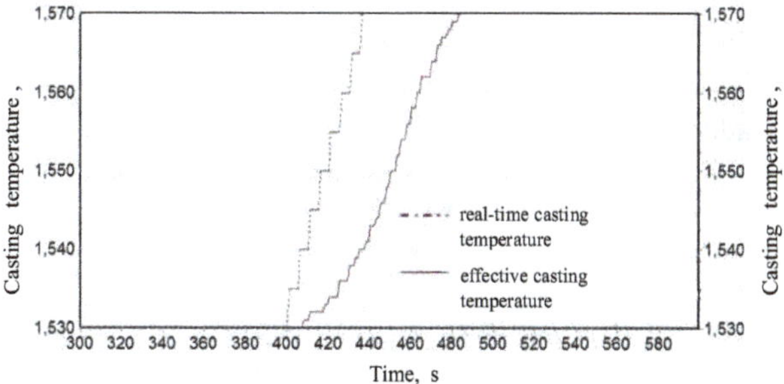

Figure 10. Effective casting temperature fluctuation in the foot-rollers cooling zone with a fluctuation of the pouring temperature

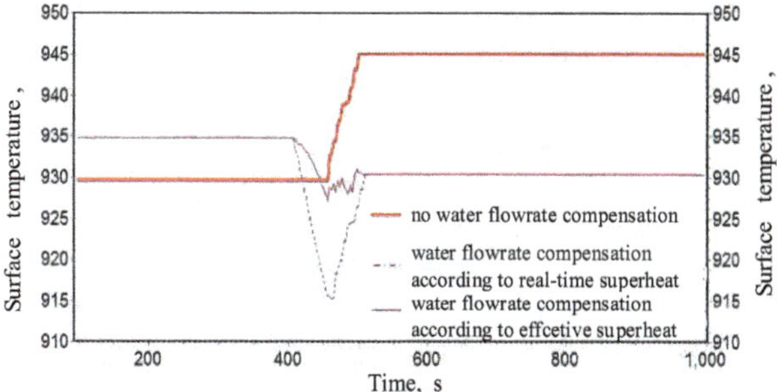

Figure 11. Comparison of the surface temperature fluctuations at the end of the foot-rollers cooling zone(1.2 m from the meniscus) under three modes of water flow rate compensation(Section is 220 mm x 1600 mm, withdraw speed is 1.0 m/min, peritectic steel)

The various control models mentioned above have different characteristics. By integrating them, a new synthesized secondary cooling control method called "synthetical model dynamic control method based on online temperature measurement" can be deduced, as shown in equation (22):

$$Q_i = f_1(V_{ei}) + f_2(\Delta T_{ei}) + f_3(\Delta T_{fi}) \tag{22}$$

Where $f_1(V_{ei})$ is the water flow rate determined by effective speed; $f_2(\Delta T_{ei})$, the water flow rate determined by effective superheat; and $f_3(\Delta T_{fi})$, the water flow rate determined by the deviation value between the target surface temperature and the calculated surface temperature.

In this control model, the surface temperature is controlled through setting the feed-forward water flow rate with effective speed and effective superheat, and carefully adjusting it with the deviational values of the target temperature, using the adjusting pattern of the PID control algorithm. In addition, this control system with an online thermometer can modify the online thermal model with time when casting conditions change. The control logic is shown in Fig.12.

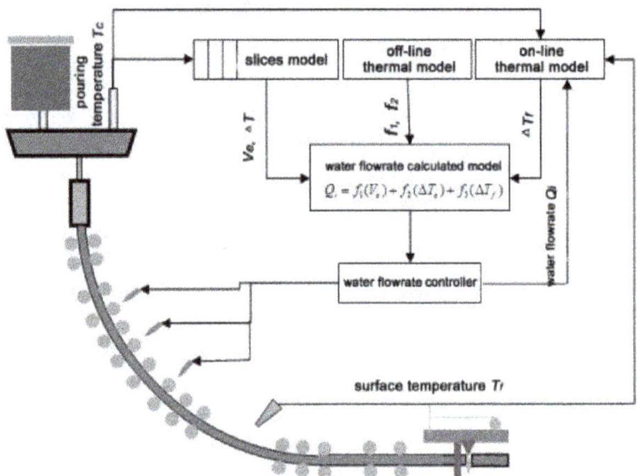

Figure 12. Control logic of synthetical model dynamic control method based on online temperature measurement

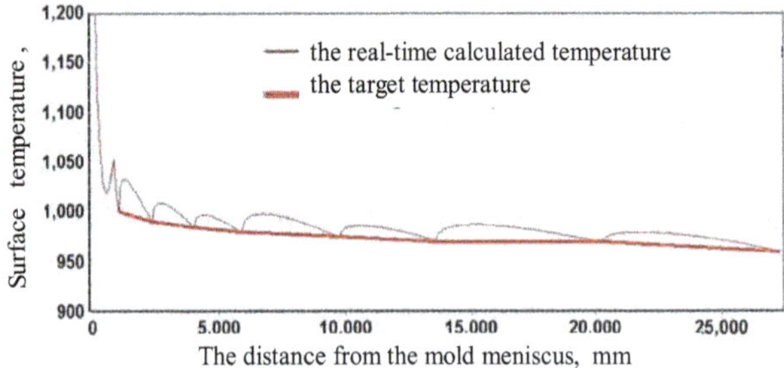

Figure 13. Comparison of real-time calculated surface temperature and the target surface temperature

Fig.13 shows the center surface temperature profile of peritectic steel whose cross section is 1600 mm × 220 mm at a withdraw speed of 1.2 m/min, and superheat 1818 K.

It reveals that in this control method, the surface temperature is well controlled around the target temperature.

This study summarizes the principles and characteristics of several popular models of secondary cooling control and furthermore, puts forward the concept of the effective superheat and an improved model called "synthetical model dynamic control based on online temperature measurement." This new control method demonstrates good control of the slab's surface temperature. As the requirements on slab quality continue to rise, the secondary cooling control system will play an important role in the casting process. Many new technologies such as dynamic soft reduction, the quality of online evaluation and forecasting and the direct rolling process, are based on an advanced control system of secondary cooling as the pre-condition. A secondary cooling control system not only needs to ensure a smooth slab surface temperature distribution, but also provides real-time information of the slab's temperature profile and the end of the liquid pool. In addition, the rapid development of information technology will also push the secondary cooling control to the level of intelligent and full automation. From work presented here, we can conclude that the subject of secondary cooling control systems needs further research and development from the following aspects:

1. The operation conditions in special periods, such as at the start or end of the casting or at the hot exchange of the tundishes, should be taken into account in the control model, in order to guarantee the slab quality at these points and improve the recovery ratio of metal.

2. Durable, accurate, online surface temperature measuring sensors should be developed to provide continuous, accurate feedback data for the secondary cooling control system, and achieve dynamically precise cooling control of the slab.

3. For further improvement of the simulation models for continuous casting processes, thermal-mechanical coupling should be introduced into the online calculation model, so that the models can not only provide real-time temperature profiles, but also provide the stress field, shell shrinkage, and a function for online crack forecasting.

4. Influence of nozzle layouts on the secondary cooling effect of slabs

The effect of spray water on heat transfer of slab surface depends on the performance of the nozzle. Therefore, in order to analyze cold characteristics of the nozzle, improve the slab quality by the optimization of secondary cooling system, and improve the continuous casting productivity, a series of experiments should be carried out.

Taking the CCM2 at the No.3 steelmaking plant of Hansteel for example, flat type air mist nozzles are used in the segments, with three nozzles arranged in each row. The distance between adjacent nozzles is 450mm, and the height from the slab surface to a nozzle is 380mm. As the spraying angle is 110°the water sprayed from the nozzles appears to be triple overlaid on the center surface of the slab, which causes water accumulation in the region. In addition, the presence of excessive water is at the corner region. Uneven cooling along the width direction of the slab can easily lead to slab cracks and other defects.

Figure 14. Experimental device of cold characteristics test for cooling nozzles

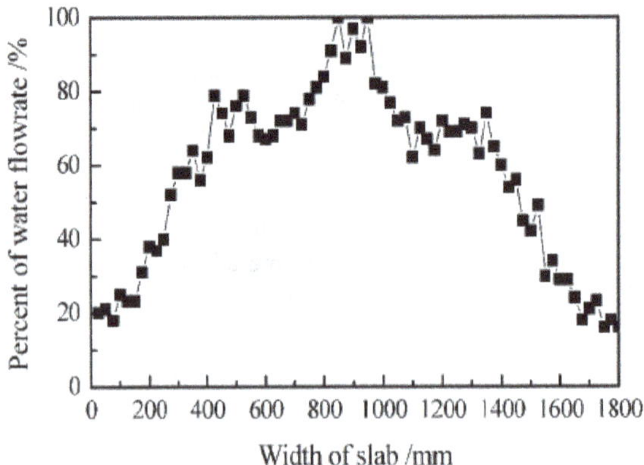

Figure 15. Water distribution along slab width direction before optimisation(water pressure, 0.2 MPa; air pressure, 0.2 MPa)

4.1. Influence of spray water distribute on secondary cooling effect of slabs

Based on the mathematical model, the stress and strain fields of the slab were also studied under specific casting conditions using the finite element software ANSYS. Considering the symmetry of a slab cross-section, half of the slab cross-section was taken as the research object. Under the current arrangement of nozzles, the distribution of water flowrate in the slab width direction was measured as shown in Fig.15. Before optimization at the straightening zone, the temperature profile of the slab surface in the slab width direction was as shown in Fig.16.

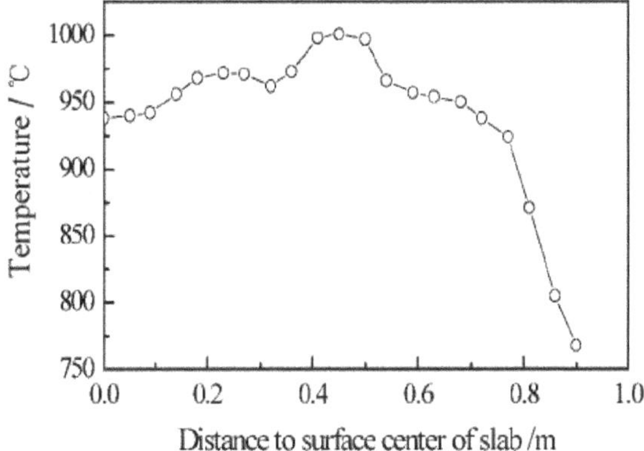

Figure 16. Temperature behavior of slab surface in slab width direction before optimization(Half section, casting speed, 0.9 m min^{-1}; superheat, 27℃, 18 m below meniscus)

As can be seen from the Fig.16, due to the poor spray cooling pattern, there is an uneven surface temperature distribution in the slab width direction. The temperature at the surface center of slab is only 938℃, while the highest temperature value of the slab surface is 1001℃, which is near the quarter of the whole slab width. Moreover, the lowest temperature of 768℃ is at the slab corner.

This chapter analyses fully the stress field of the slab in the straightening region, between 15.86 and 20.24m below the meniscus. Because the slab is not fully solidified when the slab enters into the straightening zone, the temperature of central region of slab is still above the liquidus temperature. In order to simplify the model, the equivalent stress analysis is only focused on the solidified shell. The temperature field before optimization is set as the initial condition; meanwhile, corresponding ferrostatic pressure is imposed on the solidifying front of the slab for stress analysis. The ferrostatic pressure can be expressed as equation (23).

$$\Delta P = \rho g H \tag{23}$$

The equivalent stress field of the slab is simulated under the action of the straightening roller along the casting direction, as shown in Fig. 17.

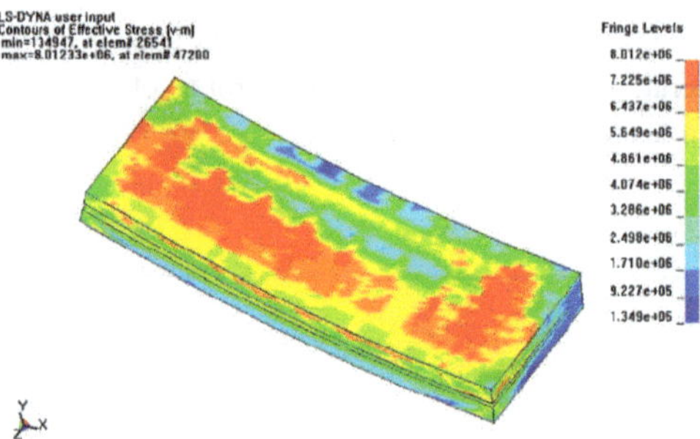

Figure 17. Equivalent stress field of slab at straightening segment before optimization(steel grade, Q420B; section, 1800×220 mm; casting speed, 0.9 m min21; superheat, 27℃)

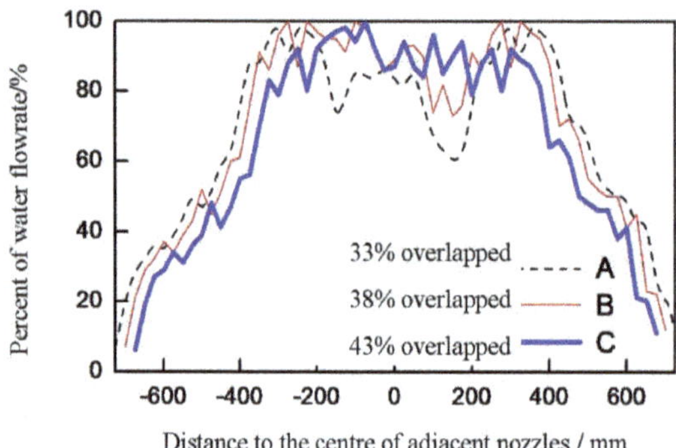

Figure 18. Water distribution with two nozzles along the slab width directionwater flowrate :3.9 L min^{-1};injection height :300mm

The figure shows clearly that the maximum equivalent stress on the slab reaches 8.012 MPa in the straightening zone under the direction of the slab, which results in a high temperature gradient in the slab. Hence, the corresponding equivalent stresses in these regions are larger than those of the other regions, which can generate easily slab defects. Uneven cooling usually appears in the width direction of a slab because of its large width. As an additional factor, the heat transfer occurs on two directions at the corner of a slab. Thus, the design scheme for a secondary cooling system should obey the rules of a homogeneous cooling distribution in the width direction and a gradual decrease in the cooling range along the width direction from the top to bottom of the caster; this should prevent defects caused by undercooling in the corner region of the slab. Based on the temperature and stress analysis of the slab, combined with cold test performance data of the nozzles, a new scheme for the secondary cooling system is proposed.

4.2. Influence of nozzle layouts on the spray water distribution

The principle of nozzles arrangement is to make spray water distribute evenly in the width direction of slab surface. Through a series of test for combined nozzles on the platform of nozzle automatic testing, the relationship between spraying overlap degree of adjacent nozzles and the uniformity of water distribution in slab width direction is analyzed from three aspects such as nozzle flow rate, injection height, water pressure and air pressure.

As can be seen from Fig.18water distribution of scheme C whose spray overlap degree of adjacent nozzles is 43%, is more even in slab width direction when the water flow rate is 3.9 L min^{-1} and injection height is 300 mm.

After the optimization, the distribution of the water flowrate in the slab width direction is improved significantly, as shown in Fig. 19.

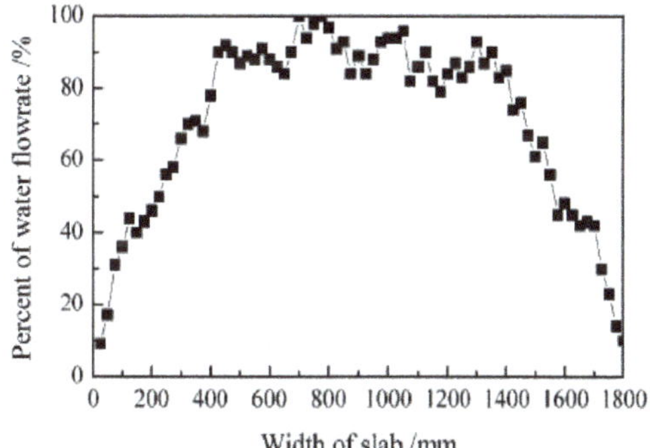

Figure 19. Water distribution along slab width direction after optimization(Water pressure: 0.2 MPa, air pressure: 0.2 MPa)

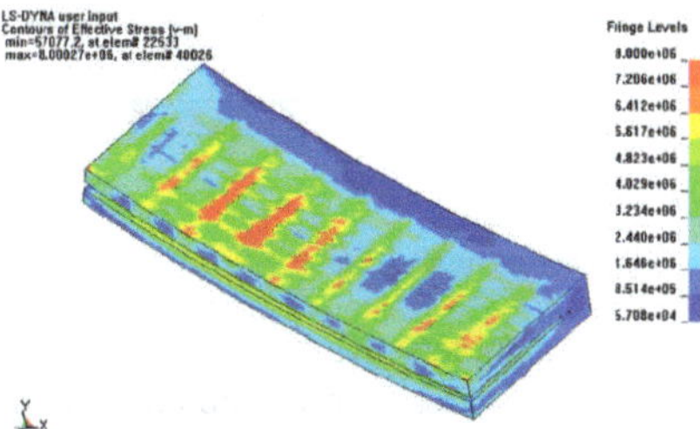

Figure 20. Equivalent stress field of slab at straightening segment after optimization (Steel grade, Q420B; section, 1800×220 mm; casting speed, 0.9 m min^{-1}; superheat, 27℃)

On the basis of the optimization of the temperature field, the stress field of a slab at the straightening zone was analyzed. The simulation results for the equivalent stress field in the straightening zone of the slab after optimization are shown in Fig. 20.

Comparison of Figs. 17and 20 shows that although the maximum values of equivalent stress decrease from 8.012 to 8.000 MPa after optimization (only reduced by 0.012 MPa), the stress concentration has almost disappeared. The larger stresses shown in Fig. 13 exist only where the slab and the rollers are in contact because of the ferrostatic pressure of the molten steel. However, a wide range of slab surface was under the state of large equivalent stress before optimization, which was harmful to the surface quality of the slab. In the software simulation, the temperature and stress fields were both greatly improved, which was useful to improve the quality of the slab.

Through the experimental studies of the flat type nozzle, nozzles arrangements have a major impact on spray water distribution, not only due to the distance of adjacent nozzles and the height of nozzles, but also due to the degree of flat type nozzle bias. As is shown in Fig.21, if the water is sprayed in a straight line at each row with same nozzle type, water pressure and air pressure, and so on.

Figure 21. Nozzles distributed in a straight lineA,B is the center of adjacent two nozzles,1,2,3 is spray area for three flat type nozzles respectively)

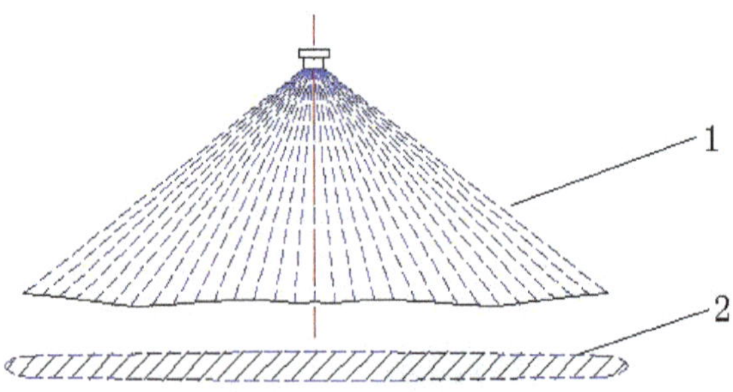

Figure 22. Zonal zones of spray water for flat type nozzle1 is jet stream of flat nozzle, 2 is zonal zones of spray water

As shown in Fig.21, a lot of water droplets will collide crosswise and vertically down between adjacent nozzles. It will lead to water concentration in some area. As shown in Fig.23, The peak phenomenon occurs in water distribution results, and the position of the distance to the edge of slab is 650mm and 1100mm (position of A and B as shown in Fig.21). The experiment proves the validity of the theoretical analysis.

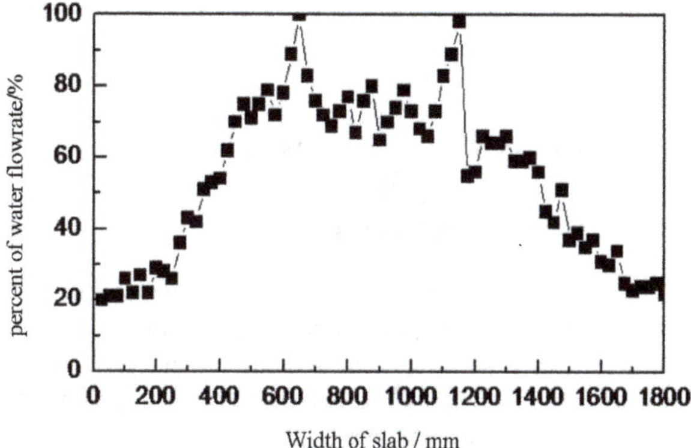

Figure 23. The result of water distributioninjection height: 380mm

In order to avoid water concentration in continuous casting process, it is helpful to prevent water spraying on the same straight line. This will effectively avoid water droplets collide leading to uneven water distribution.

The arrangement of the nozzles is shown in Fig. 24 (a). Considering the restriction of space between rolls, it suggests that the nozzle angle offset is about 5° in this study to make less collide of water spray above the slab if possible. In addition, it can also use the scheme as Fig. 24(b). The length direction of water injection for flat nozzle 1,2,3 are all perpendicular to casting direction. Nozzles of 1, 3 on the edge are located in the same plane. And the nozzle 2 in center is deviated from others a certain distance. This arrangement can also avoid spray water collide between adjacent nozzle.

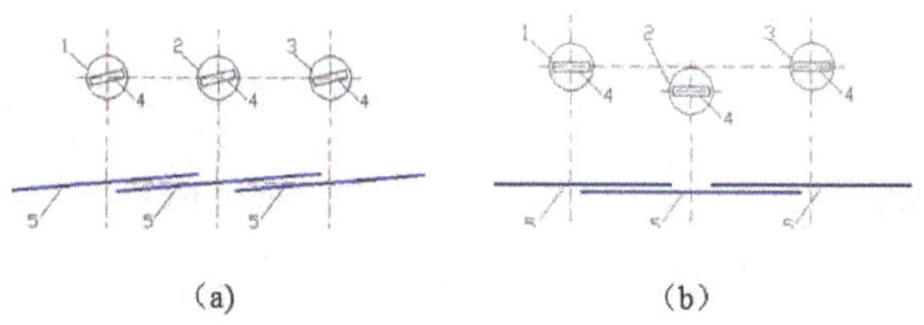

Figure 24. Arrangement of nozzles1,2,3 is the nozzle respectively4 is flat water injection5 is the spray area

After a series of experiments, it shows that the water distribution is more even and will get even cooling in the width direction of slab with above nozzles arrangements.

The control technology of continuous casting of steel not only lies in the fine process model, but also depends on the reasonable production plan. The process of continuous casting should be considered as the linking process between steelmaking process and rolling process. The research on casting production plan should be connected with the charge plan and the rolling plan.

5. Planning for the process of steelmaking-rolling

5.1. Models and solutions of planning for the process of steelmaking-rolling

Planning for the process of steelmaking-rolling is programming and decision-making of inventory or contract oriented production by considering the constraints of steel grades, specifics and due dates based on facilities and resources. The production plan could be obtained by integrating optimal charge plan with optimal casting plan and optimal rolling plan. A description of methods and procedures on planning for the process of steelmaking-rolling is described as follows.

5.1.1. The production plan compiling

Generally speaking, the charge (or heat) is the basic unit for the steelmaking process, and a charge represents the whole process. The constraints of steel grades, dimensions and due dates are all considered in the planning process. Therefore, the charge planning problem is a complicated combinatorial optimization problem subjected to several constraints. The optimal charge plan model was developed based on the objective function of the lowest penalties shown as equation (24).

$$\min z = \sum_{j=1}^{n}\left(P_j^1 + P_j^2\right) + \sum_{j=1}^{n}\sum_{p=1}^{0} P_{pj}^3 \cdot X_{pj}$$
$$x_{ik} = \begin{cases} 1\, \textit{Contract product I exists in charge k} \\ 0 \textit{else} \end{cases} \tag{24}$$

In equation (24): n - number of contract products; i,j - serial number of contract products, i, j = 1, 2,..., n; k - serial number of charges, k = 1, 2,..., m;P_k^1 - penalties for contract products unselected into any of the charges,¥; P_k^2- penalties for open order in charge k,; P_{ik}^3- penalties for differences in contract products' due dates in charge k,¥.

Genetic algorithms could be used to solve the model, the basic parameters of the algorithm's evolution algebra, search methods, population size and penalty coefficient should be set firstlythen the iteration calculation should be started with the individual evaluation index of fitness which could be described by penalties, the iteration should be calculated generation by generation until that the smallest value of penalty was obtained, the optimal charge plan could be worked out,then.

The optimal charge plan is obtained by solving the charge plan model with genetic algorithm. The optimal charge plan cannot organize the production independently, the charge plans must be grouped into multiple casting plans. The casting plan is the rational combination and reasonable sort of the charge plans.

The optimal casting plan model could be established based on two objective functions: minimum total value of penalties for all the casting plans consist of n charge plans shown as equation (25) and maximum average operating rate of continuous casting machines shown as equation (26).

$$\min z_1 = \sum_{i=1}^{m}\sum_{j=1}^{n}\sum_{s=1}^{n}\left(C_{js}^1 + C_{js}^2 + C_{js}^3\right)\times X_{ij}\times X_{is} \tag{25}$$

$$\max z_2 = \frac{\sum_{i=1}^{m} n_i, t_{i,CCM}}{t^{iE} - t^{iS}} \tag{26}$$

C_{js}^1- penalties for production with difference of steel grade between Heat j and Heat s;

C_{js}^2- penalties for production with difference of specification between Heat j and Heat s

C_{js}^3- penalties for production with difference of due date between Heat j and Heat s

$$X_{i,j} = \begin{cases} 1 & \text{Contract product I exists in charge k} \\ 0 & \text{else} \end{cases}$$
$$i = 1,2,\ldots,n; k = 1,2,\ldots,m \tag{27}$$

The optimal casting plan could be obtained by solving this casting plan model with heuristic rules by following steps:

Grouping the charge plans into casts:

The casts could be grouped by steel grade, specification of the items in the charge plans. If the total number of heats in the group exceeds the allowed number of heats of continuous casting machine, the excess heats should be considered as surplus heats.

Sorting the casts by following rules:

Rule 1: the producing steel grade of each cast on the continuous casting machine should be determined by the casts obtained by step (1);

Rule 2: the steel grades and specifications of adjacent casts should be the same as possible;

Rule 3: the casts should be sorted by due date of the order.

Based on the model and calculation above, the optimal grouped casts and optimal casting plan could be obtained.

The optimal rolling plan could be obtained by grouping and sorting the casts in optimal casting plan by the specification ranges and capacity of different mills. The planning process from charge plan to rolling plan is shown in Fig.25.

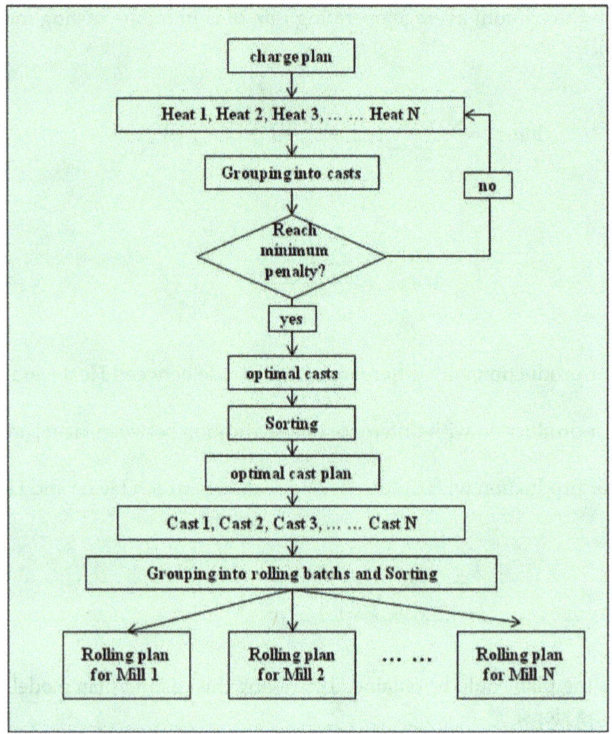

Figure 25. The planning process from charge plan to rolling plan

5.1.2. Buffer and coordination on the plan execution

In order to ensure the stability of the execution of production plan, the rules below should be used when production conflicts and equipment malfunction happened.

1. Rule 1: If the molten steel tapping of some charge delayed, and the delayed time was in the scope of allowed buffer time in refining process, the refining process could take the delayed time as buffer time by prolonging the heating time, and the operating time of continuous casting process and subsequent charges should not be changed;

2. Rule 2: If the molten steel tapping of some charge delayed, and the delayed time was exceed the scope of allowed buffer time in refining process, the refining process could take a part of the delayed time as buffer time by prolonging the heating time, and the

continuous casting process could take the rest of the delayed time as buffer time by lowering the casting speed;

3. Rule 3: If the molten steel tapping of some charge delayed, and the delayed time was exceed the scope of total allowed buffer time in refining process and continuous casting process, the continuous casting machine should stop working;
4. Rule 4: If the continuous casting production of some cast delayed, and the delayed time was in the scope of allowed buffer time in reheating process, the reheating process could take the delayed time as buffer time by changing the heating time and intensity, and the operating time of rolling process should not be changed;
5. Rule 5: If the continuous casting production of some cast delayed, and the delayed time was exceed the scope of allowed buffer time in reheating process, the rolling process should use the intermediate inventory for production.

In the dynamic execution of planning, production conflicts and equipment malfunction could be regulated dynamically by the proposed rules in order to ensure the implementation of the production plan running steadily.

5.2. Examples of planning for the process of steelmaking-rolling

5.2.1. Rule based planning system on the process of steelmaking-rolling in plant A

Based on the analysis of the process of steelmaking-rolling in plant A, a series of rules on planning were proposed, the optimal rule-based plan model was built up. The system was practiced in real productive process. The production mode of plant A is shown as Fig.26, and the window of management of casting plan in the planning system is shown in Fig.27.

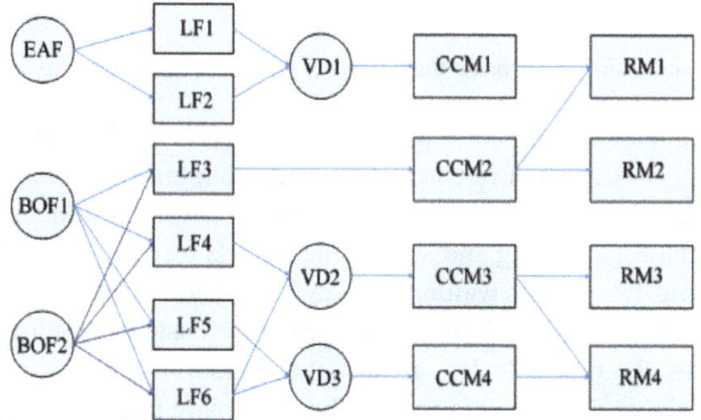

Figure 26. The production mode of plant A

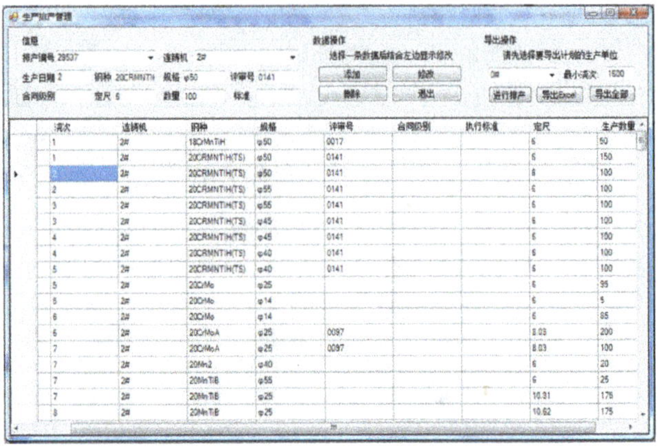

Figure 27. The window of management of casting plan in the planning system

This system optimized the planning process combined with SAP management software of plant A, which has the following characteristics:

Firstly, the storage module of planning system closely combined with the third class storage management subsystem of SAP software. The system could eliminate the existed storage production from the requirement list and drop the redundant production to the minimum.

Secondly, the system could analyze the regulation of mass flow of the special steel producing procedure and make arrangements for plan so that the plans would be able to satisfy the mass and energy balance of the productive process.

Thirdly, the system took the productive costs minimum as the goal function, set up the optimized models for each procedure, and so the factory would be able to achieve the goals of reducing costs, improving production quality, shortening producing cycle and delivering goods on time.

5.2.2. Planning and scheduling system on the process of steelmaking-continuous casting in plant B

In this example, planning and scheduling were considered as a whole process to ensure the integrality and systematization of research. The production lines in steelmaking-continuous casting process were optimized on the basis of process analysis, and the optimized production mode was obtained. Planning and scheduling models were established to optimize the total production time, the process waiting time and the operating rate of steelmaking furnaces. Meanwhile, the rule base which was composed of the basic scheduling rules, the equipment selecting rules, the time calculating rules, the manual intervening rules and the real-time adjusting rules, was developed to acquire the scheduling plan shown by Gantt chart (as shown in Fig.28, Gantt chart illustrates the start and finish time of the of each productive mission), and then the dynamic scheduling strategy is explored to cope with the dynamic events in the production process to ensure the feasibility of the scheduling plan and the stability of the production.

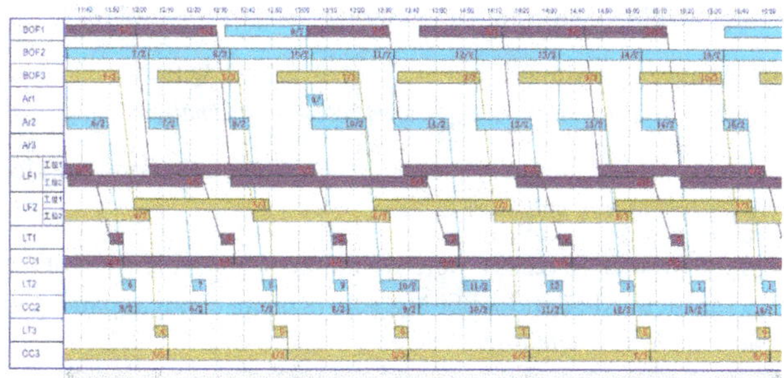

Figure 28. Gantt chart of the process of steelmaking-continuous casting in the system

The planning and scheduling system was established to integrate production planning and scheduling subsystems. The contracts could be converted to the scientific and reasonable scheduling plan by the system, and the conflicts of time or facilities could be eliminated by the dynamic scheduling. The planning and scheduling system showed the efficiency and simplicity of the model and algorithm, the scheduling plan could be obtained within acceptable time, and the proposed solutions could have a great influence on the research of planning and scheduling in steel plants.

Author details

Qing Liu, Xiaofeng Zhang, Bin Wang and Bao Wang

State Key Laboratory of Advanced Metallurgy (University of Science and Technology Beijing; (2)School of Metallurgical and Ecological Engineering, University of Science and Technology Beijing, China

References

[1] Mintz , B. (1999). The influence of composition on the hot ductility of steels and the problem of transverse cracking [J]. *ISIJ International*, 39(9), 833-855.

[2] Yong, Wang, Chunxia, Su, & Yuanpeng, Zhang. (2006). High Temperature Mechanical properties of alloy steel 0.29C-1.20Cr-0.30Ni-0.35Mo [J]. Special Steel (, 27(6), 32-33.

[3] Xiaoqun, He. (2008). Applied Regression Analysis [M]. *Beijing ChinaHigher Education Press.*

[4] Qiaodan, Lu, Zhiyuan, Zhu, Wanjun, Wang, et al. (2002). Influence of calcium treatment on hot ductility of steel for container [J]. *Iron and Steel,* 37(6), 35-38.

[5] Wang, Xin, Wang, Xianyong, Wang, Bao, et al. (2011). Differential Calculation Model for Liquidus Temperature of Steel [J]. *steel research international,* 82(3), 164-168.

[6] Wu, Y. J., Jiang, Z. H., Liang, L. K., et al. (2002). Calculation on Liquidus temperature of Steel [J]. *Journal of Iron and Steel Research,* 14(6), 6-9.

[7] Dong, H. B., & Brooks, R. (2005). Determination of Liquid Journal of Chongqing Univeus temperature in Al-Si and Al-Si-Mg Alloys using a Single-pan Scanning Calorimeter [J]. *Materials Science and Engineering A,* 413, 480-484.

[8] Miettinen, J. (1997). Calculation of solidification-related thermophysical properties for steel [J]. *Metallurgical and Materials Transactions B,* 28B(2), 281-297.

[9] Mills, K. C. (2004). Equations for the Calculation of the Thermophysical Properties of Stainless Steel [J]. *ISIJ International,* 44(10), 1661-1668.

[10] Zhichao, Dou, Liu, Qing, Wang, Bao, et al. (2011). Evolution of Control Models for Secondary Cooling in Continuous Casting Process of Steel [J]. *steel research international,* 82(10), 1220-1227.

[11] Zhao, Jiagui, Qu, Xiuli, Cai, Kaike, et al. (2000). A secondary spray water cooling control model for slab caster and its application [J]. *Metallurgical Industry Automation,* 24(3), 34-36.

[12] Liu, Qing, Wang, Liangzhou, Zhang, Liqiang, et al. (2008). Mathematical model of heat transfer for bloom continuous casting [J]. *Journal of University of Science and Technology Beijing,* 15(1), 17-23.

[13] Chen, Dengfu, Li, Hongliang, Niu, Hongbo, et al. (2007). New Model for Spraying Water of Nozzles in Secondary Cooling of Billet Continuous Casting [J]. *Journal of Chongqing University (Natural Science Edition),* 30(6), 61-64.

[14] Han, Peng, & Zhang, Xingzhong. (2002). Non-Steady Control of Secondary Cooling Used for Continuous Casting Slab [J]. *Journal of Iron and Steel Research,* 14(4), 73-76.

[15] Gilles, Herbert L. (2003). Primary and Secondary Cooling Control [M] . *The casting volume of the 11th edition of the making, shaping and treating of steel ,AISE,* 33-44.

[16] Kondo, Osamu, Hamada, Katushige, Kuribayashi, Takashi, et al. (1993). New dynamic spray control system for secondary cooling zone of continuous casting machine [M]. *Steelmaking Conference ProceedingsDallas,* 309-314.

[17] Kawasaki, S., Arita, H., Kikunaga, M., Chida, Y., et al. (1984). On the Secondary Cooling Control Technology for the Continuous Casting [J]. *Direct Rolling ProcessNippon Steel Tech-nical Report,* 23, 69-76.

[18] Morita, T., Konishi, M., Kitamura, A., et al. (1986). Control Method of Secondary-Cooling Water for Bloom Continuous Casting. *Kobelco Technical Bulletin*, 1109, 1-5.

[19] Barozzi, P., Fontana, P., & Pragliola, P. (1986). Computer Control and Optimization of Secondary Cooling During Continuous Casting. *Iron and Steel Engineer*, 63(11), 21-26.

[20] Liu, Weitao, Bai, Jubing, Qian, Liang, et al. (2008). Application of Dynamic Secondary Cooling in Continuous Casting of Steel Slab [J]. *Foundry Technology* [12], 1651-1654.

[21] Guo, Liangliang, Tian, Yong, yao, Man, et al. (2009). Temperature distribution and dynamic control of secondary cooling in slab continuous casting [J]. *International Journal of Minerals, Metallurgy and Materials*, 16(6), 626-631.

[22] Zhang, Zheng Shan, Chai, Tianyou, Wang, Wei, et al. (1999). Secondary cooling control for alloy steel billet continuous caster in Fushun Special Steel Co Ltd [J]. Metallurgical Industry Automation, , , 23(5), 32-35.

[23] Takawa, T., Takahashi, R., & Tatsuwaki, M. (1987). Mathematical Model and Control System of Cooling Process. *The Sumitomo Search*, 34, 79-87.

[24] Morwald, K., Dittenberger, K., & Ives, K. D. (1998). Dynacs cooling system-features and operational results [J] . *Ironmaking & Steelmaking*, 25(4), 323-327.

[25] Hardin, Richarda, Liu, Kai, Kapoor, Atul, et al. (2003). A transient simulation and dynamic spray cooling control model for continuous steel casting. *Metallurgical and Materials Transactions B*, 34B(3), 297-306.

[26] Okuno, K., Naruwa, H., & Kuribayashi, T. (1987). Dynacs spray cooling control system for continuous casting [J]. *Iron and Steel Engineer*, 64(4), 34-38.

[27] Spitzer, K.H., Harste, K., Weber, B., et al. (1992). Mathematical model for thermal tracking and on-line control in continuous casting [J]. *ISIJ International*, 32(7), 848-856.

[28] Camisani-Calzolari, F.R., Craig, I.K., & Pistorius, P.C. (2000). Speed disturbance compensation in the secondary cooling zone in continuous casting [J]. *ISIJ International*, 40(5), 469-477.

[29] Liu, Wen-kai, Xie, Zhi, Ji, Zhen-ping, et al. (2008). Dynamic water modeling and application of billet continuous casting [J]. *Journal of Iron and Steel Research International*, 15(2), 14-17.

[30] Cai, Ji, Zhang, Shuyan, zhao, Qi, et al. (2005). Study and implement of real-timperature field calculation and secondary cooling control model for slab caster with dynamic soft reduction [M]. *CSM 2005 Annual Meeting Proceedings, Beijing*, 3, 340-345.

[31] Xianyong, Wang, Qing, Liu, & Zhigang, Hu. (2010). Influence of nozzle layouts on the secondary cooling effect ofmedium thickness slabs i n continuous casting [J]. *Journal of University of Science and Technology Beijing*, 32(8).

[32] Wang, Xianyong, Liu, Qing, Wang, Xin, et al. (2011). Optimal control of secondary cooling for medium-thickness slab continuous casting. *Ironmaking & Steelmaking [J]*, 38(7), 552-560.

[33] Qing, Liu. (2002). Research on mode optimization of modern BOF steelmaking workshop for long products-the effect of high efficiency continuous casting technology on running control for modern BOF steelmaking workshop. *Beijing: University of Science and Technology Beijing.*

[34] Qing, Liu, Bai, Suhong, Lu, Junhui, et al. (2008). Production plan schedule for the casting-rolling process in BOF special steel plants [J]. *Journal of University of Science and Technology Beijing*, 30(5), 566-570.

[35] Bin, Wang. (2009). Research on planning and dispatching of steelmaking-rolling process at Shijiazhuang iron and sSteel corporation [D]. *Beijing: University of Science and Technology Beijing.*

[36] Chuang, Wang. (2012). Research on production planning and scheduling system for steelmaking-continuous casting process in the special steel workshop [D]. *Beijing: University of Science and Technology Beijing.*

[37] Gantt, H. L. (1910). Work, Wages, Profit. The management of projects [M]. *The Engineering Magazine, New York.*

Section 3

Evaluation of Castings

Chapter 8

Accuracy Improving Methods in Estimation of Graphite Nodularity of Ductile Cast Iron by Measurement of Ultrasonic Velocity

Minoru Hatate, Tohru Nobuki and
Shinichiro Komatsu

Additional information is available at the end of the chapter

1. Introduction

Ductile cast iron is one of the very useful and economical engineering materials and it is often used as the material for the members that are required good mechanical properties such as high tensile strength and high elongation. The microstructure of ductile cast iron basically consists of two kinds of basic components: the metallic matrix and many spheroidal graphite nodules dispersed among the matrix. As the bonding strength at the boundaries between the matrix and the graphite nodules is considered to be very little or nothing at all and also as the mechanical properties of the graphite nodules themselves are considered to be much less than those of the matrix, the tensile properties of ductile cast iron are considered to depend mostly upon the two kinds of conditions of the matrix. One of the conditions of the matrix is its microstructure (such as ferrite, pearlite and others), and the other one is the graphite nodularity which determines the continuity condition of the matrix. The latter is usually expressed by a kind of comparison number called "graphite nodularity" or "graphite spheroidizing ratio" which indicates how much the outer shapes of the graphite nodules are close to those of perfect spheres.

In a case of a ductile cast iron product under a tensile load the graphite nodules are considered to act as voids or cavities. This means that the presence of graphite nodules produces a kind of discontinuity effect to the matrix. Therefore, in the case of 100 % in graphite nodularity, which means that the outer shapes of the graphite nodules are almost perfect spheres, the continuity condition of the matrix produced by these graphite nodules is considered to be the best, and good mechanical properties of ductile cast iron can be expected. However, when the outer shapes of graphite nodules collapse from perfect spheres to the CV (compacted vermicular) side, the continuity condition of the matrix becomes worse and it results in some decreasing in mechanical properties of the ductile cast iron, especially in tensile strength, elongation and fatigue strength. Therefore, many users of ductile cast iron products tend to demand higher graphite nodularity values for the producers of the ductile cast iron products.

One of the authors belongs to a research group of nondestructive evaluation of the properties of castings in The Japan Foundry Engineering Society, and one time they held a round robin test of ultrasonic velocity for several ductile cast iron specimens with some ten different laboratories of companies and research institutes. Although they used same specimens in all the laboratories they found that the reported values of the ultrasonic velocity of a same specimen varied considerably at each laboratory. Then, they held a round robin test once again with the information of the thickness of each specimen. This time each group member measured the ultrasonic velocity of each specimen with a same value for its specimen thickness. As the results they could obtain the ultrasonic velocities which are very close to each other regardless the testing laboratories. This experience informed us that the small error in measurement of specimen thickness may result in a considerable error in the measurement values of ultrasonic velocity and thus in the graphite nodularity estimated. There may be many kinds of approaching methods to improve the accuracy in the measurement of ultrasonic velocity of ductile cast iron, but in this study we tried to approach it from the view point of measurement error in specimen thickness and the other matters related to it.

2. Relations between graphite nodularity and ultrasonic velocity in ductile cast iron

Figure 1 shows the various shapes of graphite nodules which are introduced in one the standards to measure the graphite nodularity (or graphite spheroidizing ratio) of ductile cast iron [1]. Although the graphite shape of the type of the lower far right, which is the closest to perfect spheres among the five shapes shown, is ideal to obtain the best mechanical properties in ductile cast iron from the view point of graphite shape. But sometimes we may happen to have the graphite nodules which have modified to the shapes of the other types shown in the figure due to the fading phenomenon of graphite nodules or others. So, in a case of a duct cast iron whose graphite shapes correspond to plural types of graphite shape shown in figure 1 we usually use a "graphite nodularity" value" to express numerically the average shape of the graphite nodules of the iron. In the case of the standard (JIS:G5502) which classifies the graphite shapes according to figure 1 the nodularity values of the graphite shapes are stipulated to be 0 %, 30 %, 70 %, 90 % and 100 % from the upper far left to the lower far right in the figure 1 respectively [1]. And in a case of a ductile cast iron whose graphite shapes correspond to plural types of them its graphite nodularity value is obtained from the calculation of the weighted average of the "graphite nodularity" and "number of nodules" of each type which are observed and counted through a microscope.

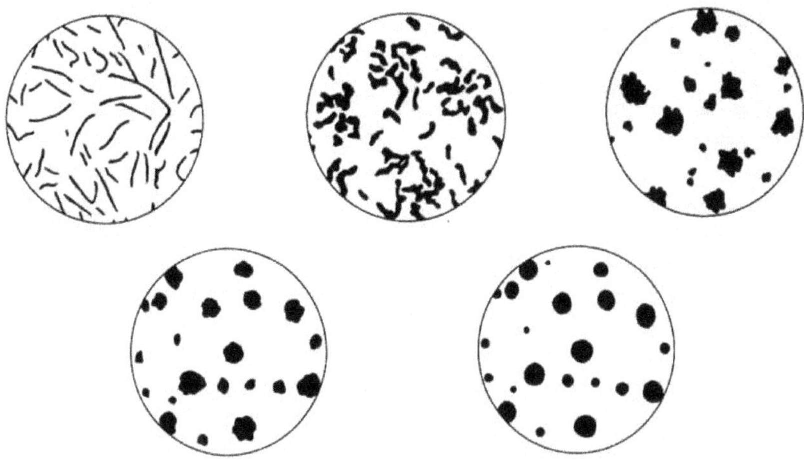

Figure 1. Various shapes of graphite nodules introduced in JIS:G5502 (Nodularity is 0%, 30%, 70%, 90% and 100% from the upper far left to the lower far right respectively) [1].

As described in the section of introduction the graphite nodularity determines the continuity condition of the matrix, and it relates largely not only to the mechanical properties but also to the ultrasonic velocity in ductile cast irons. Figure 2 is an example which shows the relation between graphite nodularity and ultrasonic velocity of ductile cast irons [2]. Although a little bit wide band is seen along the main thick line, the figure indicates that the ultrasonic velocity is largely related to the graphite nodularity in ductile cast iron. The width of the band is considered to be mainly related to the other factors beside graphite nodularity such as the amount of graphite (which may correspond to the carbon content of the iron), the sizes of graphite nodules (which may correspond to casting thickness), kind of heat treatments if any, and others. However, in a case of mass production of the products of one kind this band may become narrower and a more precise relation line is considered to be obtained.

Figure 2 indicates that the measurement of ultrasonic velocity is a useful method to evaluate or estimate nondestructively and also simply the graphite nodularity of ductile cast iron. However we find that the difference in ultrasonic velocity caused by the difference in graphite nodularity is not so large comparing to the ultrasonic velocity value itself, i.e. the difference in ultrasonic velocity which corresponds to the difference in graphite nodularity between 60 % and 80 % is merely some 130 m/s among the some 5630 m/s which is the velocity at the point of 80 % in graphite nodularity. As the range of graphite nodularity between 60 % and 80 % is the range subjected mainly to be inspected at a common commercial foundry factory, and as the difference of the ultrasonic velocity in this range is very small comparing to some 5630 m/s which is the velocity at the point of 80 %, the measuring methods of ultrasonic velocity are considered to require very high accuracy in many points.

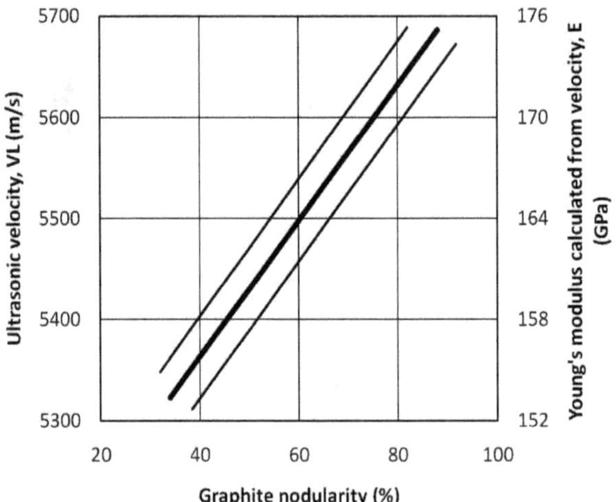

Figure 2. An example of the relation between graphite nodularity and ultrasonic velocity in ductile cast irons [2].

The ratio (R) of the difference of ultrasonic velocity (=130 m/s) between 60 % and 80 % in graphite nodularity to 5600 m/s (the standard ultrasonic velocity of ductile cast iron) is calculated as follows and is found to be relatively small.

$$R = 130(\mathrm{m} / \mathrm{s}) \;\; / \;\; 5600(\mathrm{m} / \mathrm{s}) \; = \; 0.0232 \; = \; 2.32 \; \%$$

As the difference of 20% in graphite nodularity corresponds to this difference of 2.32% in ultrasonic velocity, the difference of 1 % in graphite nodularity corresponds to the difference of only 0.116 % in ultrasonic velocity. Figure 2 and the detailed values introduced above are the data of one example and these may be somewhat different in details in other cases of experiments, but they are considered not to be so different from this figure within the material field of ductile cast irons. Therefore we should know that the measurement of ultrasonic velocity should be conducted with very high accuracy all the way.

Figure 3 shows another example which indicates a practical usage of ultrasonic velocity in ductile cast iron. This figure shows a relation between the tensile strength of ductile cast iron and its product of ultrasonic velocity and Brinell hardness [3]. This figure shows a very good correlation between them, and the reason for this can be explained as follows. The tensile strength of an ordinary ductile cast iron (ferrite/pearlite mixed in matrix microstructure) is considered to be determined mainly by two basic factors concerning to its microstructure. The one factor is the graphite nodularity which determines the continuity condition of its matrix and thus the ultrasonic velocity. The other factor is the microstructure condition (mainly the volume fractions of ferrite and pearlite) of its matrix, which is considered to be largely related to its Brinell hardness. As these two factors of the microstructure are considered to determine synergistically the basic tensile strength of ductile cast iron, the product of the ultrasonic velocity value and the Brinell hardness value is considered to become a kind of index value to estimate roughly the tensile strength of an unknown ductile cast iron.

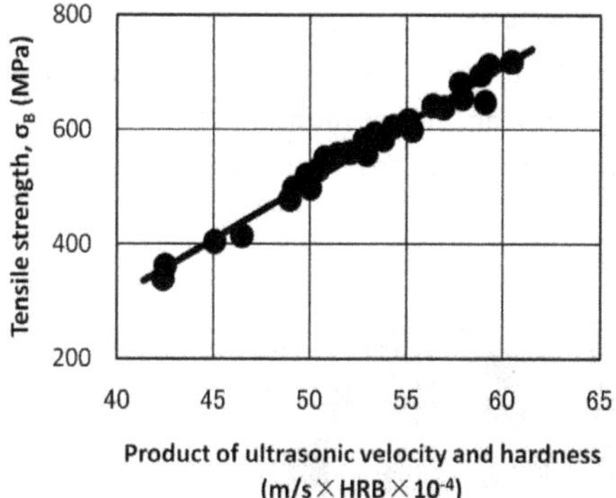

Figure 3. An example of the relation between tensile strength and product of Brinell hardness and ultrasonic velocity in ductile cast iron (ferrite/pearlite matrix) [3].

The close relationship between ultrasonic velocity and graphite nodularity in ductile cast iron is considered to be produced from the close correlation between modulus of elasticity and ultrasonic velocity in solid materials, which is shown in equation 1.

$$V = \sqrt{\frac{E(1-\nu)}{\rho(1+\nu)(1-2\nu)}} \tag{1}$$

where, V is ultrasonic velocity, E is modulus of elasticity, ρ is density and ν is Poisson's ratio.

The modulus of elasticity of ductile cast iron is affected largely by the continuity condition of the matrix, and the continuity condition of the matrix is determined by graphite nodularity. In other words, the graphite nodularity determines the continuity condition of the matrix, and the continuity condition of the matrix determines the modulus of elasticity, and the modulus of elasticity determines the ultrasonic velocity.

Shiota and Komatsu (one of the authors of this paper) once reported a paper on the close relationship between graphite nodularity and tensile strength of ductile and CV cast irons from the view point of "the effective sectional area ratio" of matrix which is determined by graphite nodularity [5]. The continuity condition of matrix mentioned above is similar to the view point of that paper.

3. Experimental procedures

The measuring equipments of ultrasonic velocity used in this study were a NDT tester of type "AD-3212/3212A" made by A&D and a transducer named "V110" made by "PANA-

METRICS" in Japan. The frequency used was 10 MHz. Figures 3 and 4 show the NDT tester and the transducer used in this study respectively. Several kinds and groups of the specimens were prepared in order to investigate the influence of several kinds of errors concerning to specimen thickness. The details of the specimens used will be introduced at each section in the Results and Discussions. The words "specimen thickness" used in this paper means the distance for the propagation of ultrasonic waves in a specimen and it is sometimes referred as "specimen length" or "specimen width" in some literatures.

Figure 4. Ultrasonic velocity tester used.

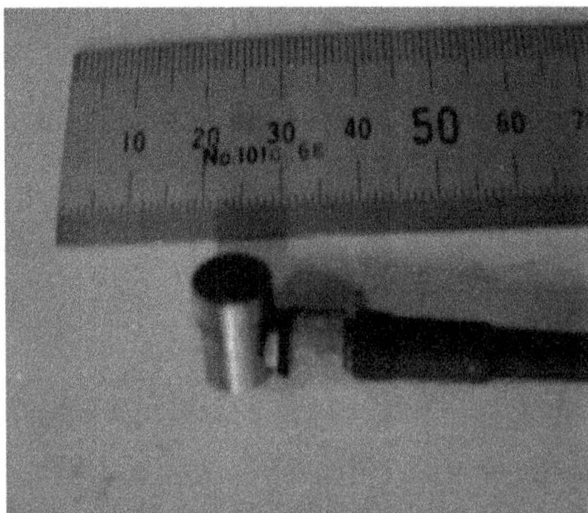

Figure 5. Transducer used.

4. Results and discussions

4.1. Affection of measurement error in specimen thickness

As mentioned in the introduction the measurement error in specimen thickness is considered to be produced easily by the difference in inspector and/or time of the measurement. Then, in order to see a sample case of this kind of error we asked several students who had no measurement experience at first to measure the ultrasonic velocity of a carbon steel (S35C) bar specimen with 32 mm and 40 mm in diameter and length respectively. Figure 6 shows the results of the measurements. This figure indicates that the measurement results varied largely depending on inspector and time of the measurements even in a case of a same specimen. As we found that these large errors had been produced mainly from the rough measurement manners of the specimen thickness by the amateur students we discussed on the scale (magnitude) of the errors in ultrasonic velocity which can be created by the measurement error of specimen thickness more precisely.

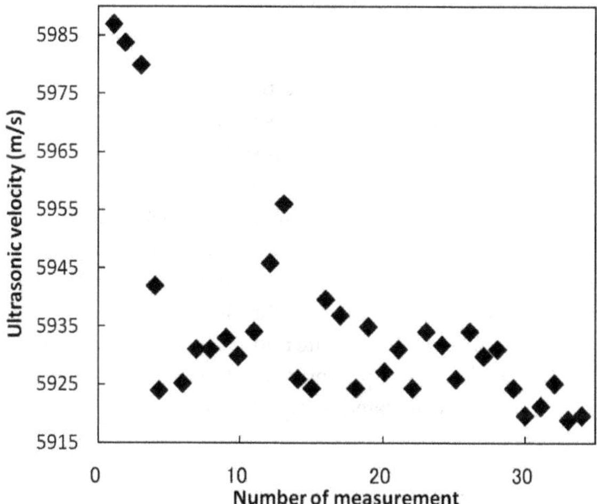

Figure 6. An example of measurement error in ultrasonic velocity for a same specimen caused by the difference of inspectors and time.

Usually, the measurement of ultrasonic velocity of a specimen is conducted by contacting a transducer to one surface of a specimen, and the ultrasonic velocity of the specimen is obtained by the calculation from the two measured values. The first one is the specimen thickness which corresponds to the distance between the surface facing to the transducer and the surface that reflects the ultrasonic waves, and the second one is the round-trip propagating time of ultrasonic waves between these two surfaces. The ultrasonic velocity of a specimen is calculated by using the equation 2.

$$V = \frac{2L}{T} \tag{2}$$

where, V is ultrasonic velocity, L is specimen thickness and T is round-trip propagating time of ultrasonic waves. The "2" in the numerator corresponds to the round trip propagating distance of the ultrasonic waves. In many cases of ductile cast iron products their thickness (L) to be objected for the measurement of ultrasonic velocity are up to some 100 mm or so and they are usually measured by using vernier calipers (slide calipers). The minimum measurement unit by a conventional vernier caliper is 0.05 mm. Following to the equation 2, we tried to figure out how much velocity error (E_V) would be created in the case that a measurement error of 0.05 mm was made in the specimen thickness of 50 mm. And also we tried to calculate how much graphite nodularity error (E_N) would be created in this case assuming that the difference of 130 m/s in ultrasonic velocity corresponds to the difference of 20 % in graphite nodularity which was mentioned in Figure 2.

$$\text{Velocity error } (E_V) = \frac{2 \times 0.05(\text{mm})}{50(\text{mm})} \times 5600(\text{m / s}) = 11.2(\text{m / s})$$

$$\text{Nodularity error } (E_N) = \frac{11.2(\text{m / s})}{130(\text{m / s})} \times 20(\%) = 1.72(\%)$$

These results indicate that a small measurement error of 0.05 mm in specimen thickness, which is usually paid very small attention because of its being the minimum measuring unit of vernier caliper, may result in a considerable error in ultrasonic velocity measured and thus graphite nodularity estimated. These values shown above are for the case of 50 mm in specimen thickness but these values are considered to increase in the case of specimen thickness less than 50 mm and to decrease in the case of specimen thickness larger than 50 mm. Then, we discussed to see the co-relationships among the four valuables relating to each other deeply ; true specimen thickness (L), measurement error in specimen thickness (E_L), measurement error in ultrasonic velocity (E_V) and measurement error in graphite nodularity (E_N). The measurement error in ultrasonic velocity may be calculated by equation 3, and the measurement error in graphite nodularity estimated from E_L may be calculated by equation 4.

$$E_V(\text{m / s}) = \frac{2 \times E_L}{L} \times 5600(\text{m / s}) \tag{3}$$

$$E_N(\%) = \frac{E_V(\text{m / s})}{130(\text{m / s})} \times 20(\%) \tag{4}$$

Figure 7 shows the relations of Equation 3 which is the relation between ultrasonic velocity error (E_V) and specimen thickness (L) in the cases of measurement errors (E_L) of 0.05 mm, 0.1 mm, 0.2 mm and 0.5 mm in specimen thickness. This figure indicates that even a small measurement error of 0.05 mm which is the minimum measurement unit of a conventional vernier caliper creates considerably larger ultrasonic velocity errors, especially in the range of relatively smaller values in specimen thickness. These results may indicate that we need to use a micrometer instead of a vernier caliper and to measure the thickness for several times precisely in order to obtain good measurement results especially in the case of a relatively smaller specimen thickness.

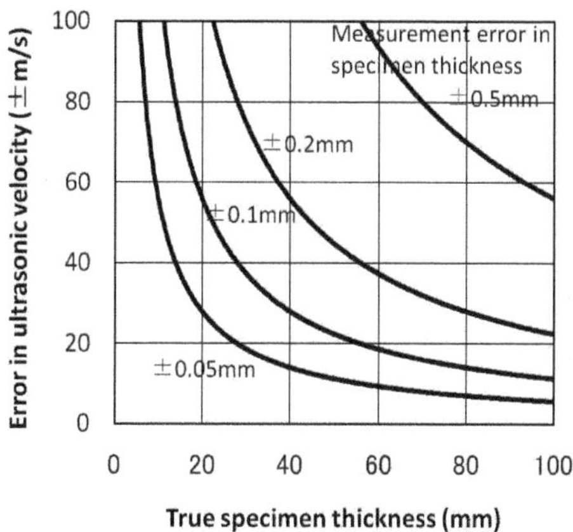

Figure 7. Relations between the specimen thickness and the ultrasonic velocity error at various measurement errors in specimen thickness.

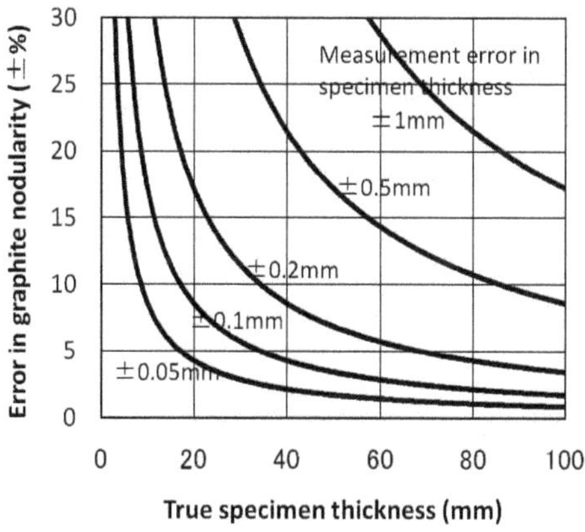

Figure 8. Relations between the specimen thickness and the graphite nodularity error at various measurement errors in specimen thickness.

Figure 8 shows the relation of Equation 4 which is the relation between graphite nodularity error (E_N) and specimen thickness (L) in the cases of measurement errors (E_L) of 0.05 mm, 0.1 mm, 0.2 mm and 0.5 mm in specimen thickness. This figure indicates the similar tendencies to those of Figure 7, because E_N is in direct proportion to E_L as shown by Equation 4. These equations and figures indicate that the measurement error of 0.05 mm in specimen thickness of 20 mm creates some 4.3 % in the error of graphite nodularity, which can't be ignored easily in actual production of ductile cast iron products. These results are considered also to recommend the inspectors to use a micrometer instead of a vernier caliper and to make more precise measurements in specimen thickness especially in the case of a relatively small specimen thickness.

4.2. Affection of specimen temperature

As the previous section indicated that a small amount of measurement error in specimen thickness may result in a considerable error in graphite nodularity especially in the case of relatively smaller thickness, the authors tried to investigate on the affection of specimen temperature which might be considered to produce some measurement error in specimen thickness because the thermal expansion effect is expected to be produced by the difference of room temperature. Figure 9 shows the measurement results of ultrasonic velocity in the case that a S35C specimen bar with 32 mm in diameter and 80 mm in length was subjected to change its specimen temperature variously between 18 and 26 ℃.

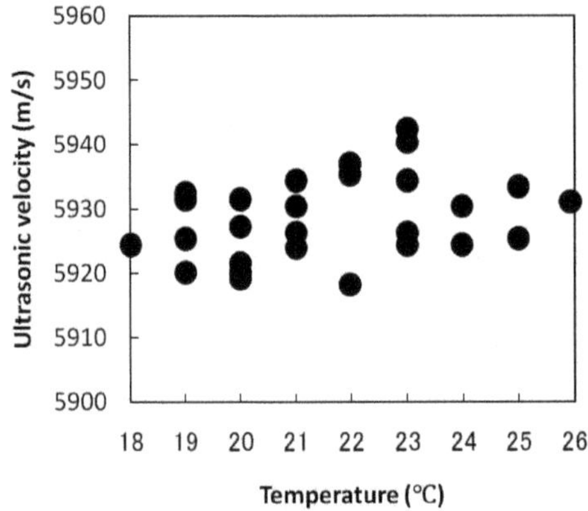

Figure 9. A measurement example of the relation between specimen temperature and ultrasonic velocity (S35C bar).

This figure indicates no significant tendency of the affection of specimen temperature to ultrasonic velocity measured. We also tried to calculate the affection of specimen temperature from the view point of thermal expansion effect. The followings are the calculations for the measurement errors in ultrasonic velocity (E_V) and graphite nodularity (E_N) which may be created by the temperature difference (t) of 20 ℃ in the case that only the specimen thickness becomes larger by the thermal expansion effect and all the other variables are constant. Here, V_0 and L_0 are the ultrasonic velocity and the specimen thickness respectively of the specimen in the original state, and V_1 and L_1 are those in the specimen in the state of 20 ℃

above the original state, and T is the propagating time which is assumed to be constant. The coefficient of linear expansion (α) and the basic value of ultrasonic velocity (V_0) were supposed to be 12 x 10^{-6} (1/℃) and 5600 (m/s) respectively for ductile cast iron.

$$E_V = V_1 - V_0 = L_1 / T - L_0 / T = (1 + at) L_0 / T - L_0 / T = at\ L_0 / T = at\ V_0$$
$$= 12 \times 10^{-6} (1 / \square) \times 20(\square) \times 5600(\mathrm{m/s}) = 1.34(\mathrm{m/s})$$
$$E_N = E_V \times 20(\%) / 130(\mathrm{m/s}) = 1.34(\mathrm{m/s}) \times 20(\%) / 130(\mathrm{m/s}) = 0.206(\%)$$

These calculation results are considered also to indicate that the difference in specimen temperature caused by the difference of room temperature within some 20 ℃ gives only a little affection to the measurement error in ultrasonic velocity of ductile cast iron.

4.3. Affection of surface finishing

In many cases at foundry factories the ultrasonic velocity measurement of ductile cast iron products is performed by placing a transducer to the surface of one surface which has been machine-finished to be skin-free and flat by means of using a grinder or a similar equipment, but in many cases the surface of the other end which is subjected to reflect the ultrasonic waves has been left in the condition of as-cast or shot-blasted. In this section we tried to investigate the affection of surface finishing on the measurement errors of ultrasonic velocity of ductile cast iron products.

Figure 10. The specimens with three kinds of surface conditions (as-cast, machined and shot-blasted: 100 mm square, continuously cast bar).

Figure 10 shows the square shaped specimens of ductile cast iron we used in this section. The material of these specimens is a continuously cast square bar of ductile cast iron with 100 mm in two sides, and the heights of the specimens are varied into four kinds from 12.5 mm to 75 mm. This material of continuously cast ductile iron bar was selected for the reason of expectation for better homogeneity in the microstructures of all the specimens with various kinds of thickness because of its uniform solidification rate comparing to those of the materials cast by the other kinds of methods. In order to investigate the affection of surface condition of specimen three kinds of specimens whose surface conditions varied for three kinds were prepared. The surfaces which are subjected to contact to a transducer are machine-finished surfaces in all the specimens but the surfaces which are subjected to reflect the ultrasonic waves were varied for three conditions which are "as-cast", "machine-finished" and "shot-blasted", which means that the three kinds of specimens, whose combinations of surface conditions are (a) machined surface / as-cast surface, (b) machined surface / machined surface, and (c) machined surface / shot-blasted surface, were prepared and conducted to the measurement of ultrasonic velocity.

Figure 11 shows the test results of ultrasonic velocity of the specimens whose thickness is 12.5 mm which is the smallest among all the specimens we made, and Figure 12 is the test results of the specimens whose thickness is 75 mm which is the largest among all the specimens.

For each specimen the measurement of ultrasonic velocity was carried out for five times, and the five results of each specimen are illustrated in the figures with a line which connects all of them. So in these figures, the lines indicate only that the data connected by each line are of one specimen, and the order of the data along the X-axis and the inclinations of the lines have no significant meaning at all. However this kind of illustration method is considered to be useful to examine the difference between the averages of the data of each group with comparison to the scale (size) of variation among the data within each group.

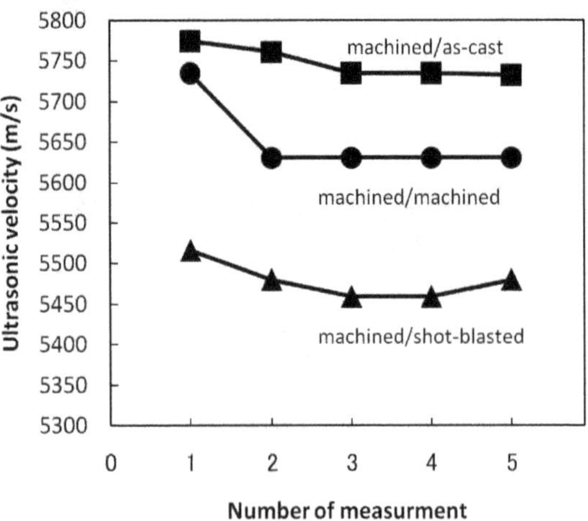

Figure 11. Comparison of the ultrasonic velocity measured by the specimens with various combinations of surfaces (Specimens 12.5 mm thick).

Figure 11 for the specimens 12.5 mm thick shows that the difference between the largest data and the smallest data in each specimen looks much larger than those in the figure 12 for the specimens 75 mm thick. This is considered to be due to the tendency that the measurement error in specimen thickness results in a larger measurement error in ultrasonic velocity in the case of a specimen with a smaller thickness than in the case of a specimen with a larger thickness, which was described in the former section of this paper.

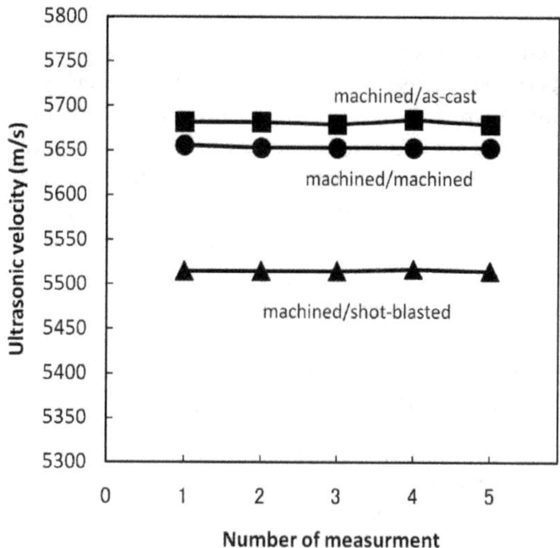

Figure 12. Comparison of the ultrasonic velocity measured by the specimens with various combinations of surfaces (Specimens 75 mm thick).

When we compare the order of the average values of the ultrasonic velocities of these three kinds of specimens in both figures, we can see that the ultrasonic velocity becomes somewhat larger in the case of "machined/as-cast" than in the case of "machined/machined", and also we can see that it becomes somewhat smaller in the case of "machined/shot-blasted" than in the case of "machined/machined". This means that the order of the values of ultrasonic velocities measured by these three kinds of specimens becomes the order shown below. (Note that the surface for the transducer is always a machined surface.)

machined/as-cast > machined/machined > machined/shot-blasted

Among all of the three kinds of the specimens the specimen of machined/machined is considered to be the best to measure the true ultrasonic velocity of the material because its microstructure is more uniform through all the entire length (thickness) of the specimen than the other two specimens with the skins of different kind. This may also be confirmed from the test results that the average ultrasonic velocity of the specimen 12.5 mm thick with machined surfaces is very similar to that of the specimen 75 mm thick with machined surfaces. However, in the case of the specimens with an as-cast surface or a shot-blasted surface the microstructure of the layers beneath these surfaces of reflection is considered to be somewhat different from that of the original material and some measurement error in ultrasonic velocity might be created by this inhomogeneity in microstructure.

Some reasons may be considered for the tendency of becoming somewhat larger of ultrasonic velocity when it is measured by specimens with as-cast surface. From the view point of the measurement error in specimen thickness, the undulation in the as-cast surface, which may be found more often and/or more largely than in the case of machined surface, may cause the specimen thickness to be measured somewhat larger than the true thickness (average of the undulation) of the specimen because the vernier caliper always touches the top of the highest peak in the undulation. As the transmission time is measured by the ultrasonic waves which may be considered to reflect at the cutting section of the average thickness (mid height of undulation) of the specimen, the ultrasonic velocity calculated from equation 2 is considered to pretend to be somewhat larger than the true velocity. Another reason may be considered from the view point of microstructure. Usually the microstructure of the layer of the as-cast surface differs from that of the material inside due to its rapid solidification, oxidization by air, chemical reactions with molds and others. Right now we haven't examined precisely yet, but we estimate that the ultrasonic velocity may be a little bit larger in the as-cast layer than in the material portion beneath the layer because of its finer and harder microstructure and smaller amount and size of graphite nodules due to its rapid solidification.

When we compare the two ultrasonic velocities of the specimens with as-cast surfaces in figures 11 and 12, we find that the ultrasonic velocity measured by the specimen 75 mm thick is a little bit smaller than that by the specimen 12.5 mm thick, and also it is closer to that of the specimens with machined surfaces. This is because the amount of the errors created from the undulation and the difference in microstructure of the layer of as-cast surface is constant regardless the specimen thickness, and its affection to the ultrasonic velocity calculated by equation 2 becomes smaller in the case of a specimen with a larger thickness. In the case of the ductile cast iron specimens or the products with the as-cast surfaces made by sand molds we may need to pay more precise attention to the influence of the measurement error created by the undulation or conditions of their as-cast surfaces. In the case of the as-cast surfaces made by sand molds the measurement error in ultrasonic velocity is considered to be made not only from the undulation of the surfaces. The surface which is used for the measurement of specimen thickness by a vernier caliper may be located at the peaks of the roughness made by sand particles, but the surface which mainly reflects the ultrasonic waves may be located at the bottom of the roughness made by sand particles, and the distance between these two surfaces is considered to give a considerable error in ultrasonic velocity measured. Therefore, in the case of measurement of a ductile cast iron product made by sand molds we are recommended to remove the as-cast skins totally not only from the transducer side but also from the reflection side of ultrasonic waves.

Figures 11 and 12 also show that the ultrasonic velocity measured by the specimens with shot-blasted surfaces tends to become a little bit smaller than the true velocity value which is measured by the specimens with machined surfaces. This is considered to be resulted from the affection of the surface layer which has been deformed plastically by shot blasting. As the modulus of elasticity of a ferritic material becomes smaller in a plastically deformed condition than in an elastic condition, the ultrasonic velocity of a ferritic material also becomes smaller in a plastically deformed condition than in an elastic condition because the ultrasonic velocity is in direct proportion to the modulus of elasticity as shown in equation 1. As the ultrasonic waves takes more time to propagate through the surface layer which has been plastically deformed by shot-blasting, the total length of propagation time becomes a little bit longer than in the case of the specimen without the shot-blasted layer. And this seemingly longer time results in a seemingly smaller ultrasonic velocity through equation 1. Therefore, in the case of measurement of the castings with shot-blasted surfaces, we are recommended to remove the plastically deformed layer totally in measurement or to revise the measured values with the results from another experiment using similar materials.

4.4. Affection of non-parallelism of specimens

In an actual measurement of ultrasonic velocity of a specimen we measure the distance and the propagation time between two surfaces: the surface facing to a transducer and the surface for the reflection of ultrasonic waves. In many cases we check the parallelism of these two surfaces by means of the visual checks with inspector's eyes. However our results shown in previous sections have indicated that even a small measurement error in specimen thickness may result in a considerable measurement error in ultrasonic velocity and thus graphite nodularity, and a small non-parallelism in the two surfaces is considered to produce somewhat errors in these measurement results. In this section we investigated on the affection of the non-parallelism of these two surfaces of specimens.

Figure 13. Specimens whose non-parallelism angles are differed variously.

Figure 13 shows the specimens which we used in this section. They are made from a material of continuously cast ductile iron, and the size of each specimen is approximately 30 x 30 x 50 mm.

The surface for reflection of ultrasonic waves of each specimen was made to have an angle of non-parallelism between 0 and 10 degree against to the surface for a transducer. Therefore, even in one specimen the thickness value is not constant. Figure 14 shows an example of a specimen in the group. The ultrasonic velocity of a specimen was measured at the three points marked A, B and C shown in figure 14. The point A corresponds to the largest, the point B corresponds to the average (middle) and the point C corresponds to the smallest among these three points in specimen thickness. The ultrasonic velocity at each point was calculated from the propagating time and the thickness which were measured at each point. For example, the thickness at point A was obtained by measuring the distance from point A to the intersection point of the straight line of the surface for reflection of ultrasonic waves and the straight line which is perpendicular to the surface for the transducer at point A. Although we measured all the specimens shown in figure 13, we found that the affection of non-parallelism between the two surfaces is extraordinarily large, therefore we introduce only about the case of 1 degree which was the smallest among all the specimens we made.

Figure 14. Three measuring points of a specimen (A: largest, B: middle, C: smallest in thickness).

Figure 15 shows the results of the ultrasonic velocity measurements at the three points (A, B and C) of the specimen whose non-parallelism angle is 1 degree. At each measuring point we measured velocity for five times, and the results of them were illustrated with a connecting line in the figure. Figure 15 indicates that the largest ultrasonic velocity was measured at point A which is the largest point in specimen thickness among all the three points. The smallest ultrasonic velocity was measured at point C which is the smallest point in specimen thickness among the three, and the ultrasonic velocity at point C which is the middle point between points A and B was measured approximately to be the average of the two at the points A and B. The reason for the relatively larger ultrasonic velocity at point A is considered as follows. Although the ultrasonic waves are sent off at point A they reflect not only at the opponent point (the point where the thickness was measured from) of the point A but also from the points (place) which are smaller than the thickness at the point A because the ultrasonic waves reflect not at one point but at a face with a some kind of area in the surface of reflection. Then, the ultrasonic waves which were used to read the propagation time may contain the affections of the waves reflected at the points which are a little bit smaller in specimen thickness than at the point A, and it results in a little bit shorter time in propagation, and then a little bit larger ultrasonic velocity was measured from equation 2. On the other hand, in the case of the relatively smaller ultrasonic velocity at the point C is also considered to be resulted from the opposite reason to that at the point A mentioned above. Figure 15 also shows that the difference in ultrasonic velocity between the two points A and B is as large as some 170 m/s, which corresponds to the measurement error of some 26% in graphite nodularity. This means that we are strongly recommended to pay much more attention for the non-parallelism of the surfaces for reflection of ultrasonic waves in order to obtain good measurement results.

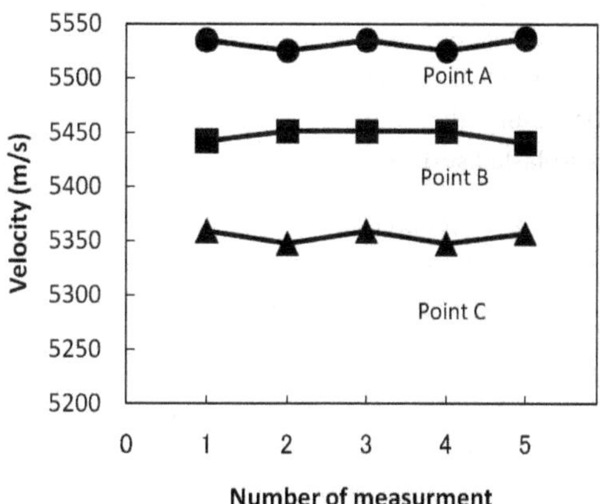

Figure 15. An example of measurement error in ultrasonic velocity created by1 degree of non-parallelism angle between the surface for transducer and the surface for reflection.

In this study we tried to measure the graphite nodularity of a specimen by two steps; measuring the ultrasonic velocity at first and then converting it into graphite nodularity by a relation figure between graphite nodularity and ultrasonic velocity. Recently we see some practical measuring apparatuses which convert the ultrasonic velocity into graphite nodularity automatically and show graphite nodularity directly [6], but even in the case of using those equipments we also need to know the various affections mentioned above and pay much careful attentions for precise measurements in specimen thickness.

5. Conclusions

We investigated on the sources and the scales (magnitude) of the errors in estimation of graphite nodularity of ductile cast iron by measuring ultrasonic velocity from several points of view such as the specimen thickness measurement error, the surface finishing methods, the error of non-parallelism in two surfaces and others, and obtained the following conclusions.

1. The difference of 20 % in graphite nodularity was found to correspond to the difference of some 130 m/s or only 2.3% of some 5600 m/s which is the standard ultrasonic velocity of ductile cast iron. This means the necessity for precise measurements for good measurement results.

2. As the specimen thickness are usually so small comparing to the standard ultrasonic velocity of some 5600 m/s, a measurement error in specimen thickness is expanded largely in the calculation result of ultrasonic velocity. For example, 0.05 mm error which is the minimum unit of conventional vernier caliper may result in the errors of some 11.2 m/s in ultrasonic velocity and some 1.72 % in graphite nodularity in the case of 50 mm in specimen thickness.

3. The ultrasonic velocity and graphite nodularity are measured to be larger than their true values in the case of measurement by the specimens with as-cast surfaces, and they are measured to be smaller than their true values in the case of measurement by the specimens with shot-blasted surfaces, although their skin thickness seems very small.

4. Even a small angle such as 1 degree of non-parallelism between the surface for transducer and the surface for reflection of ultrasonic waves is recognized to create extraordinary large measurement errors in ultrasonic velocity and graphite nodularity.

5. In order to improve the accuracy in estimation of graphite nodularity of ductile cast iron by measurement of ultrasonic velocity we are recommended to know and recognize the amount of the affections of their sources and minimize the errors from such view points as measurement error in specimen thickness, error of non-parallelism of reflection surface, error by existence of the layer of as-cast or shot-blast and others.

Author details

Minoru Hatate, Tohru Nobuki and Shinichiro Komatsu*

*Address all correspondence to: komatsu@hiro.kindai.ac.jp

Kinki University, School of Engineering, Higashihiroshima, Japan

References

[1] Japan Industrial Standards - JIS:G5502.

[2] Japan Foundry Engineering Society. (2004). Investigation on the Nondestructive Evaluation of Properties of Castings. *Research Report* [94], 22.

[3] The Japanese Society for Non-Destructive Inspection. (1991). The Investigation on the Standardization of Non-Destructive Evaluation Technology. *The Report of the year of Heisei* 3, 22.

[4] Japan Foundry Engineering Society. (2004). Investigation on the Nondestructive Evaluation of Properties of Castings. *Research Report* [94], 24.

[5] Toshio SHIOTA and Shinichiro KOMATSU. (1977). The relations between effective sectional area and tensile strength of cast irons. *IMONO (Journal of Japan Foundry-men's Society)*, 49, 602-607.

[6] Dakota Japan Co. Ltd. (2012). *The catalogue for graphite nodularity measuring equip-ments,* http://www.dakotajapan.com/autoscan/point.html.

Chapter 9

Fracture Toughness of Metal Castings

M. Srinivasan and S. Seetharamu

Additional information is available at the end of the chapter

1. Introduction

From the continuum mechanics point of view, fracture toughness of a material may be defined as the critical value of the stress intensity factor, the latter depending on a combination of the stress at the crack tip and the crack size resulting in a critical value.

The local stress σ_{local}, shown in Figure 1, scales as $\sigma\sqrt{(\pi c)}$ for a given value of r, where $\sigma$is the remotely applied stress, σ_{local} is the stress in the vicinity of the crack at a distance r from the tip and c is the crack length; this combination is called the Stress Intensity Factor (K). For the type of load shown (tensile load) K is denoted as K_I. Thus,

$$K_I = Y\sigma\sqrt{(\pi c)} \tag{1}$$

where Y is a dimensionless constant to account for the crack geometry. K_I has units of MPa $m^{1/2}$. The material fractures in a brittle manner when K_I reaches a critical value, denoted by K_{IC}; if there is significant crack tip plasticity, instability occurs at this critical value, leading to fracture. A simpler view of the fracture toughness is that it is a measure of the resistance of the material to separate under load when a near-atomistically sharp crack is present.

The stress intensity factors are usually identified by the subscript I for "opening mode", II for "shear mode" and III for "tearing mode". The opening mode is the one that has been investigated widely and hereafter only K_I will be considered.

Under load, a metallic material first undergoes elastic deformation and plastically deforms when its yield stress is exceeded. Fracture occurs when the ability to plastically deform under load is exhausted. The chief cause of the plastic deformation is the movement of dislocations and the resistance to its movement causes increased plastic flow stress and abetment of fracture.

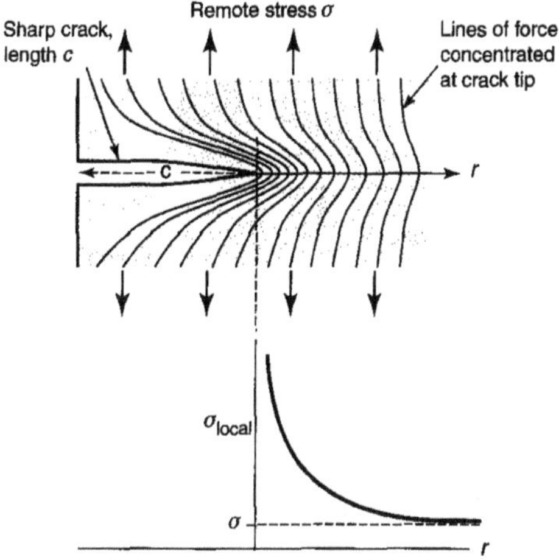

Figure 1. Lines of force and local stress variation from a body with sharp crack. Source: M.F. Ashby, et al [1]

The term "metal casting" represents an umbrella consisting of many variants, as will be briefly described later. The composition of the metal (alloy) will usually depend on the variant. The common factor among the variants is that they are all products of liquid-to-solid transformation, usually termed "solidification". Solidification of castings involves nucleation and growth of solid. Casting alloys usually consist of more than one phase. The simplest solidification occurs in a pure metal or an isomorphous system wherein the solid consists of only one phase. As the complexity increases, an eutectic system consisting of two solid phases may be formed, which may be totally different in properties. In low carbon steel castings, a high temperature reaction known as peritectic reaction will occur, which have some influence on the room temperature microstructure. Adding to this complexity, solid-to-solid transformations may occur, as for example, the eutectoid reaction in cast iron and steel. The phase diagrams will at best give useful guidance on the development of microstructure as they are based on equilibrium, but most castings solidify under nonequilibrium conditions resulting in departures from the phase diagram predictions. Commercial castings invariably contain various impurities that may affect the microstructure. Certain aspects of the casting microstructure have a fundamental influence on the fracture resistance [2]. It is therefore pertinent to consider the influence of valence electrons on the fracture behavior. Covalent bonds have shared electrons and the limited mobility of the electrons impedes plastic flow resulting in brittle fracture. Though metal castings in general have metallic bonds, they may contain covalent compounds such as nitrides, carbides and others as inclusions. Silicon, an important constituent in Al-Si casting alloys also has covalent bond. Ionic bonds permit better electron mobility than covalent bonds, but may cause brittle behavior when like poles interact while slipping. Metallic bonds offer least restriction to electron mobility, but as stated earlier, only a few commercial castings are made of pure metals or isomorphous alloys. Another important factor that needs attention is the dislocation dynamics as affected by the casting microstructure, despite the fact that the dislocation density in castings is much lower than in cold worked materials, in the as-cast state; this difference may get less under stress in castings. Unfortunately little quantitative information is available on the significance of dislocation dynamics on fracture in castings. Some casting alloys have compositions suitable for heat treatment involving solid state transformations. The microstructure is substantially changed after heat treatment and thus, in heat treated castings, the fracture behavior will usually be different as compared to the as-cast counterparts.

From the microstructural point of view, the route to increase the fracture toughness of castings would involve conflict in increasing both the fracture toughness and the yield strength. The factors of importance are [3]: improved alloy chemistry and melting practice to remove or make innocuous impurity elements that degrade fracture toughness; development of microstructures and phase distributions to maximize fracture toughness. through proper choice of composition and process variables; microstructural refinement through solidification control.

Thus it is clear that continuum mechanics provides the theoretical basis for designing against fracture in castings, but a thorough understanding of the microstructure and its effects is essential to fine-tune the final design against fracture. As noted by Ashby [4],

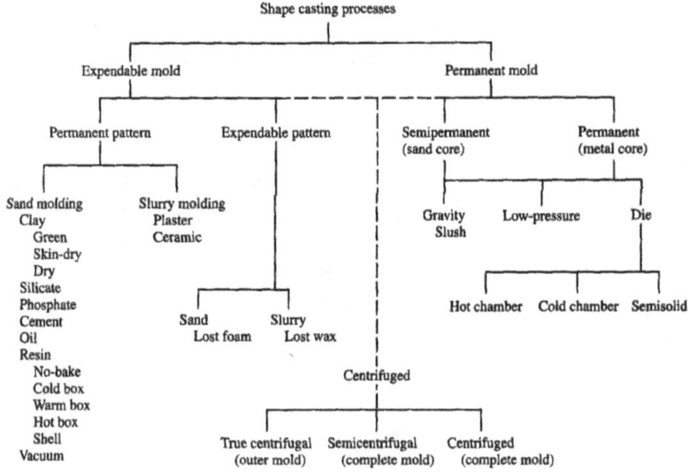

Figure 2. Classification of Metal Casting Processes Source: J.A. Schey, Introduction to Manufacturing Processes [6]

the real value of a well-functioning product is easy to assess, but the value of a failed product eludes evaluation until the extent of damage is known. Such knowledge can often fall under the category of "too little, too late".

In what follows, the process variables of the different members of the family of castings will be briefly considered with a view to differentiate the type of microstructure that is developed in the castings. The principles and evaluation methods of fracture toughness will then be briefly described. Selected papers from the literature will next be analyzed with a view to highlight the role of the microstructure in determining the fracture toughness of the castings. The effect of common castings defects on fracture toughness will then be very briefly considered. The use of fracture toughness - yield strength bubble charts for design against fracture [5], based on continuum mechanics, will be indicated.

2. Casting Processes

The umbrella covering the casting processes is shown in the figure below.

It is clear from Figure 2 that there are many avenues for making a casting, depending on the type of pattern, the type of mold and whether pressure is used for assistance in filling the mold. Not all processes are suitable for all the casting alloys. Investment casting (ceramic slurry, lost wax) is perhaps the most accommodative process for most alloys and others have limitations based on resistance to high temperature, chemical reaction and other factors. It is therefore customary to choose the casting process with due regard to the casting alloy. A recent addition to the umbrella is the squeeze casting process which is somewhat analogous to transfer molding of polymers. The microstructure of the casting is strongly affected by the process used for making it. The explanation is given below.

As stated earlier, casting is the product of solidification, which consists of nucleation and growth of solid from the liquid metal alloy). The final microstructure is decided by the composition of the alloy, the solidification rate and any melt treatment used. The alloys, based on their phase diagram may be of long-freezing range or short-freezing range type. The solidification rate is governed by the rate at which the mold is able dissipate the latent heat and superheat of the metal poured into the mold. Permanent molds like metal and graphite molds have higher thermal conductivity than disposable molds like sand and ceramic shells and therefore provide higher solidification rates. If there is no melt treatment, finer scale microstructure can be expected when these higher conductivity molds are used. Melt treatment however, can change this picture. The object of this treatment is to refine the microstructure and the treatment is variously termed as grain refinement (in the case of single phase alloys), modification or inoculation (in the case of second phase alteration of binary alloys). The application of continuous pressure as in squeeze casting may also substantially affect the microstructure. Long-freezing range alloys cooled at a relatively slow rate, as for instance in a sand mold, tend to solidify in a "mushy" or "pasty" manner. During the progress of solidification, there will be three distinct zones: liquid, liquid+solid, solid in most cases. The liquid+solid zone is the mushy zone. If this zone has large width, the final microstructure will consist of large amount of distributed interdendritic shrinkage areas, as any feed metal from the riser will find it difficult to access many of these areas due to tortuous path involved. The width of the mushy zone is reduced as the cooling rate increases, as in metal mold castings, with consequent reduction in distributed shrinkage. When the mushy zone is absent or too small, the solidification is termed "skin-forming" and the feed metal from a properly designed riser will have good access to the solidifying areas, thus minimizing distributed shrinkage. The shrinkage under these conditions can be totally eliminated that the feed metal has access to the final solidifying area. The application of Chroninov's rule, which states that the solidification time is proportional to the square of the volume-to-surface area of the casting and the riser or its modifications to account for the shape, will be helpful in this regard. The basic idea is to design the riser such that its solidification is more than that of the casting and its feeding distance is appropriate to reach the last solidifying zone of the casting. In long freezing range alloys solidifying in a mushy fashion, hot tear or hot crack can develop near above the solidus temperature when the network of solid crystals is unable to sustain any thermal stress gradients, particularly when the feed metal is unable to reach these locations. These cracks are usually sharp, capable of rapid propagation. Another important consideration in castings is the porosity caused by gas liberation during solidification. Gases like hydrogen are easily soluble in the liquid state but the solubility is substantially reduced in the solid state. This may result in pores of various sizes in the solid or even microcracks when there is significant resistance to the escape of the gases. It is therefore desirable to degas the liquid metal prior to pouring in the mold. A useful law in this context is Sievert's law which states that the solubility of a dissolved gas is proportional to the square root of its partial pressure. Using this law, degassing in the liquid state can be achieved by applying vacuum (difficult and expensive) or purging with an inert gas which serves the dual purpose of lowering the partial pressure of the dissolved gas and acting as a carrier for the escape of the dissolved gas, thus reducing the harmful effect of gas porosity in the solid.

As microstructure is the key to fine-tuning of the fracture toughness of castings, the influence of casting process factors on the microstructure must be well understood, if such fine-tuning is attempted. Needless to say, metallurgical knowledge such as phase diagram and the effect of non equilibrium cooling rate on it, nucleation and growth of the different phases in the microstructure, evolution of defects through impurities and interaction of the molten metal with melting atmosphere, the furnace lining, the mold, etc., will be very useful in this regard. Heat treatment can substantially affect the microstructure and therefore, knowledge of kinetics of solid state transformations is also important to understand the effect of the particular heat treatment on the microstructure.

3. Basics of fracture toughness testing

3.1. Linear Elastic Fracture Mechanics [LEFM]Approach

Linear Elastic fraction mechanics approach may be defined as a method of analysis of fracture that can determine the stress required to unstable fracture in a component.[7] The following assumptions are made in applying LEFM to predict failure in components [8].

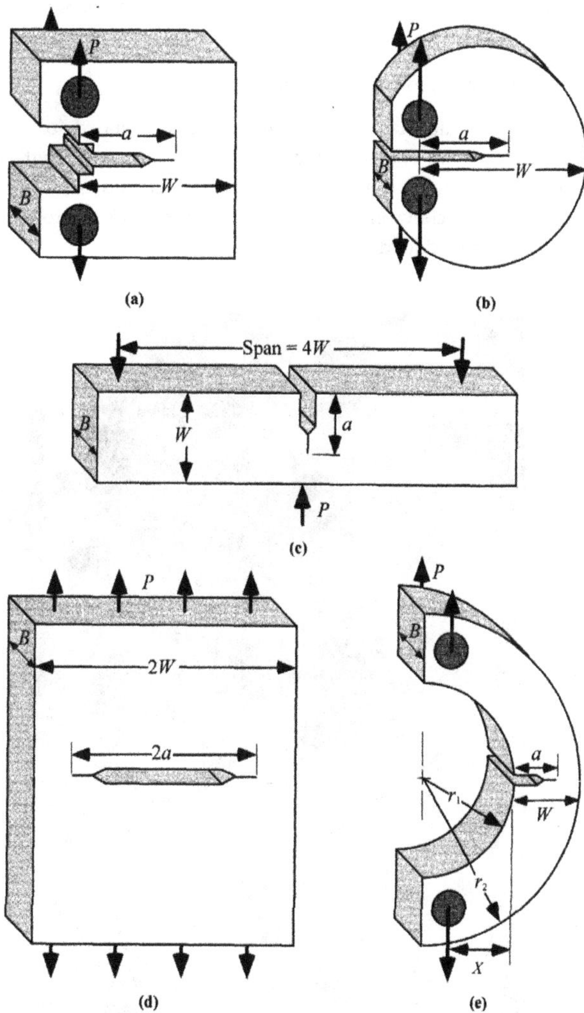

Figure 3. Standardized fracture mechanics test specimens: (a) compact tension (CT) specimen, (b) disk-shaped compact tension specimen, (c) single-edge-notched bend (SEB) specimen, (d) middle tension (MT) specimen and (e) arc-shaped tension specimen. Source: T.L. Anderson [9]

1. A sharp crack or flaw of similar nature already exists; the analysis deals with the propagation of the crack from the early stages.
2. The material is linearly elastic.
3. The material is isotropic.

4. The size of the plastic zone near the crack tip is small compared to the dimensions of the crack.

5. The analysis is applicable to near-tip region.

Figure 3 below shows standardized test specimens recommended for LEFM testing. Each specimen has three important characteristic dimensions: the crack length (a), the specimen thickness (B) and the specimen width (W). In general, W=2B and a/w = 0.5 with some exceptions For brittle materials, a chevron-notch is milled in the crack slot to ensure that the crack runs orthogonal to the applied load.

Figure 4. A typical view of the test set up for fracture toughness testing Source: Seetharamu [10]

In most cases fracture toughness tests are performed using either CT specimen or SEB specimen. The CT specimen is pin-loaded using special clevises. The standard span for SEB specimen is 4W maximum; the span can be reduced by moving the supporting rollers symmetrically inwards.

It is to be noted that the tip of the machined notch will be too smooth to conform to an "infinitely sharp" tip. As such, it is customary to introduce a sharp crack at the tip of the machined notch. Fatigue precracking is the most efficient method of introducing a sharp crack. Care must be taken to see that the following two conditions are met by the precracking procedure: the crack-tip radius at failure must be much larger than the initial radius of the precrack and, the plastic zone produced after precracking must be small compared to the plastic zone at fracture. This is particularly necessary for metal castings as many exhibit plasticity; a notable exception is flake graphite cast iron castings made in sand molds.

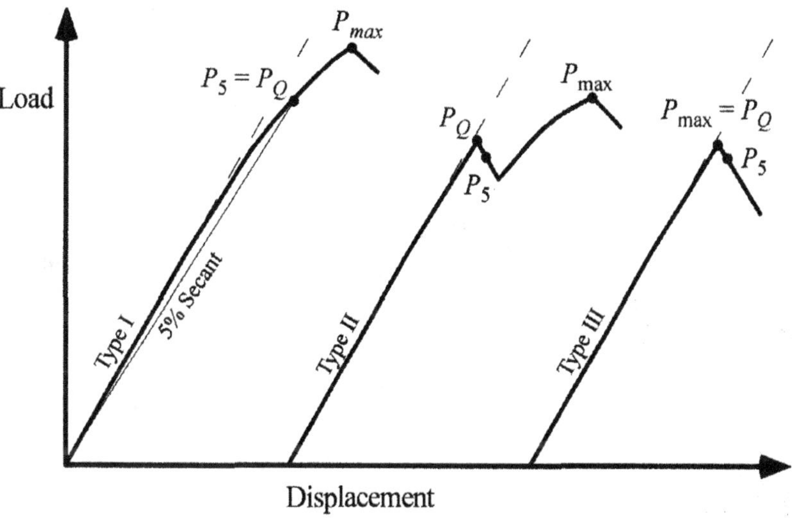

Figure 5. Type I, Type II or Type III behavior in LEFM test Source: T.L. Anderson [12]

LEFM tests are conducted as per ASTM E 399 [11]. A typical test set up is shown in Figure 4. All except the MT specimen noted in Figure are permitted to be used as per this standard. The ratio of 'a' as defined in each figure to the width W should be between 0.45 and 0.55.

The load-displacement behavior that can be obtained in a LEFM test, depending on the material, can be one of three types as shown in the Figure 5 below.

First a conditional stress intensity factor K_Q is determined from the particular curve obtained using

$$K_Q = \frac{P_Q}{B\sqrt{W}} f(a/W) \tag{2}$$

where f (a/W) is a dimensionless factor of a/w.

The conditional stress intensity factor K_Q is the critical stress intensity factor if

$$B,\ a \geq 2.5\left(\frac{K_Q}{\sigma_{ys}}\right)^2 \tag{3}$$

where σ_{ys} is the yield stress of the material.

If this is not the case, the result is invalid, most likely because of significant crack tip plasticity. This would imply that triaxial state of stress required to ensure plane strain condition at the crack tip is not achieved and any determined stress intensity factor at fracture as per ASTM E399 would be an overestimate of the resistance to crack growth. Use of such values in design would be dangerous. In such cases, an elastic-plastic fracture mechanics (EPFM) method must be employed to determine the specimen's resistance to the propagation of a sharp crack.

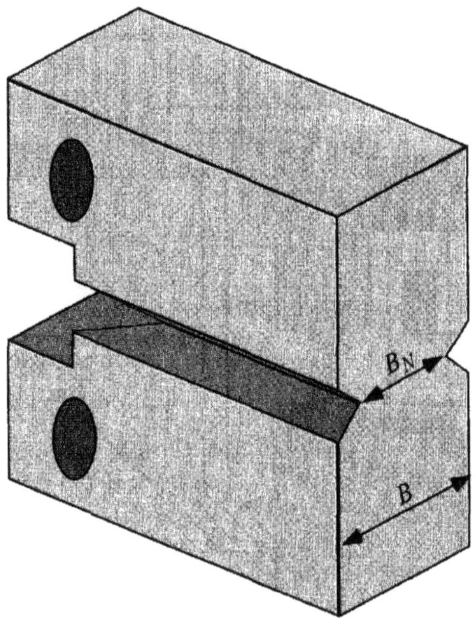

Figure 6. Side-grooved Fracture Toughness Test Specimen Source: T.L. Anderson [14]

3.2. Elastic Plastic Fracture Mechanics (EPFM) Approach

Among the different methods available to determine the sharp crack growth resistance in specimens with significant plasticity at the crack tip (much less than what is required to cause total plastic collapse) the J-integral method and the Crack-tip Opening Displacement (CTOD) have been more widely adopted. The recent trend however, is to use the J-integral approach and only this method will be briefly described here. ASTM E 1820 [13] gives two alternative methods: the basic procedure and the resistance curve procedure. The basic procedure normally requires multiple specimens, while the resistance curve test method requires that crack growth be monitored throughout the test. The main disadvantage of this method is the additional instrumentation and skill are required. Though this method has the advantage of using a single specimen, making of multiple specimens as nearly externally identical-looking castings is not a major problem; any inconsistent results among the different specimens will give an opportunity to see if the casting microstructure is properly controlled. Therefore only the basic test procedure will be considered here.

3.3. The Basic Test Procedure and J_{IC} Measurements

The ASTM standard that covers J-integral testing is E 1820 [13]. The first step is to generate a J resistance curve. To ensure that the crack front is straight the use of a side grooved specimen as shown in Figure 6, is recommended.

A series of nominally identical specimens are loaded to various level and then unloaded The crack growth in each sample, which will be different is carefully marked by heat tinting or fatigue cracking after the test. The load-displacement curve for each sample is recorded. Each specimen broken open and the crack growth in each specimen is measured.

J is divided into elastic and plastic components, by using

(4)

$$J = J_{el} + J_{pl} \tag{5}$$

$$J_{el} = \frac{K^2(1-\upsilon^2)}{E} \tag{6}$$

$$K = \frac{P}{\sqrt{BB_N W}} f(a/W) \tag{7}$$

$$J_{pl} = \frac{\eta A_{pl}}{B_N b_0} \tag{8}$$

η is a dimensionless quantity given by

$$\eta = 2 + 0.522(b_0/W) \tag{9}$$

In equations (6) and (7) b_0 is the initial ligament length.

A_{pl} is the plastic energy absorbed by the specimen determined from Figure 7.

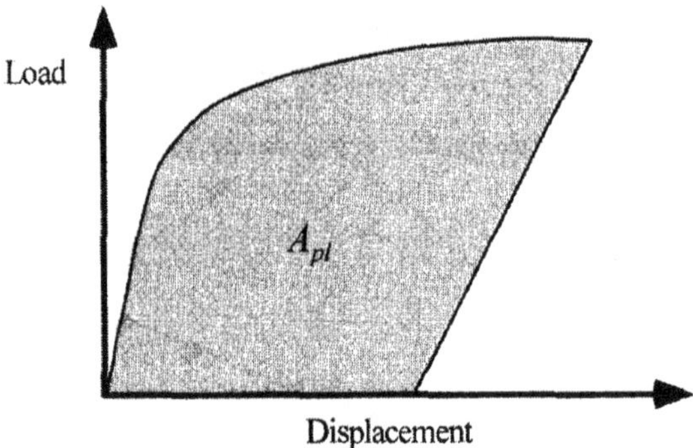

Figure 7. Plastic energy absorbed during J-integral test Source: T.L. Anderson [15]

The J values obtained from equation (3) are plotted against the crack extension Δa for each successive specimen to obtain J-R curve shown in Figure 8.

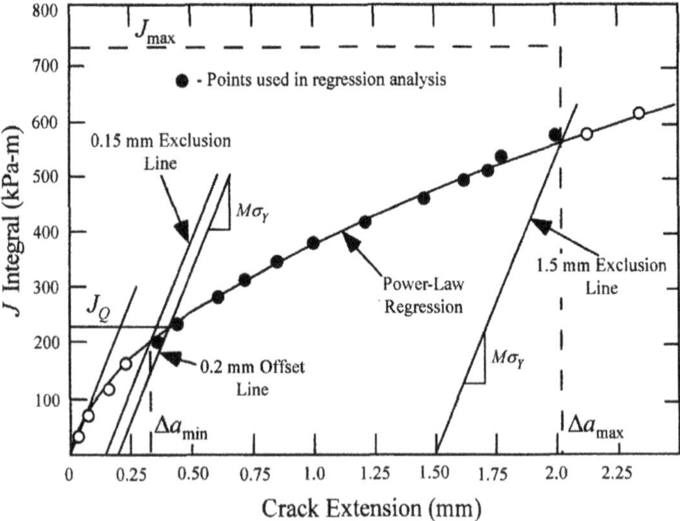

Figure 8. Determination of J_Q as per ASTM E 1820 Source: T.L. Anderson [16]

M value in the figure is related to crack blunting and the default value is 2. As seen in the figure, the provisional critical value J_Q is obtained from the intersection of the J-R curve with the line $M\sigma_Y$ where σ_Y is the flow stress given by the average of the tensile and yield stresses.

The provisional J_Q is taken as the critical value J_{Ic}if the condition:

$$B, b_0 \geq \frac{25 J_Q}{\sigma_Y} \tag{10}$$

is satisfied.

The equivalence between J_{IC} and $K_{IC}$is given by:

$$J_{IC} = \frac{(K_{IC})^2}{E}(1-\upsilon^2) \tag{11}$$

where E is the elastic modulus and υ is the Poisson's ratio.

If it is assumed that a steel sample has a yield strength of 350 MPa, tensile strength of 450 MPa and Young's modulus (E) of 207 GPa and fracture toughness of 200 MPa-$\sqrt{m}$, it can be shown that E 399 thickness requirement for validity is 0.816 m, while the E 1820 thickness requirement for validity, based on the equivalence shown in equation (6) is only 11 mm. The advantage of E1820 approach over E399 approach in regard to valid specimen thickness requirement is thus obvious.

4. Fracture toughness of metal (alloy) castings

In what follows reported fracture values of various castings will be presented and discussed.

4.1. Aluminum alloy castings

In recent times, the most widely studied nonferrous casting alloys for fracture behavior are aluminum casting alloys. Among them, aluminum silicon alloys have attracted the most attention as they are widely used because of good castability and high strength-to-weight ratio. The microstructure of aluminum silicon alloys can be significantly affected by changes in the process variables as typically shown below in Figure 9 for aluminum-5% silicon alloy. The figure shows the variation of microstructure with cooling rate. Figure 9(a) refers to a sand casting where the cooling rate is the lowest among the three, sand cast, permanent mold cast and die cast. The dendrite cells are large, the silicon flakes (dark) are coarse and iron-silicon-aluminum intermetallics (light grey) are seen. The resistance to crack propagation will be the lowest with this type of microstructure. Figure 9(b) refers to a permanent mold casting where the cooling rate is higher than in a sand castings. It is seen that there is refinement in both primary aluminum and eutectic silicon as well as the intermetallics. The resistance to crack growth will be higher than in sand castings. Figure 9(c) shows the microstructure of a die casting of the alloy where high degree of refining of dendrite cells and eutectic silicon are seen. Other things being equal, the resistance to crack growth will be maximum in this type of microstructure. However other things will not be equal in general, the main factor being the yield strength of the casting. Thus crack tip plasticity will be high in the sand casting, intermediate in the permanent mold casting and lowest in die casting. Thus the fracture toughness increase in the casting will not be in direction proportion to the reduction in cooling rate. The factors favoring increase in fracture toughness would be decreased dendritic cell size and refinement of the covalent bonded silicon and mixed bonded intermetallics. The opposing factor would be reduced plasticity due to increase in yield strength, both due to primary cell and eutectic refinement.

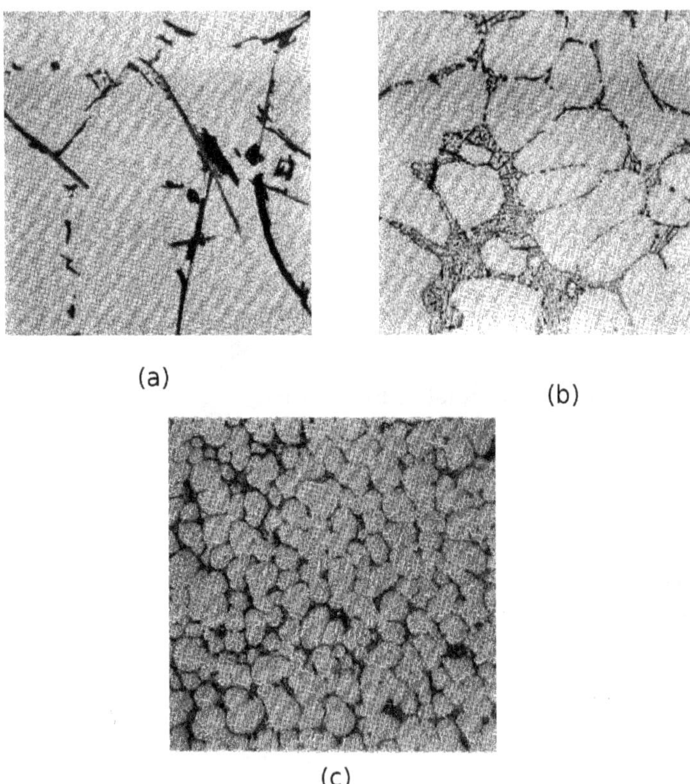

Figure 9. Microstructure of aluminum casting alloy 443 (Al-5%Si) (a) Alloy 443-F, as sand cast, (b) Alloy B443-F, as permanent mold cast, (c) Alloy C443-F, as die cast All were etched with 0.5% hydrofluoric acid and photographed at 500 X. Source: W.F. Smith [17]

Figure 10 refers to the microstructural variations brought by heat treating alloy 356- Al-7% Si-0.3%Mg (sand cast, constant cooling rate). Figure 10(a) refers the microstructure after artificial aging. The coarse dark platelets are silicon, black script is Mg_2Si and the light scripts are intermetallics of iron-silicon-aluminum and iron-magnesium-silicon-iron. The dendrite cell is coarse. Figure 10(b) refers to alloy 356-F as sand cast, which is modified with 0.25% sodium. The cells are still coarse but the silicon particles are refined and in the form of inter-dendritic network. Figure 10(c) refers to alloy 356-T7 which is modified with 0.025% sodium, solution treated and stabilized. The microstructure shows rounded interdendritic silicon and iron-silicon-aluminum intermetallics. Here again the opposing factors discussed above will come into play but the plasticity in primary aluminum will be around the same and the dual causes for increase in yield strength as in the previous case will not be present.

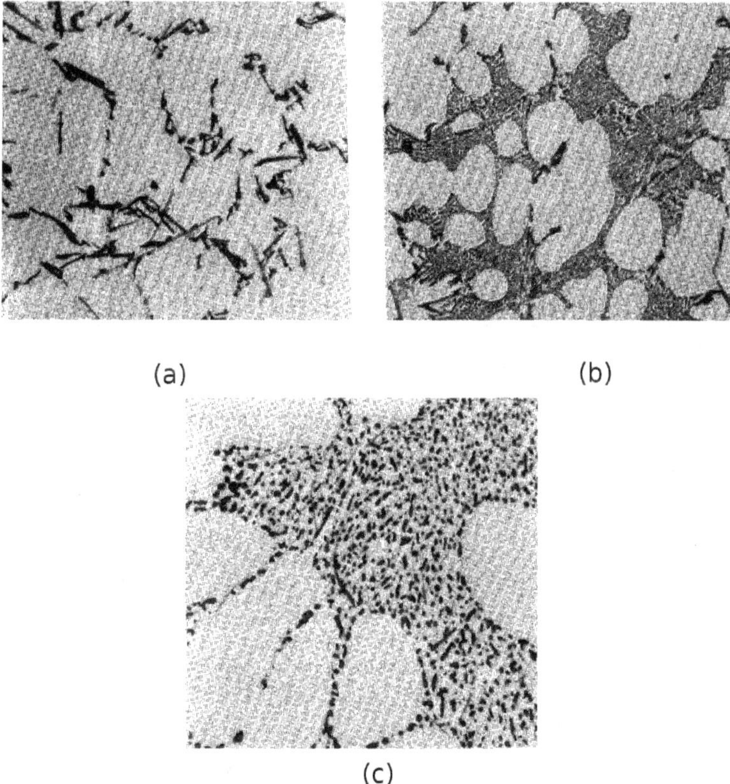

Figure 10. Microstructure of alloy 356 sand cast and heat-treated in different conditions (a) alloy 356-T51: sand cast, artificially aged, (b) alloy 356-F: as sand cast, modified with 0.025% sodium, (c) alloy 356 T7: sand cast, modified by sodium addition, solution treated and stabilized. All were etched with 0.5% hydrofluoric acid and photographed at 250 X. Source: W.F. Smith [18]

Having noted these factors in affecting the fracture toughness, some recent papers on fracture toughness of aluminum alloys will now be examined.

Hafiz and Kobayashi [19] studied the fracture toughness of a series of aluminum silicon eutectic alloy castings made in graphite and steel molds. The microstructure was varied by treating with different amounts of strontium. J-R curve obtained from multiple specimens was used to determine J_Q values. Extensive microstructural and SEM fractographic studies were made. They defined a ratio (λ / DE_{Si}) where λ is the silicon particle spacing and $(DE)_{Si}$ is the equivalent silicon particle diameter. They also defined the void growth parameter as $VGP = \sigma_y(\lambda / DE)_{Si}$ They found that the equation $J_Q = -9.94 + 0.38(VGP)$ is obtained in their samples, with J_Q varying in a straight line fashion from about 7 kJm^{-2} to about 78 kJm^{-2} when the VGP varied from 50 MPa to 200 MPa. Their main conclusion is that in eutectic Al-Si alloy castings, greater the refinement of eutectic silicon, higher will be the fracture toughness.

Kumai, et al, [20] on the other hand focused on the dendrite arm spacing of alloy A356, (which is hypoeutectic) permanent mold and direct chill (semi continuous) cast tear test samples in their work. The area under the load-displacement curve was determined as the total energy and was divided into energy for initiation and propagation. It was found that in direct chill casting, both initiation and propagation energies increased with decrease in the dendrite arm spacing (DAS); decrease in DAS resulted only in increase of propagation energy in permanent mold casting. The fracture surface was perpendicular to the load in permanent mold castings while it was slanted in DC casting indicating higher energy absorption during the fracture process. This test could at best be qualitative in determining the fracture behavior.

Tirakiyoglu [21] has examined the fracture toughness potential of cast Al-7%Si-Mg alloys. He has reported that based on Speidel's data [22] a relationship of the form:

$$K_{IC}(\text{int}) = 37.50 - 0.058\sigma_{ys} \tag{12}$$

can be developed between the maximum (intrinsic) fracture toughness and yield strength of this alloy. However, as suggested by Staley [23] there are several extrinsic factors such as porosity, oxides and inclusions that tend to lower the fracture toughness. If these extrinsic factors are eliminated the intrinsic fracture toughness can be higher, given by:

$$K_{IC}(\text{int}) = 50.0 - 0.073\sigma_{ys} \tag{13}$$

Equation (11) gives the potential maximum fracture toughness of the Al-7%Si-Mg cast alloy in the absence of defects. A nice feature of this paper is the listing of dendrite arm spacing of different types of aluminum-silicon-magnesium alloy castings.

Tohgo and Oka [24] have studied the influence of coarsening treatment on fracture toughness of aluminum-silicon-magnesium alloy castings. The alloy: Al-7%si-0.4%Mg was cast in permanent mold and solution treated for 6 hr at 803 K followed by aging for 6 hr at 433 K. One batch was tested in this condition while a second batch was further given a coarsening treatment at 808 K for 50 hr, 100 hr, 150 hr and 200 hr. J-R curves were constructed using 5 specimens and J_Q values were determined. The fracture toughness increased to 27 $MPam^{1/2}$ after coarsening of silicon, as compared to 20.8 $MPam^{1/2}$ for uncoarsened sample. The authors attribute the improvement to the increased plastic deformation of α-Al owing to more uniform distribution of silicon particles, energy dissipation due to damage of silicon particles around a crack and the rough fracture path in the coarsened sample.

Kwon, et al [25] have investigated the effect of microstructure on fracture toughness of rheo-cast and cast-forged A356-T6 alloy. Interdendritic silicon was observed in the microstructure of rheo-cast sample while there was alignment of cells in the cast-forged sample along with more uniform dispersion of silicon particles. Fractographs of fracture toughness specimens indicated cleavage type fracture in the rheo-cast sample while there was fibrous fracture in the cast-forged sample. As to be expected the fracture toughness of the rheo-cast sample was 20.6 $MPam^{1/2}$ while the cast-forged sample showed a fracture toughness of 24.6 $MPam^{1/2}$.

Alexopoulos and Tirayakioglu [26] have determined the fracture toughness of A357 cast aluminum alloys with a few minor chemical modification. The raw stock for further machining required for studies was continuously cast with intent to keep porosity and inclusions at a minimum level. The continuous casting process is the patented SOPHIA process capable of providing cooling rates of up to 700 K/min. As compared to an investment cast sample, the dendrite arm spacing in the SOPHIA-cast sample would be lower by about 33%. The fracture toughness values, determined from CTOD measurements, ranged from about 18 $MPam^{1/2}$ to about 29 $MPam^{1/2}$, depending on the composition and the heat treatment. The higher value was obtained in the plain A357 cast by SOPHIA process and subjected to solution treatment for 22 hr at 538 C and aged for 20 hr at 155 C. The main aim of these authors was to establish correlation between tensile properties and fracture toughness and the major part of the paper deals with evaluation of tensile behavior under different conditions.

Lee, et al [27] have investigated the effect of eutectic silicon particles on the fracture toughness of A356 alloy cast using three different methods: low pressure casting (LPC), casting-forging (CF) and squeeze casting (SC). They used ASTM E 399 procedure and as to be expected, got invalid fracture toughness results (sample thickness was 10 mm). They also conducted in-situ SEM studies on crack morphology, where plane stress was present. Thus only qualitative comparisons can be made on the influence of the three different casting processes on the fracture toughness. A notable observation is that significant shrinkage pores were present in LPC samples, while they disappeared in CF and SC samples, evidently due to the higher pressures applied. The eutectic cell size was the least in SC samples while it was similar in size in PC and CF samples. SEM fractographs from all the three samples showed fibrous fracture, with LPC samples showing the additional effect of stress concentration at the edges of shrinkage cavity. Though the SC sample had the most refined microstructure, the apparent fracture toughness was the lowest on account of reduced spacing between the eutectic silicon particles that apparently encouraged fracture initiation.

Tirakiyoglu and Campbell [28] have analyzed the fracture toughness of Al-Cu-Mg-Ag (A201) alloy from data on premium quality castings. When molten metal is poured into a mold, the Reynolds number is invariably in the turbulent flow region to facilitate proper filling of the mold. In aluminum alloys, the surface oxide that forms as a result becomes folded into the bulk of the melt. These oxide "bifilms" have neutral buoyancy, unlike in say, steel castings and tend to travel with the melt into the mold cavity. As they do not bond with the liquid, the solidified casting will have the bifilms remaining as cracks due to the discontinuity. Also, the layer of air in the folded bifilms can grow into a pore or remain as a crack in the casting. The authors point out that in aluminum (and other drossing alloys) this is perhaps the most ignored defect as far as plans for elimination of defects are concerned. This extrinsic defect will result in the intrinsic fracture toughness not being attained. As per the authors, the intrinsic fracture toughness in A301 casting can be represented by:

$$K_{IC}=\left\{\ln\left[1+\frac{\exp(-0.0032\sigma_{ys})}{100}\right]\right\}^{3/2}\left(\frac{2kE'\sigma_{ys}}{3}\right)^{1/2} \tag{14}$$

Here,

$$E'=\frac{E}{1-\upsilon^2} \tag{15}$$

The intrinsic value of K_{IC} can exceed 45 MPa m$^{1/2}$ if the yield strength is around

350 MPa.

4.2. Steel Castings

Jackson [29] has published a comprehensive paper on the fracture toughness of steel castings. He has considered that steel is susceptible to ductile-brittle transition and has reported the fracture toughness for lower shelf using LEFM and for the upper shelf using EPFM. While the LEFM method he used was the same as ASTM E399, use of CTOD was more in vogue in England at the time he wrote the paper and therefore either the critical CTOD (δ_C) or the equivalent J_{IC} have been reported in the paper, using the relation:

$$\delta_C=\frac{J_{IC}}{\sigma_{ys}} \tag{16}$$

Steel	K_{IC}(MPa m$^{1/2}$)	σ_{ys}(MPa)	Critical flaw size (mm) Surface Embedded
1. 0.5%C, 1% Cr	46	480	3.7 4.4
2. 1.5%Ni-Cr-Mo	86	740	5.4 6.2
3. 1.5%Ni-Cr-Mo	104	1280	2.6 3.2

Table 1. Table 1. The fracture toughness, yield strength and chosen values of critical flaw size for three cases are shown in Table 1.

Steel 3 was vacuum melted while the other two were air melted, showing that a stronger steel has the disadvantage of lower critical flaw size (elliptical flaw, ratio of major-to-minor axis is 8-to-1).

One important point made by Jackson is that the chemical composition effects on fracture toughness may be masked by those of features such as shrinkage. Though it is known that increasing sulfur and phosphorus leads to decrease in fracture toughness, in the researcher's experiments, shrinkage masked this expected effect. Shrinkage encountered in the crack path may cause multiple crack fronts deviating from the main path resulting in increased fracture toughness to be observed; this overestimates the intrinsic fracture toughness and may cause problems when applied in design. The best remedy is therefore is to minimize shrinkage using proper feeding techniques.

As reported by Jackson, in the case of a 0.5%Mo, 0.33% V steel casting the lower shelf fracture toughness is about 55 MPa $m^{1/2}$ (temperature < 60C) while the upper shelf value increases to about 180 MPa $m^{1/2}$ (temperature > 110 C). This behavior is inherent in BCC alloys like steel and should be considered in equipment where there is a wide difference between the cold start temperature and operational temperature. The problem then is to avoid brittle fracture during cold start and onset of plastic instability at normal operating temperatures.

Barnhurst and Gruzleski [30] have investigated the fracture toughness of high purity cast carbon and low alloy steels. A notable feature of this work was that only blocks that were found to be radiographically sound were used for the preparation of fracture toughness specimens. The inclusion level in all the castings were low enough to classify them as extremely clean. The steel compositions were according to AISI/SAE 1030, 1527, 1536, 2330, 2517 for low carbon steels, 1040,5140,1552,5046, 2345 for medium carbon steels and 1055,5155,3450, 52100 for high carbon steel. Other than carbon, each grade no other element or one alloying element, with impurities being kept to a minimum. All castings were austenitized in the range of 840 C- 900 C depending on the alloy, for 4 hr, oil quenched, held mostly at 650 C for 2 hr (with two exceptions: two samples directly air cooled from 900 C soak, one sample held at 300 C after oil quenching and then air cooled. The fracture mode in most castings was ductile, with only a few showing cleavage or mixed ductile/cleavage fracture. The K_{IC} values determined from J_{IC} ranged from 41.6 MPa $m^{1/2}$ (1.0 C, 1.61 Cr) to 247.8 MPa $m^{1/2}$ (0.25 C, 4.60 Ni). The conclusions drawn were that under carefully controlled composition, heat treatment, inclusion and impurity content, exceptional fracture toughness values at room temperature can be obtained, at the expense of tensile properties. The critical flaw sizes would exceed the section thickness of most designs. Under normal production conditions where attainment of such high purity is impractical, this study does provide the guidelines that the influence of alloying elements like nickel, chromium and manganese is relatively small at medium carbon levels and that heat treatment, additions of molybdenum and silicon may have significant influence on room temperature fracture toughness.

Chen, et al [31] have studied random fracture toughness values of China Railway Grade B cast steel wheels using LEFM approach. The wheel was first stress relieved, and then the rim was quenched and tempered, while the hub was shot peened. K_Q values reported range from 50.52 MPa $m^{1/2}$ to 63.77 MPa $m^{1/2}$ in the wheel hub and, 60.70 MPa $m^{1/2}$ to 76.40 MPa

$m^{1/2}$ in the wheel rim. Only the specimen thickness (~25 mm) has been indicated but the yield strength values have not been provided: it is therefore difficult to say whether these values are valid or not. Narrative description of the fracture surface using SEM indicates the predominance of cleavage with little evidence of fibrous rupture.

Kim, et al, [32] have evaluated the fracture toughness of centrifugally cast high speed steel rolls. The carbon equivalent, defined as C + 1/3 Si was in the rang if 1.89 to 2.28 and the tungsten equivalent, defined as W + 2 Mo was in the range of 9.82 to 13.34. Vanadium content was varied between 3.95 and 6.26 and the chromium content was kept constant in the range of 4.0-6.0. Precracking presented difficulties and therefore the authors used 30-50 μm machined notch. Tests were made otherwise as per ASTM 399. K_Q values were in the range of 21.4 MPa $m^{1/2}$ to 28.2 MP a $m^{1/2}$. They have concluded that the fracture toughness is determined by the total fraction of carbides, characteristics of the tempered martensitic matrix, distribution and fraction of intercellular carbides and fraction of cleavage and fibrous mode on the fracture surface. The best fracture toughness as obtained when a small amount of intercellular carbides was distributed in a relatively ductile matrix of lath martensite.

James and Mills [33] have investigated the fracture toughness of two popular as cast stainless steels, CF8 and CF*M. Toughness tests were conducted at 24 C, 371 C, 427 C and 482 C using multiple specimen J-R curve method. Exceptionally high J_{IC} values, in the range of 1397 kJ/m^2 at 24 C to 416 kJ/m^2 at 482 C demonstrated that fracture control is not a concern in unirradiated condition. However, neutron irradiation reduces J_{IC} by an order of magnitude and therefore fracture control becomes essential.

4.3. Cast Iron

The metallurgy of cast iron is among the most complex of all alloys. Cast iron shows metastability anomaly. Under certain conditions of composition and cooling rate the eutectic formed upon solidification consists of austenite and graphite. Under certain other conditions an eutectic of austenite and iron carbide is formed. The former is known as graphitic cast iron, while the latter as white cast iron. In low sulfur and oxygen cast iron melts, if magnesium is added so that its residual amount is 0.05% or above (but not too high) the graphite formed will be nodular rather than the flake form found in untreated graphitic cast iron melts. In the latter the flake may be of undercooled type (Type D- when the sulfur content is low in sand or investment castings or with normal sulfur when the cooling rate corresponds to that in permanent molds); it will be in the interdendritic form, with branching). Under normal conditions found in commercial sand castings, the graphite will be a part of the eutectic cell formed with austenite, graphite having a loose "cabbage" shape with the interleaf region occupied by eutectic austenite. Adding a silicon-bearing inoculant will increase the number of eutectic cells in flake cast iron and nodule count in nodular (or, ductile) iron. The white cast iron forms graphite in the solid state when heat treated (and is called malleable iron), but the melt-formed graphite in the other two types of cast iron will be largely unaffected by any solid state transformation. In recent times another type called compacted graphite cast iron has been developed where the residual magnesium is lower than in ductile iron. All types of cast iron noted above are governed by eutectoid decomposition, which means that the matrix may consist of various combinations of ferrite and pearlite under near-equilibrium conditions. These irons are also affected by isothermal or continuous cooling transformations at nonequilibrium rates giving rise to bainitic or martensitic or tempered martensitic cast irons. In recent times, a bainitic ductile iron known as austempered ductile iron (ADI) has become popular in industrial applications. In what follows, investigations on the fracture toughness of some of these cast irons will be briefly discussed.

The exact reasons for the formation of different types of graphite in cast iron have been a matter of debate for many years. The type of graphite found in commercial cast irons may have one or more of the following types: flake (Type A), undercooled (Type D), coral, compacted, nodular. A generalized view, based on the growth of graphite (in the liquid state) is presented in Figure 11 below.

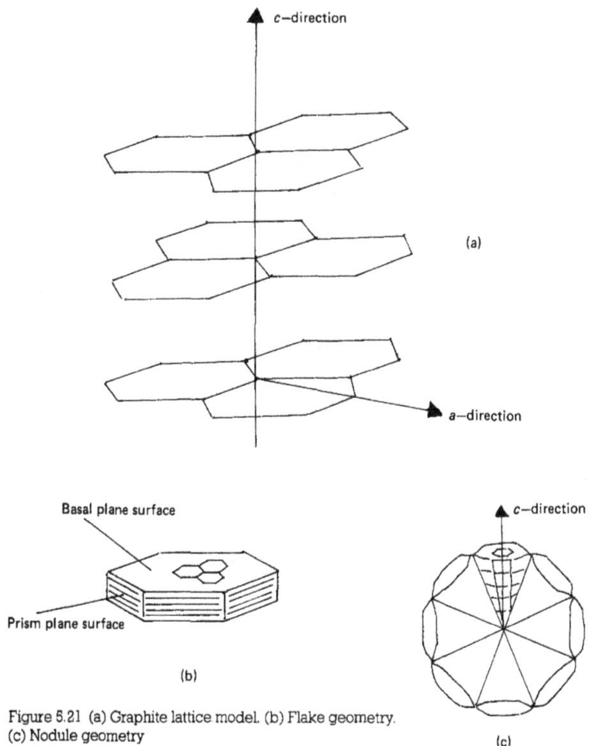

Figure 11. Extremities in the growth of graphite in the liquid Source: Elliott [34]

Graphite has a layered hexagonal lattice structure (a), with strong covalent bonds in the hexagonal chains, with the layers bonded by weak secondary bonds. The hexagonal plane is called the basal plane and the edge of the block formed by bonding of layers with weak bonds is called the prism plane. The basal planes tend to grow in the "a" direction and the prism planes in the "c" direction. When growth in the "a" direction is dominant, flake form is obtained, the thickness being determined by the growth rate and the graphite source; slower the growth rate and lower the number of eutectic cells, the thicker would be the flake. When "a" growth is suppressed and "c" growth is fully encouraged, nodular form results. Intermediate forms like Type D, coral or vermicular forms result when there is progressive resistance to the formation of Type A, or alternatively, decreasing encouragement to the nodular shape formation. It is to be realized that the exact reasons for these resistances or discouragements may be due to the interaction of fine-scale multiple activities, often at the atomic scale, related to both nucleation and growth. Thus, at this time one has to accept that these different forms of graphite, which curiously are relatively stable forms during the life of a component, do exist and there is need to understand, for instance, the details of how the crack propagation is affected by their interactions with the fine-scale features of the neighborhood of the crack..

The fracture toughness of graphitic cast iron is determined by the type of graphite, the type of matrix and the interaction between the graphite and the matrix. In view of numerous combinations possible, the fracture toughness could be expected to vary over a wide range: this is indeed the case. Once again it follows that to fine-tune the fracture toughness of cast iron the microstructural features should be analyzed and examined if corrective measures can be taken, consistent with cost-benefit analysis.

The fracture toughness of flake cast iron ranges from 11-19 MPa $m^{1/2}$ [35]. Whether these are valid results as per ASTM E399 is subject to the acceptance of the tensile strength instead of the yield strength for validity criterion, as flake cast iron has non linear elastic part in the stress-strain curve and 0.2% offset method can not be applied to determine the yield strength. Thus the above noted values may be cited by some as K_Q and by others as K_{IC}. In critical applications these low value force the assumption of a high factor of safety. A pertinent observation with respect to flake cast iron is that the ductile-brittle transition temperature is well above the room temperature and therefore the fracture toughness at normal or below-normal operating temperatures seems to be unaffected by the temperature.

Because of the steel-like mechanical behavior of nodular graphite cast iron, the fracture toughness of this iron has been vastly studied. The fracture toughness values range from about 25 MPa $m^{1/2}$ in an iron with yield strength of about 450 MPa to nearly 60 MPa $m^{1/2}$ in an iron of yield strength of 370 MPa [36]. It is possible that the intrinsic fracture toughness of nodular iron would be higher if the inherent shrinkage, among the highest in cast irons, is reduced. A particular grade, D7003 (quenched and tempered) posses both good fracture toughness and high yield strength. Salzbrenner [37] evaluated the fracture toughness of samples of different compositions, but adopted a constant heat treatment with intent to have a ferritic matrix. The heat treatment involved solutionizing at 900 C for 4 hr, followed by slow furnace cool (at 10 C/hr) to 700 C and holding at this temperature for 24 hr followed by slow cooling. He followed EPFM approach and obtained fracture toughness values ranging from a high of 79 MPa $m^{1/2}$ (with small, well distributed nodules) to as low as 25 MPa $m^{1/2}$ in a sample with non-spherical nodules. The better fracture toughness of nodular iron in relation to flake cast iron is often attributed to the relatively smooth graphite-matrix interface in the former. This statement may however, be an oversimplification as there are factors such the diversion of the crack and ability to absorb energy in the interlayer regions of the nodule to be considered.

Doong, et al [38] have investigated the influence of pearlite fraction on fracture toughness of nodular iron and their results show that when the pearlite fraction is 4% or 27% the fracture toughness shows a decreasing trend in the range of -75 C to 75 C, while the fracture toughness of samples with 67% and 97% pearlite show an increasing trend in the same temperature range. The nodularity in all these castings was 95% or better..

Nodular iron castings are generally made in sand molds but the present authors investigated the fracture toughness of permanent mold-cast magnesium-treated iron. A hypereutectic composition with a high silicon percentage (3-3.4%) was used to avoid the formation of iron carbide in the as-cast state. The graphite consisted of overlapping nodules, possibly as a result of high thermal convection in permanent molds. It is also possible that inoculation was needed to provide more nucleation of graphite and reduce the possibility of overlapping, by rapid austenitic shell formation around the nodules.

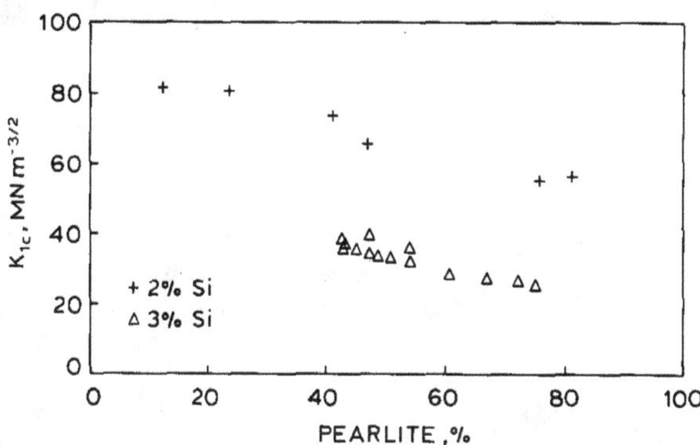

Figure 12. Effect of pearlite content on fracture toughness of permanent mold ductile iron Source: Bradley and Srinivasan [39]

Figure 12 seen above shows the effect of pearlite content in two types of permanent mold ductile iron. The top curve refers to a melt with 2% silicon, which led to a chilled casting and was soaked at 900 C and cooled at different rates to obtain different combinations of ferrite and pearlite in the matrix. The lower curve refers to a set of chill-fee castings, obtained by solidifying castings with 3% silicon. Different pearlite/ferrite combinations were produced by varying the casting thickness. It is seen that increase in silicon significantly lowers the fracture toughness. A possible reason is that on increasing the silicon level to 3%, the ductile-brittle transition temperature is raised to well above the room temperature. It is also possible that any residual stress present in the as-cast state is minimized in the heat treated state. In any case the differences in the modes of fracture in the two cases are clearly seen in Figure. 13 and Figure 14 shown below.

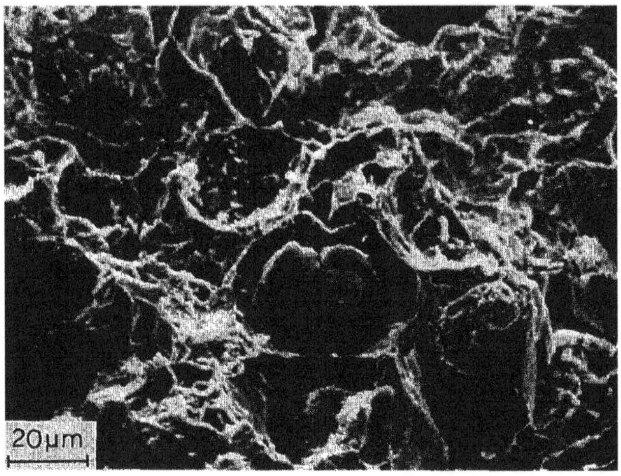

Figure 13. Fibrous fracture in the stable crack growth region of 2% Si casting Source: S. Seetharamu [10]

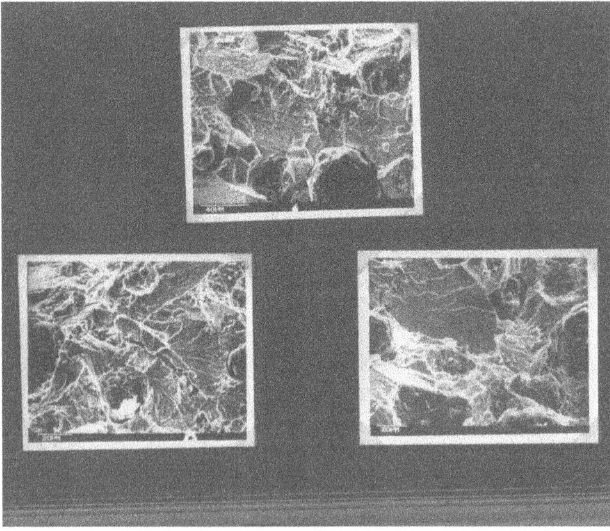

Figure 14. Transgranular cleavage in crack growth region of 3% Si casting Source: S. Seetharamu [10]

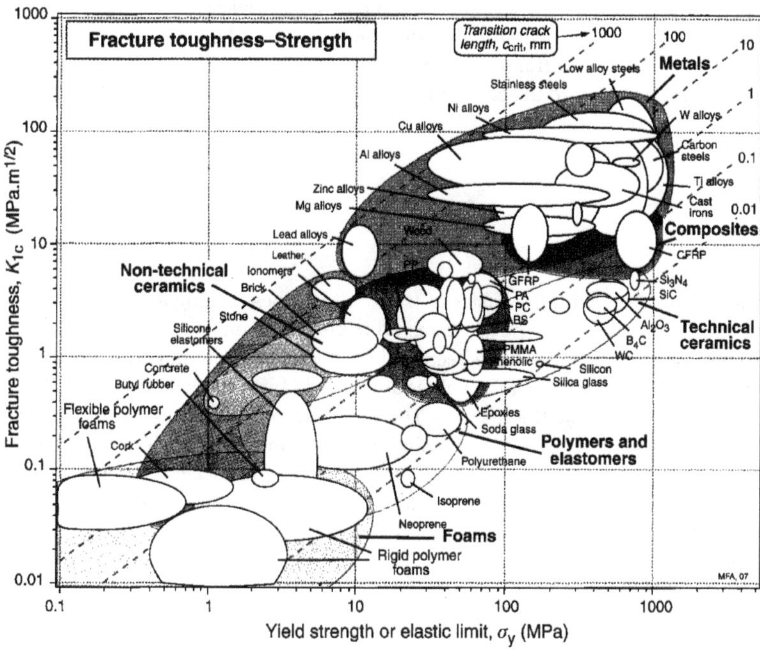

Figure 15. Fracture toughness, yield strength and transition crack length of materials. Source: M.F. Ashby, et al [5]

A relatively new development in the field of ductile irons is Austempered Ductile Iron (ADI) which is commercially available in different grades [40]. The fracture toughness of ADI can be in the range of about 59-86 MPam$^{1/2}$ and therefore exceeds the fracture toughness of most other ductile iron grades, except Ni-resist. The fracture toughness values are best determined using EPFM. However, Lee, et al [41] have used ASTM E399, which seems to be justified as the ratio 2.5 $(K_{IC}/\sigma_{ys})^2$ is below the test sample thickness of 25 mm; as the authors have not reported the yield strength, but only the Brinell hardness, the yield strength (MPa) is assumed to be 3.3 times the Brinell hardness, for the purpose of making this statement..

It is important to realize that both stress and crack size should be within limits for safe use of any casting. When the failure mode is brittle, the critical flaw size is given by equation (1) when K_I reaches a critical value $K_{IC.}$ When there is significant crack tip plasticity the transition from stable crack growth to unstable mode occurs at a length given by

$$C_{crit} = \left(\frac{K_{IC}}{\sigma_{ys}}\right)^2 \frac{1}{\sqrt{\pi}} \tag{17}$$

In Figure 15 is shown a plot of fracture toughness versus yield strength with the transition crack length (mm), based on equation 14, of different values shown as parallel broken lines. All materials cut by a given transition crack line will have the same transition crack length. It would be a great benefit to the casting industry if similar charts are available only for casting alloys.

5. Summary

This chapter first deals with the basics of fracture toughness testing and microstructure development in castings. Several publications on fracture toughness of aluminum alloys, steel and different types of cast iron have been reviewed with intent to note the typical values of fracture toughness and infer that the values are affected by not only the type of alloy but the processing adopted to make the castings. There is need to minimize extrinsic processing defects (for example, bifilms [28,42] in drossing alloys, shrinkage, porosity and others) so that the intrinsic fracture toughness, governed by the bond and dislocation mobility is approached, if highly fracture-resistant castings are to be produced. Of course this problem should be tackled based on cost-benefit relationship. The need for further research in this area is clearly evident.

Acknowledgements

Dr. T.L. Anderson is sincerely thanked for permission to use the numerous illustrations and equations used in this chapter. The authors also acknowledge the permission to use illustrations from other distinguished book authors referenced in this chapter.

Author details

M. Srinivasan[1] and S. Seetharamu[2]

1 Department of Mechanical Engineering, Lamar University, Beaumont, Texas, USA

2 Materials Technology Division, Central Power Research Institute, Bangalore, India

References

[1] Ashby, M. F., Shercliff, H., & Cebon, D. (2007). *Materials: engineering, science, processing and design*, Elsevier, Amsterdam, 167.

[2] Hertzberg, R. W. (1983). *Deformation and Fracture of Engineering Materials*, John Wiley and Sons, New York, 353.

[3] Hertzberg, R. W. (1983). *Deformation and Fracture of Engineering Materials*, , John Wil ey and Sons, New York, 355.

[4] Ashby, M. F., Shercliff, H., & Cebon, D. (2007). *Materials: engineering, science, processing and design*, Elsevier, Amsterdam, 164.

[5] Ashby, M. F., Shercliff, H., & Cebon, D. (2007). *Materials: engineering, science, processing and design*, Elsevier, Amsterdam, 173.

[6] Schey, J. A. (2000). *Introduction to Manufacturing Processes*, McGraw-Hill, Boston, 209.

[7] (2012). http://www.termwiki.com/EN:linear_elastic_fracture_mechanics.

[8] (2012). http://www.public.iastate.edu/~gkstarns/ME417/LEFM.pdf.

[9] Anderson, T. L. (2005). *Fracture Mechanics*, Taylor and Francis, Boca Raton, 300.

[10] Seetharamu, S. (1982). *Ph.D. thesis*, Indian Institute of Science.

[11] E399-97. (1997). *Standard Test Method for Plane Strain Fracture Toughness of Metallic Materials*, American Society for Testing and materials, Philadelphia, PA.

[12] Anderson, T. L. (2005). Fracture Mechanics Taylor and Francis, Boca Raton , 310.

[13] E1820-01,. (2001). *Standard Test Method for Measurement of Fracture of Toughness*, American Society for Testing and materials, Philadelphia, PA.

[14] Anderson, T. L. (2005). *Fracture Mechanics*, Taylor and Francis, Boca Raton, 308.

[15] Anderson, T. L. (2005). *Fracture Mechanics*, Taylor and Francis, Boca Raton, 321.

[16] Anderson, T. L. (2005). *Fracture Mechanics*, Taylor and Francis, Boca Raton, 322.

[17] Smith, W. F. (1981). *Structure and Properties of Engineering Alloys*, McGraw-Hill, Boston, 203.

[18] Smith, W. F. (1981). *Structure and Properties of Engineering Alloys*, McGraw-Hill, Boston, 208.

[19] Hafiz, M. F., & Kobayashi, T. (1996). Fracture toughness of eutectic Al-Si casting alloy with different microstructural features. *Journal of Materials Science*, 31, 6195 -6200.

[20] Kumai, S., Tanaka, T., Zhu, H., & Sato, A. (2004). Tear toughness of permanent mold cast DC A 356 aluminum alloys. *Materials Transactions*, 45(5), 1706-1713.

[21] Tirayakioglu, M. (2008). Fracture toughness potential of cast Al-7%Si-Mg alloys. *Materials Science and Engineering A.*, 497, 512-514.

[22] Speidel, M. O. (1982). *6th European Non-Ferrous Industry Colloquium of the CAEF.*, 65-78.

[23] Staley, J. T. (1976). Properties related to fracture toughness. *ASTM STP.*, 605, 71-96.

[24] Tohgo, K., & Oka, M. (2004). Influence of coarsening treatment on fatigue strength and fracture toughness of Al-Si-Mg alloy castings. *Key Engineering Materials*, 261-263, 1263-1268.

[25] Kwon, Y. N., Lee, K., & Lee, S. (2007). Fracture toughness and fracture mechanisms of cast A356 aluminum alloys, Key. *Engineering Materials*, 345-346, 633-636.

[26] Alexopoulos, N. D., & Tirayakioglu, M. (2009). Relationship between fracture toughness and tensile properties of A357 cast aluminum alloy. *Metallurgical and Materials Transactions*, 40A, 702.

[27] Lee, K., Kwon, Y. N., & Lee, S. (2008). Effects of eutectic silicon particles on tensile properties and fracture toughness of A356 aluminum alloys fabricated by low-pressure casting, casting-forging and squeeze casting processes. *Journal of Alloys and Compounds*, 461, 532-541.

[28] Tirayakioglu, M., & Campbell, J. (2009). Ductility, structural quality and fracture toughness of Al-Cu-Mg-Ag (A201) alloy castings. *Materials Science and Technology*, 25(6), 784-789.

[29] (2012). *FRACTURE TOUGHNESS IN RELATION TO STEEL*, www.sfsa.org/sfsa/pubs/misc/Fracture%20Toughness.pdf,.

[30] Barnhurst, R.J., & Gruzleski, J.E. (1985). Fracture toughness and its development in high purity cast carbon and low alloy steels. *Metallurgical Transactions A.*, 16A, 613-622.

[31] Chen, L., Zhao, Y. X., & Song, G. X. (2011). Random critical fracture toughness values of China railway Grade B cast steel wheel. *Key Engineering Materials*, 480-481, 381-386.

[32] Kim, C. K., Park, J. I., Lee, S., Kim, Y. C., Kim, N. J., & Yang, J. S. (2005). Effects of alloying elements on microstructure, hardness and fracture toughness of centrifugally cast high-speed steel rolls. *Metallurgical and Materials Transactions A.*, 36A, 87-97.

[33] James, L.A., & Mills, W.J. ,(1988). Fatigue-crack propagation and fracture toughness behavior of cast stainless steels. , Engineering Fracture Mechanics , 29(4), 423-434.

[34] Elliott, R. (1983). *Eutectic solidification processing*, Butterworths, London, 194.

[35] Walton, C.F., & Opar, T.J. (1981). *Iron castings handbook*, Iron Castings Society, 263.

[36] Walton, C.F., & Opar, T.J. (1981). *Iron castings handbook*, Iron Castings Society, 357.

[37] Salzbrenner, R. (1987). Fracture toughness behavior of ferritic ductile iron. *Journal of Materials Science*, 22, 2135-2147.

[38] Doong, J., Hwang, J., & Chen, H. (1986). The influence of pearlite fraction on fracture toughness and fatigue crack growth in nodular cast iron. *Journal of Materials Science*, 21, 871-878.

[39] Bradley, W.L., & Srinivasan, M.N. (1990). Fracture and fracture toughness of cast irons. *International Materials Reviews*, 35, 156.

[40] (2012). http://www.ductile.org/didata/section4/4intro.htm#Austempering.

[41] Lee, S., Hsu, C., Chang, C., & Feng, H. (1998). Influence of casting size and graphite nodule refinement on fracture toughness of austempered ductile iron. *Metallurgical and Materials Transactions A.*, 29A, 2511-2521.

[42] Campbell, J. (2003). *Castings*, Elsevier.

Chapter 10

Segregation of P in Sub-Rapid Solidified Steels

Na Li, Shuang Zhang, Jun Qiao, Lulu Zhai, Qian Xu,
Junwei Zhang, Shengli Li, Zhenyu Liu,
Xianghua Liu and Guodong Wang

Additional information is available at the end of the chapter

1. Introduction

Macro-segregations of different kinds and degrees exist not only in traditional continuous-cast thick blanks, but also in continuous-cast thin slabs and even thinner cast strips, even though the solidification speed has increased significantly. Atom of P is center segregated in continuous-casting blanks and slabs, while center negative segregation of P is found in strip cast samples. The macro-segregation of P is also recently found in rapidly-solidified steel droplets. The macro-segregation can not be removed during the following rolling and heat treatment and has negative impacts on product properties; therefore, it is one of the most important research subjects of steel.

There is an increasing need to create high-quality steels from steel scraps due to economic and ecological reasons. Phosphorus is one of the most notorious impurities in steel scraps, which can result in steel embrittlement. On the other hand, P is beneficial in providing a fine solidification structure by decreasing the prior-γ grain size in the cast steels, and can improve the properties such as strength and corrosion resistance as long as it is in solid solution. Therefore, if P remains finely dispersed, it is possible to overcome the poor properties to some extent, even in low grade steels [1].The weathering steels with P addition have been demonstrated to have a better corrosion resistance than carbon steels in various atmospheres [2,3]. Yoshida et al.[4,5] reported the beneficial effects of P addition on prior austenite grain refinement in low carbon steels containing high content of impurity and cooled at the cooling rates from 0.1 to 40 K s^{-1}.

2. Segregation of P in Twin-Roll Casting Strips

Near-net-shape-casting processes with high cooling rates are thought to be appropriate for forming ultra-fine grained steels from steel scraps [6]. Strip casting technique is a process to produce strip coils directly from molten steel with the conventional hot rolling process being omitted, which is a potential substitute to hot strip rolling. The progress in strip-casting technology with low production cost [7] makes it possible to increase the cooling rate and resist the equilibrium segregation of alloying elements during casting.

In the present study, low carbon steel strips with different P addition were produced using the twin roll strip casting process and the effects of P on microstructure were studied.

2.1. Experimental

Low carbon steels with different P contents were prepared by melting in a 10 kg-medium-frequency induction furnace. Steel strips with 240 mm in width and 1.2 mm in thickness were produced using a vertical type twin roll strip caster followed by air cooling.

Figure 1 shows the schematic diagram of the twin roll strip caster.

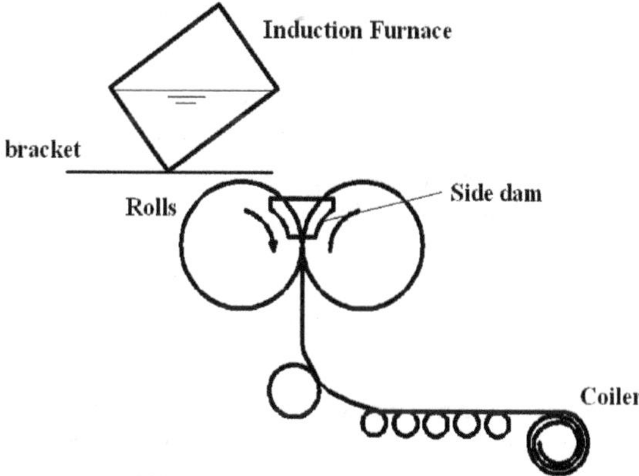

Figure 1. Schematic drawing of the operating lab-scale strip caster.

The experimental steels had a base composition of 0.16C-0.15Si-0.60Mn in mass fraction, and P content varied from 0.008% to 0.70% in weight percent. The chemical compositions of the test steels are listed in Table 1. For comparison, a sample designed as Z01 with the same composition as P01 was cast with the normal mould followed by air cooling. Samples were cut from the as-cast strip along the casting direction after casting, followed by polishing and etching with a reagent of nital at room temperature. The samples were examined using an optical microscope (OM), a scanning electron microscope (SEM) and X-ray diffraction (XRD) analysis.

Heat number	C	Mn	Si	P	Cu	Al	S
P00	0.160	0.623	0.133	0.008	0.003	0.007	0.004
P01Z01	0.158	0.607	0.140	0.100	0.301	0.009	0.006
P03	0.160	0.599	0.157	0.280	0.300	0.004	0.004
P05	0.156	0.630	0.143	0.520	0.312	0.009	0.011
P07	0.162	0.605	0.130	0.680	0.300	0.006	0.007

Table 1. Chemical compositions of test steels wt %.

2.2. Microstructure

XRD results of the as cast strips show that all the samples composed of α-ferrite and pearlite, which corresponds to the peaks of cementite in the XRD spectrum. However, XRD analysis gives less information about the amount of each phase. Figure 2 shows the microstructural observations on the cross section along the rolling direction. More α-ferrite precipitated near the surface of the samples with higher P content.

Quantitative micrographic tests were performed to decide the volume fraction of α-ferrite near the samples surfaces, as shown in Figure 3. By adding 0.68% phosphorus, there is about 90% α-ferrite precipitates in the steel. The morphology of α-ferrite transformed from widmanstaten ferrite to globular ferrite when P content was more than 0.3%. This may relate to the finer grains, which lead to the shorter diffusion distances of carbon and iron atoms.

Figure 2 also shows that the as-cast microstructures by twin-roll casting are finer than those of normal casting. Finer grains were achieved with the increase of the P content. Figure 4 shows the SEM microstructures of the cast strips. It can be seen that pearlite and bainite developed near the surfaces of the samples. The prior-γ grain size, *dr*, was also evaluated using Eq. (1) [8,9]:

$$dr = \left(A_g / n\right)^{0.5} \tag{1}$$

where A_g is the observed area and n is the number of prior-γ grains in this area. Film-like α-grains were observed along the prior-γ grain boundaries in the etched microstructure, and were used as markers for the prior-γ grain boundaries. The measured and calculated results are shown in Figure 5. The grain size decreased remarkably with P content increasing from 0.008% to 0.03%, while the decrease gradient become lower when the P content is higher than 0.3%.

Prior γ grains precipitated mainly near the surface of the strips, and there were less γ grains in the central regions. More γ grains precipitated in the surface region where reaches the austenitizing temperature more quickly and consequently offers more growing time than the center during the strip casting process. Calculation results indicated [10] that the surface region suffers a reheating process when the strip leaves the rolls, causing further growth of the γ grains.

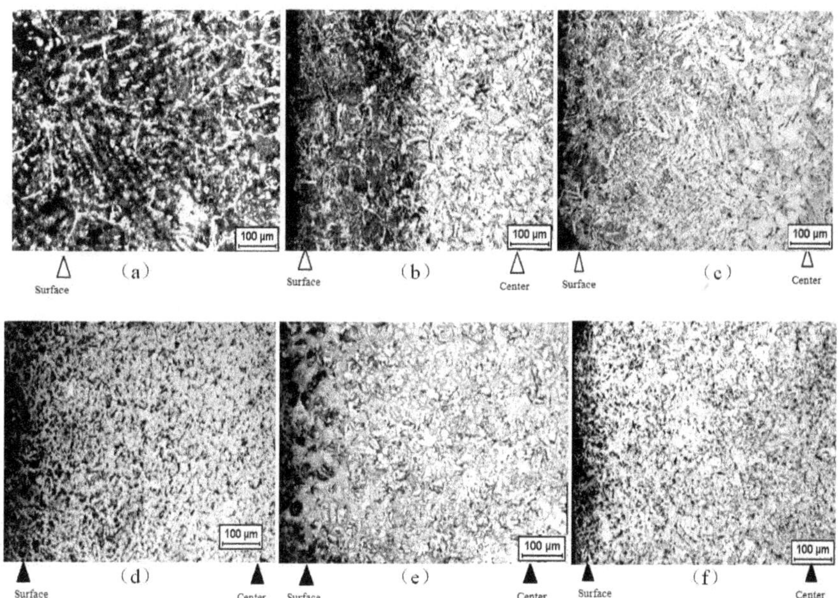

Figure 2. The microstructure of each sample: (a) Z01, (b) P00, (c) P01, (d)for P03, (e) P05 and (f) P07.

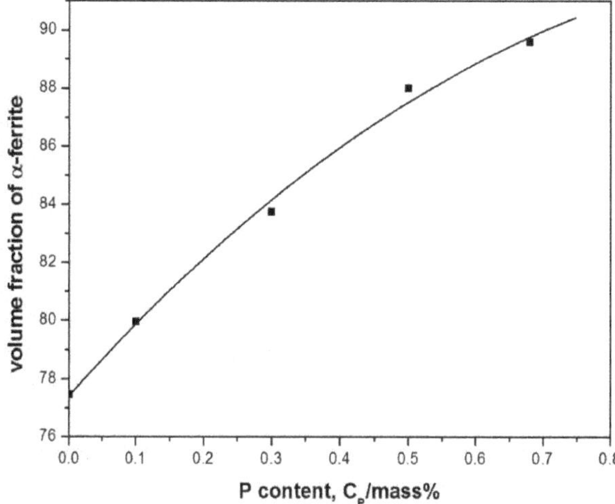

Figure 3. The volume fraction of αferrite phase with different phosphorus addition.

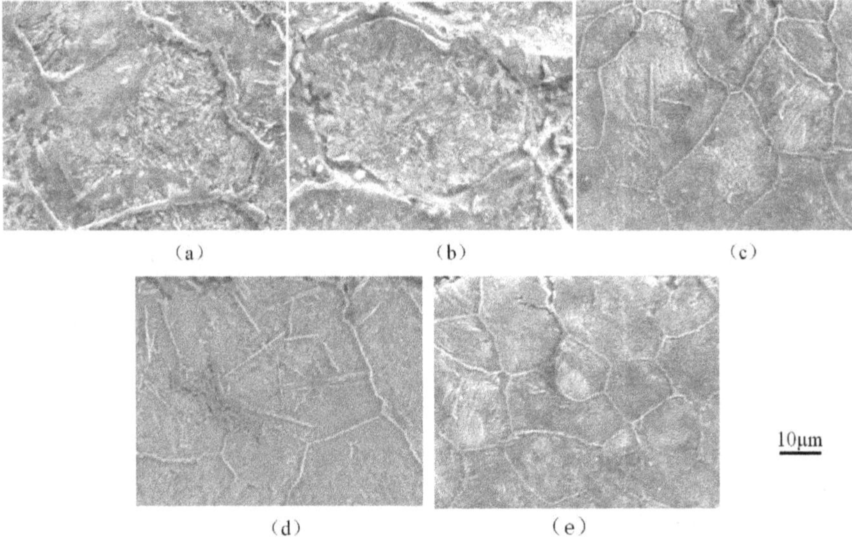

Figure 4. Microstructure near the surface of as-cast strip in the longitudinal section of (a) P00 (b) P01 (c) P03 (d) P05 and (e) P07.

2.3. P Distribution

Figure 6 shows the P distribution examined by EPMA through the whole cross-section of sample P01 and P03, P05 and P07. With P content below 0.1%, no obvious macro-segregation behavior was detected, P distributed uniformly through the thickness section. While in the steels with P content of more than around 0.3%, more P distributed near the surfaces than that at the center. The peak values may correspond to phosphide eutectics or P micro-segregation.

Fig. 7 shows the scanning maps and EDX results of P near the surface of P01 and P03. In P00 and P01, no obvious P segregation region was found by EPMA. While small phosphide eutectics can be found in sample P03, P05 and P07. Most eutectics prefer to form near the strip surfaces at the grain boundaries. This kind of phosphide eutectics must precipitate during solidification process, that is, the liquid-solid phase transformation process. While the small round phosphide showed in Fig. 8 may precipitate during the solid-solid phase transformation, and the P content in these phosphide is a little lower than that in phosphide eutectics. And the P content in the grain was also much higher than the normal value, indicating that there was much P acting as solid solution in the high-P samples.

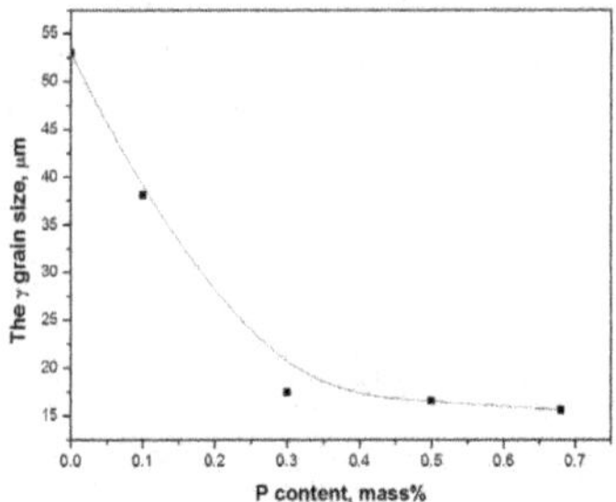

Figure 5. Effect of P content on the γ grain size.

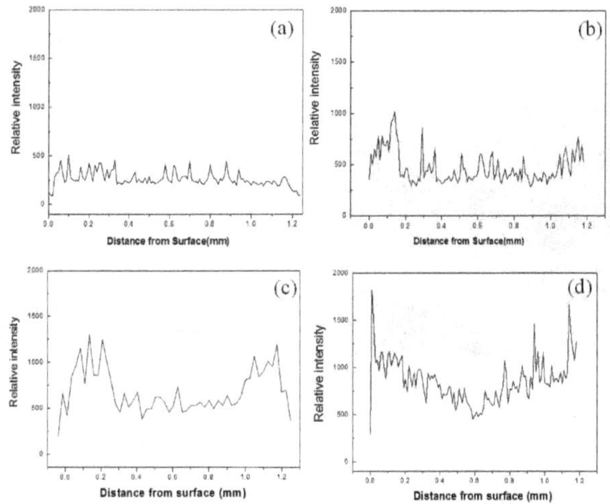

Figure 6. EMPT test results of P of sample P01(a), P03(b), P05(c), P07(d) along the thickness direction.

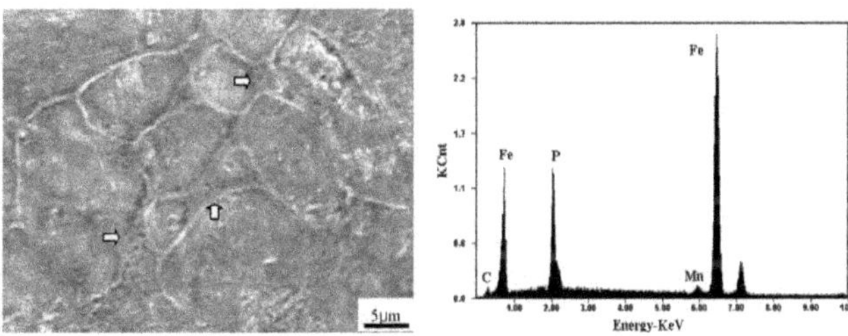

Figure 7. The morphology of phosphide eutectics and corresponding EDX in sample P03.

After the samples were annealed (at 1073K for 1h) and cooled rolled (to 0.65mm), the P distribution along the cross section of the samples also showed the same tendency as the cast samples (Fig. 9), which indicated that P distributions can be maintained until its usage stage of the steel.

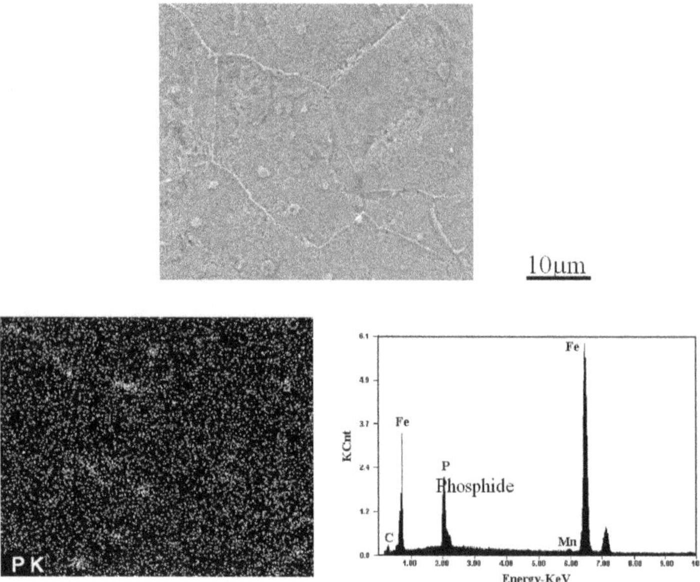

Figure 8. The mapping images of phosphorus micro-segregation in P03 by EDX.

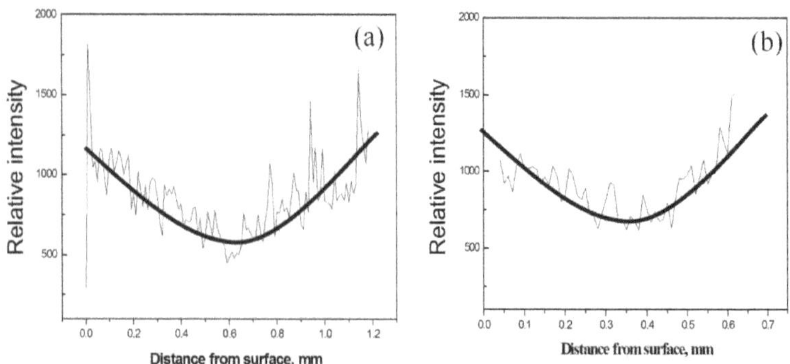

Figure 9. P distribution profiles along the thickness direction of 0.7P before (a) and after (b) annealing and cold rolling.

2.4. Properties

Micro-hardness

Test points distributed uniformly from one surface to another through the cross-section of the samples, and all test showed centro-symmetrical results. Fig. 10 shows the micro-hardness tests results from surface to center of each sample. The micro-hardness is higher near the surface than that at the center for all the samples, which corresponds to the distribution of α and γ phases in as-cast microstructure.

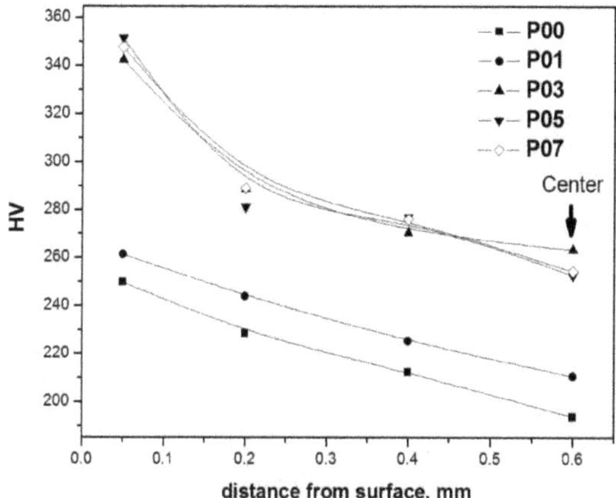

Figure 10. Micro-hardness from surface to center of each sample with different phosphorus addition.

The micro-hardness of the samples keeps increasing with the increase of P content from 0.008% to 0.68%. It is well known that P is a strong solid solution strengthening element, and the statistical micro-hardness test results also prove that most of the P atoms exist in the solid solution state in steels. The micro-hardness of P01 increased uniformly due to the well-distributed P atoms. While for sample P03, the micro-hardness near the surface is ultra high, and the relative gradient of the total curve is higher than those of the two samples with less P addition, which corresponds to the P distribution characteristic in P03 detected by EMPA. Therefore, P is an effective solution strengthening element within the studied content range and under the applied experimental conditions.

Tensile property

Tensile tests were conducted on the annealed and cold rolled samples and the results are shown in Fig. 11. The high-phosphorus steels present higher tensile strength and lower plasticity. Therefore P addition does improve the strength of the cast strips with a sacrifice of plasticity. Samples with a proper amount of about 0.01% P addition offer both high strength and elongation.

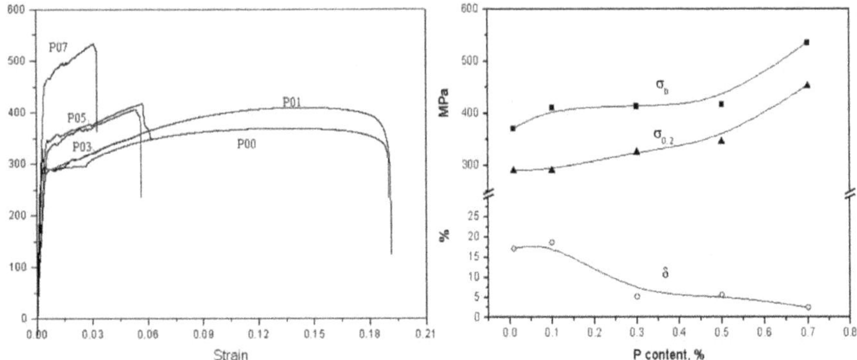

Figure 11. Tensile properties of rolled and annealed steels with different phosphorus addition.

Corrosion property

The corrosion experiments were conducted for 60 cycles with a 0.5%NaCl corrodent to simulate the see atmosphere condition. The weight increase status is shown in Fig. 12.

As shown in Fig 12, the samples with P addition exhibited high corrosion rate during the initial period, followed by rapid leveling off within 2 corrosion cycles, while the sample P00 exhibited a relatively stable corrosion rate throughout the whole corrosion process.

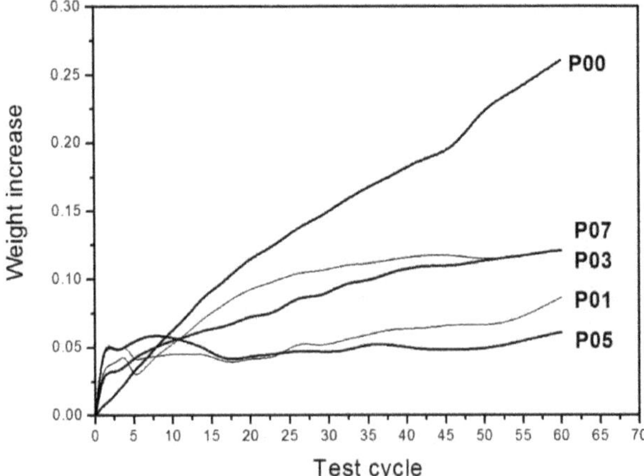

Figure 12. Corrosion properties of strip cast steels with different phosphorus addition.

The samples P01, P03, P05 and P07 show the similar corrosion properties, which indicates that P content of about 0.1% is able to increase the corrosion resistance of steel, and extra P addition can not further improve corrosion resistance.

2.5. Discussion

Phase diagram

The binary alloy containing Fe and a bcc-stabilizing element has a closed γ single phase region, or the γ-loop, in the Fe-rich side of the phase diagram [11]. Alloying elements such as P and Si can narrow the γ single phase region by decreasing the A_{e4} (δ/γ) transformation temperature and increasing the A_{e3} (γ/α) transformation temperature.

Phosphorus is a well known ferrite stabilizing element, which decreases the liquidus (T_L) and solidus (T_S) of steel and phosphorus also has a significant effect on A_{e4} and A_{e3} temperature. The addition of phosphorus in steel may lower T_L, T_S and A_{e4}, and raise A_{e3} by changing the gradients of K_L, K_S, k_{A4}, and k_{A3}, as shown in Table 2 [12], which are evaluated from the phase diagram [7] and the empirical relationship quoted by Leslie[13]. The values in the bracket in Table 2 were evaluated by thermo-dynamic calculation [5]. The A_{e4} and A_{e3} temperature for Fe-0.16C can be calculated to be 1736.5 K and 1128 K respectively, from the Fe-C phase diagram. With the gradients of transformation temperature per unit content listed in Table 2, the phase diagram of the Fe (0.16C-0.15Si-0.60Mn-0.30Cu)-P pseudo-binary system can be calculated, as shown in Fig. 13. This steel system has a typical γ-loop and a negative high gradient of δ/γ transformation temperature, or k_{A4}. The γ-loop closed at the P content of about 0.52 mass fractions. For the higher P content, no single γ phase region is formed, corresponding to the more α-ferrite volume fraction shown in Figure 3.

Element	Mn	Si	Cu	Al	P	S
K_L[K/mass%]	-4.9	-7.6	-4.7	-3.6	**-34.4**	-38
K_S[K/mass%]	-6.5	-20.5	/	-5	**-500**	-700
k_{A4}[K/mass%]	+12	-60 (-52)	/	-81	**-140 (-550)**	-160
k_{A3}[K/mass%]	-30	+44.7 (+77)	-20	+140 (+400)	**+700 (+340)**	/

The K_L and K_S are the gradient of liquidus (T_L) and solidus (T_S) of steels

The k_{A4} and k_{A3} are the gradient for A_{e4} (**δ/γ**) and A_{e3} (**γ/α**) respectively

Table 2. Effect of alloying elements on phase transformation temperature in Fe binary alloys.

P addition also has great effects on Fe-C phase diagram. Figure 14 shows the effect of P on the Fe-C phase diagram calculated by thermodynamic calculation. The mushy zone becomes wider with P addition, and the super-cooling degree increased correspondingly during solidification, which raises the amount of spontaneous nucleation and refines grains.

Figure 14 also shows that the single γ-phase region becomes smaller with an increasing P content until the tie line does not pass through the single γ-phase region at all with the P mass content of about 0.7. Therefore, γ-phase was prevented from both nucleation and growth in the high-P strips, and the finer grains can be observed in the test steels.

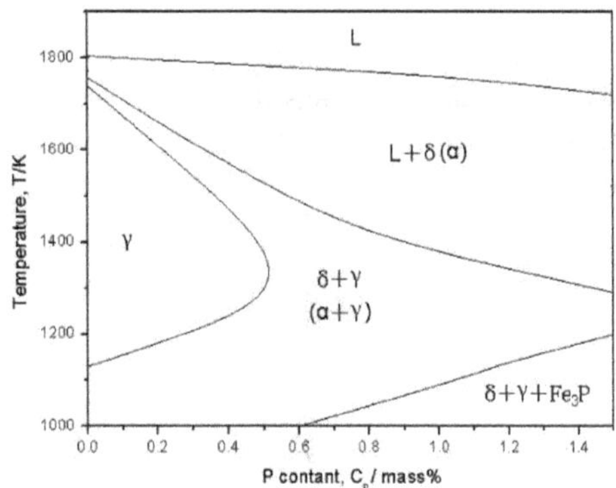

Figure 13. Phase diagram of Fe (0.16C-0.15Si-0.60Mn-0.30Cu-0.4Al-0.04S)-P pseudo-binary system.

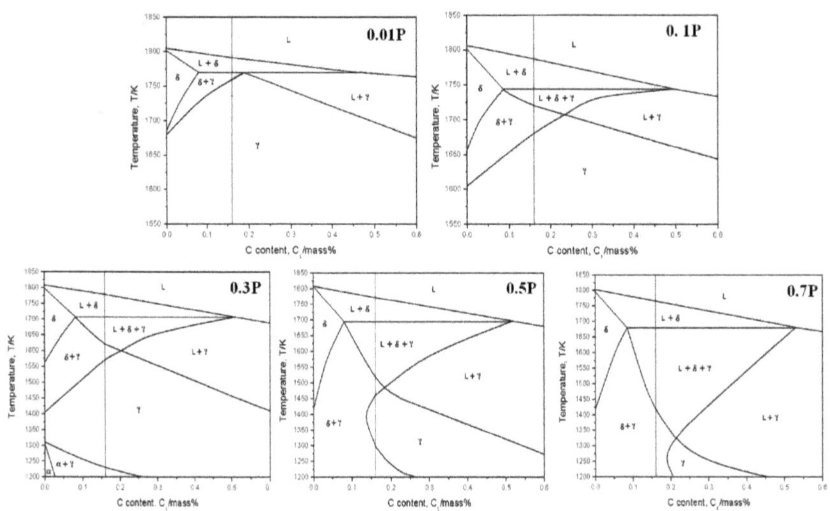

Figure 14. Phase diagrams of Fe (0.15Si-0.60Mn-0.30Cu-0.4Al-0.04S)-C pseudo-binary system with different phosphorus contents.

Solidification processing

EPMA and EDX results clearly show negative segregation at the centre of strip thickness with P content of higher than around 0.3%. Fig. 15 showed the schematic drawing of strip casting process. It is supposed that the height of liquid pool is stable during the strip cast process, and A, B and C are the schematic solidification front with different heights in the liquid pool. A solidified shell is formed near each roll surface during casting. Y.K. Shin et al [14] reported that surface inverse segregation of Mn was observed in as-cast strip resulting from the roll-separating force, and this phenomenon was not observed in the permanent mold cast strip. Y.K. Shin et al regarded [15] that as these shells were forced together and had started to be rolled, the solute enriched liquid was squeezed upwards away from the final solidification position and was extruded into prior inter-dendritic spaces, and therefore the solute content at the strip centre is consequently depleted.

However, the squeezing stress is rather low before the solidified strip is rolled, and the surfaces of the strip contract greatly at the higher cooling speed, so it is difficult for the solute enriched liquid to reach the surfaces of the strips. Meanwhile, columnar grains are significantly damaged when the solidified shell is rolled [15], and there is no obviously columnar crystals observed in the strip cast microstructures as shown in Fig. 2 and Fig. 4, there is actually no solute transmission path. Therefore, the phenomenon of high solute content near the strip surface can not be explained.

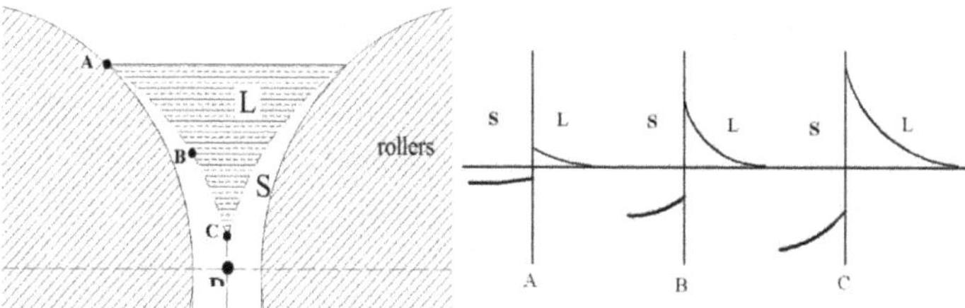

Figure 15. Schematic drawing of strip casting process.

During the strip casting process, the cooling rate near the surface is higher than that at the center of the strip, which may cause the higher partition coefficient content, that is, $K_A > K_B > K_C$ as shown in Fig. 15, and the solid solution content of P is also higher than that at the center when the P content is high enough. The coefficient K may approaches 1 when the cooling rate is high enough, implying that the solute content in solid approaches that in liquid at the solidification front.

P-rich liquid is enriched in the lower temperature regions in the melting pool [16]. Thermodynamic calculation and numerical simulation results [17] show that the lower temperature region is near the meniscus, as shown in Fig. 16. The letters A to I in Fig. 16(b) refer to the regions with corresponding temperature centigrade in the melting pool. During the strip casting process, P is redistributed in the melting pool under the stress of liquid flow and squeezing of the rolls, and more P-rich melt prefer to distributed near the meniscus, which can enter the solidified skull under the rapid solidification condition and lead to the higher solute content at solidification front, and finally form the higher content of P near the surfaces than at the centre of the strip.

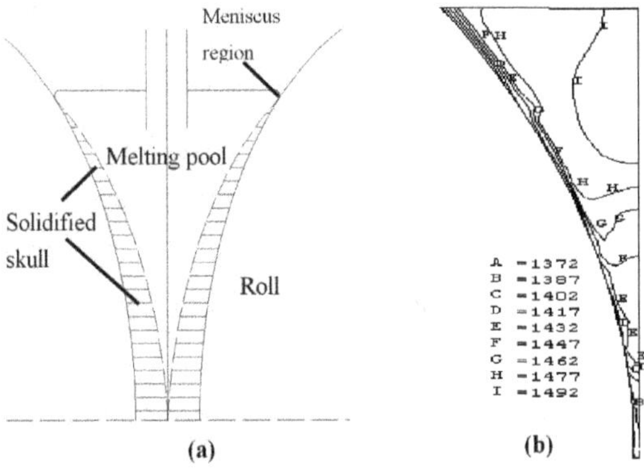

Figure 16. Schematic drawing of melting pool, (a); and typical simulated temperature fields, (b), in the middle cross section along the axial direction [18].

2.6. Effect of C on the Distribution of P

Carbon is one of the important elements in steel, and experiments also show that carbon has great affection on the segregation behavior of phosphorus [18].Carbon steels with different P and C contents were prepared using the twin roll strip caster followed by air cooling with 240 mm in width and 1.2mm, 1.4mm, and 1.8mm, in thickness, respectively.

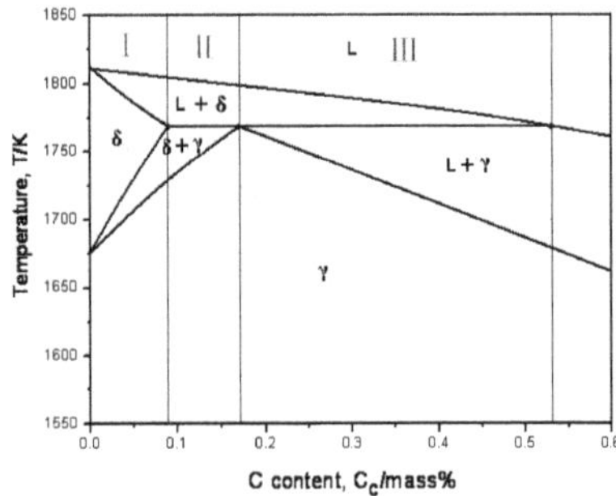

Figure 17. Part of the Fe-C phase diagram.

A portion of the Fe-C phase diagram can be divided into three parts (regions I, II, and III) at points M, N, and O as shown in Fig. 17. Three kinds of carbon steels with different carbon contents were selected in regions I, II, and III, which were designated as group I, II, and III, respectively, then different amounts of P were added by introducing the Fe-P alloy. The chemical compositions of the steels were examined and the results are shown in Table 3.

Fine grains and dendrite structure were observed in high-P steels. The distribution of P was measured by electron probe micro-analyzer (EPMA-810Q). For the steels with different carbon contents, phosphorus distribution in the thickness direction of the strip is obviously different, as shown in Fig. 18.

Group	Heat number	C	Mn	Si	P	Cu	S
I	603	**0.054**	0.121	0.334	**0.320**	0.520	0.005
	602	**0.045**	0.130	0.330	**0.600**	0.312	0.010
II	303	**0.160**	0.599	0.157	**0.280**	0.300	0.004
	307	**0.162**	0.605	0.130	**0.680**	0.300	0.006
III	206	**0.430**	0.610	0.125	**0.280**	0.300	0.011
	212	**0.440**	0.630	0.140	**0.630**	0.306	0.007

Table 3. Chemical composition of test steels wt%.

The samples taken from the head, the middle, and the end of each strip show the same results. P distribution in the thickness direction varies with different C and P contents.

It can be deduced from Fig. 17 that the higher the carbon content, the longer the length of the mushy zone (LMZ), that is, LMZI<LMZII<LMZIII. P negative segregation increases severely with increasing LMZ according to the experimental results as shown in Fig. 18.

For samples 603 and 602 of group I, the mushy zone is relatively short even though it becomes longer with more P addition [6] and it induces no obvious P segregation along the thickness direction as shown in Fig. 18. When the P content is high enough, there is a tendency of center segregation of P. The mushy zone is the longest in region III, there is obvious P negative center segregation in both strips 206 and 212. The P content near the surfaces of the strips is much higher than that near the center where the P content is relatively uniform. The negative segregation becomes severe with more P addition.

From Fig. 18, it can be observed that the segregation mode is different with different group, which may be related to the solidification outgrowth. The solidification outgrowths are δ-Fe, δ and γ phases, and single γ phase corresponding to Fig. 17, respectively. The Fe-P phase diagram shows that up to 2.8wt% P is soluble in $\delta(\alpha)$ ferrite at 1050°C and about 1wt% P is still soluble at room temperature. Therefore, P as a trace element acts as solute in δ (α) ferrite. The solubility of P in γ phase is very low and the maximum solubility at 1200°C is only about 0.28wt%, and even lower P is soluble in γ phase at the temperature below 911°C. The solubility of P in the α+γ region is between that of the above two regions and is much lower than that in ferrite [19].

The α (δ) ferrite percentage in the solidification microstructure of samples 603 and 602 is very high because of the low carbon content. The solubility of P in α (δ) ferrite is higher than the content of P added in the experimental steels, and the diffusion speed of P in α-ferrite is much higher at high temperatures, so the distribution of P is rather uniform in the thickness direction of the cast strip as shown in Fig. 18.

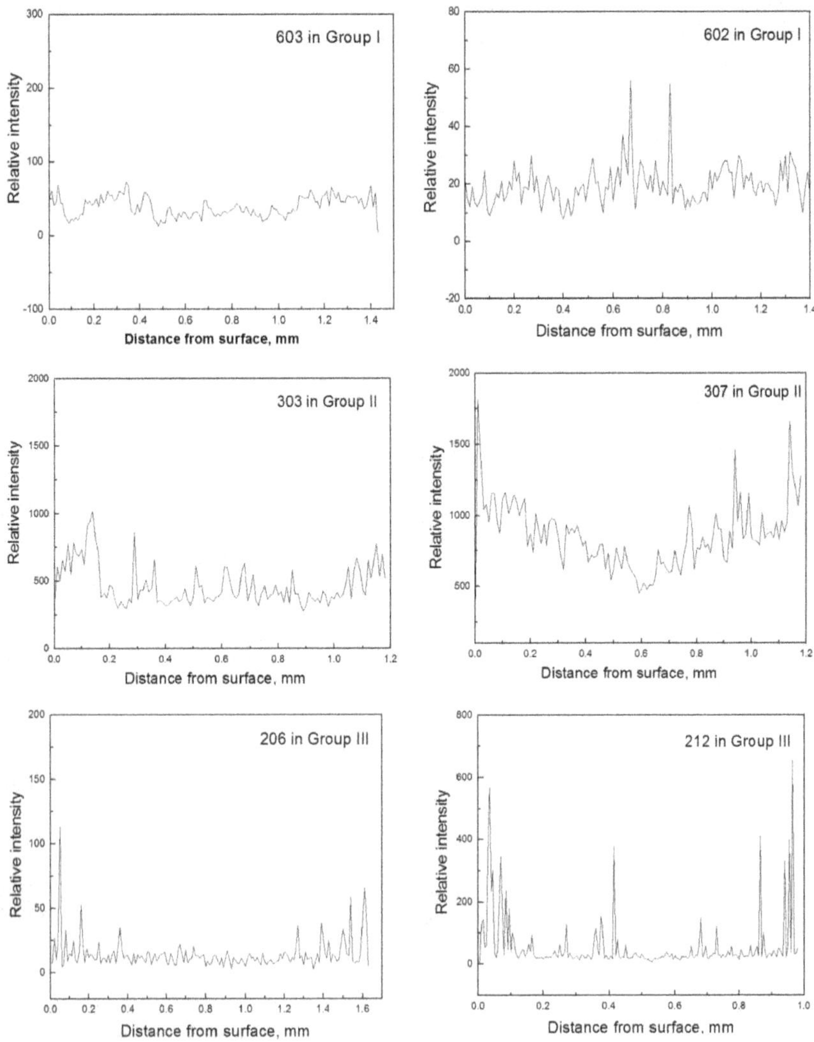

Figure 18. P distribution profiles along the cross section of each sample in different groups.

P in cast strips 206 and 212 is concentrated near the surface as shown in Fig. 18 because the solubility of P in γ austenite phase is rather low and the solidification outgrowth is only γ austenite in region III. In region II, the solidification outgrowth is $\alpha+\gamma$ and the solubility of P in $\alpha+\gamma$ is between that in α and in γ, therefore the P distribution characterization in this region is intermediate as shown in Fig. 18.

3. Segregation of P in steel droplets

Rapid solidification is a significant research subject in the field of material science and condensed physics and plays a major role in material engineering and crystal growth [20], which can remarkably increase the solid solution of alloying elements, produce fine microstructures and reduce or eliminate the segregation of alloying elements. However, the segregation of P and C was observed in rapid solidified strip-casting steel strips.

Container-less processing is an important method to realize the under-cooling and rapid solidification of materials. During container-less processing, the contact between the melt and container wall can be avoided and heterogeneous nucleation can be suppressed to some extent; hence high under-cooling and rapid solidification can be achieved. A drop tube is a special technique for investigating rapid solidification through combining high under-cooling and rapid cooling [21].

3.1. Experimental

Carbon steels with P addition and different B and C contents were prepared in a 2-kg high-frequency vacuum induction furnace, and the compositions are listed in Table 4. Small samples with the size of 2 mm × 2 mm × 2 mm (TM) and 5 mm × 5 mm × 5 mm (FM) were taken from the bulk. All the sides of the small samples were ground and then cleaned with alcohol.

The dry samples were re-melted in a suspension-type vacuum furnace and the melted droplets were then solidified in silicone oil, both the vacuum furnace and the silicone oil were placed in the vacuum drop tube. The schematic diagram of the experimental device is shown in Fig. 19. The drop heights were set to be about 0.2 m and 50m respectively.

Heat No.	C	Mn	Si	S	P	B
0	0.035	0.176	0.024	0.006	**0.005**	
1	0.035	0.179	0.036	0.004	**0.089**	--
2	0.030	0.182	0.033	0.004	**0.525**	
3	0.038	0.181	0.028	0.005	**0.097**	**0.0033**
4	**0.141**	0.178	0.033	0.004	**0.091**	

Table 4. The composition of droplet samples wt%.

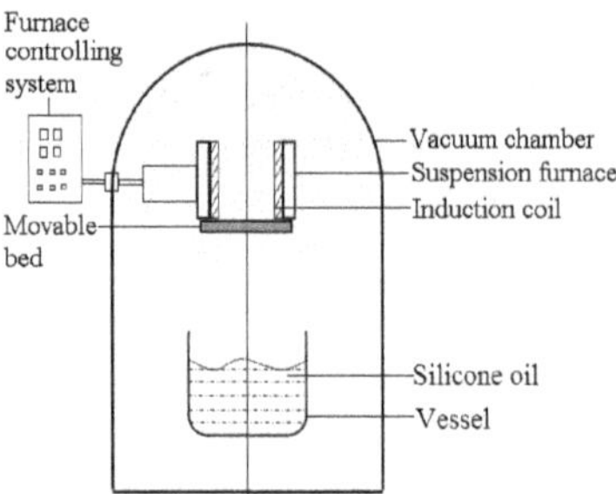

Figure 19. Schematic diagram of the vacuum drop tube.

The microstructures were observed with optical microscope, the alloying elements were detected by electron probe micro-analysis (EPMA-1610), and the micro-hardness were measured and compared with the bulk samples.

3.2. Microstructure

Microstructure at different falling height:

Fig. 20 shows the microstructure near the surface and at center of drop TM sample 1 with the falling height of 0.2m and 50m in drop tube respectively. The microstructure is fine ferrite. The convection heat transfer were ignored because of the vacuum condition in drop tube, and for the volume of the droplets were rather large, it was regarded that the solidification process happened until the liquid drops met the silicone oil even though the falling height is 50m. It can be seen from Fig. 20 that the microstructures of Fig. 20 (c), (d) were a little finer than those of Fig. 20 (a), (b). It indicated that the solidification speed was a little higher for 50m drop samples, for the decreased temperature during falling period may correspond to lower casting temperature, and the lower casting temperature can result in finer microstructure. The finer microstructures were also observed in other 50m drop samples.

By comparing Fig. 20 (a) and Fig. 20 (b), Fig. 20 (c) and Fig. 20 (d) respectively, there is not obvious difference between the microstructure near the surface and that at the center. It indicated that the solidification speed is approximately the same from the surface to the center.

Microstructure with different drop sizes:

Figures 21 and 22 show the microstructures near the surface and at the center of TM sample 1 and FM sample 1 solidified in the drop tube (all at the falling height of 0.2m). The microstructure is mainly fine ferrite. It can be seen from Figs. 21 and 22 that the microstructures in Fig. 21 are a little finer than those of Fig. 22. It may indicate that the solidification speed is a little higher for the TM drop samples. By comparing Fig. 21 (a) with Fig. 21 (b), and Fig. 22 (a) with Fig. 22 (b), it was found that there is no obvious difference between the microstructure near the surface and that at the center. It indicates that the solidification speed is approximately the same from the surface to the center for both TM and FM samples.

For sample 2 with higher carbon content, as shown in Figs. 23 and 24, the microstructures of FM sample 2 are rather finer than those of the FM sample 1 and of the TM samples 1 and 2. This is opposite to the observed results for sample 1; and the microstructures are uniform from the center to the surface. It was analyzed that, recalescence is an important phenomenon that could not be ignored during the rapid solidification process. Recalescence comes from the release of the latent heat of crystallization, which is in direct proportion to the volume of the melt. So the effect of recalescence on the FM samples is considerably greater (more than 15 times) than that on the TM samples. When the carbon content is increased in the steel, the heat transfer capability and the latent heat of crystallization are decreased gradually [22]. So the solidification speed is higher in sample 2 than that in sample 1, which leads to the finer grains. Moreover, the carbon content of sample 2 approaches the eutectoid steel, which may make the microstructure further refined during the cooling process after solidification.

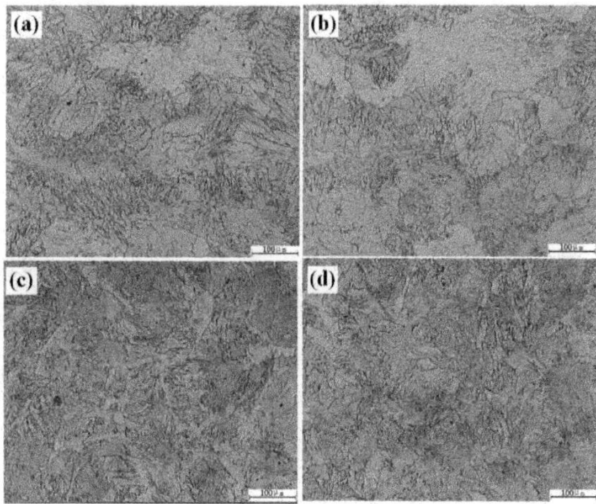

Figure 20. Microstructures of droplets TM sample 1. (a), (b) with the falling height of 0.2m: (a) near the surface, and (b) at the center; (c), (d) with the falling height of 50m: (c) near the surface, and (d) at the center.

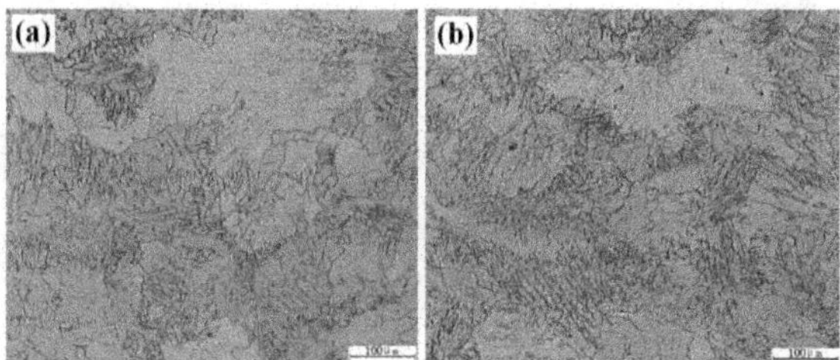

Figure 21. Microstructures of steel droplet TM sample 1: (a) at the center, (b) near the surface.

For sample 2 with higher carbon content, the microstructure of TM samples at the center (Fig. 23a) is quite different from that near the surface (Fig. 23b). There is more pearlite appearing near the surface. This means there is higher carbon content near the surface than at the center, where the microstructure presents more ferrite. The surface temperature of the droplet sample declines to γ phase zone earlier than the center during the solidification process, and the solidification speed near the surface reduces due to the release of latent heat of crystallization [23], even though it is not re-melted.

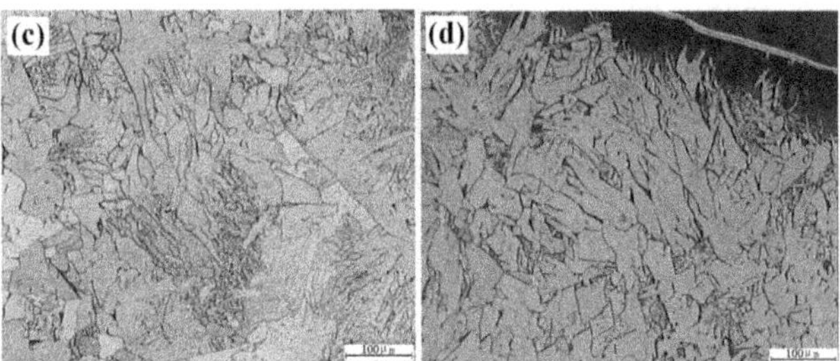

Figure 22. Microstructures of steel droplet FM sample 1: (a) at the center, (b) near the surface.

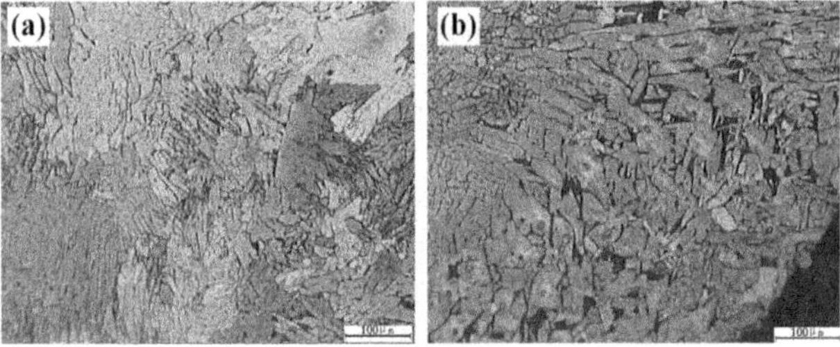

Figure 23. Microstructures of steel droplet TM sample 2: (a) at the center, (b) near the surface.

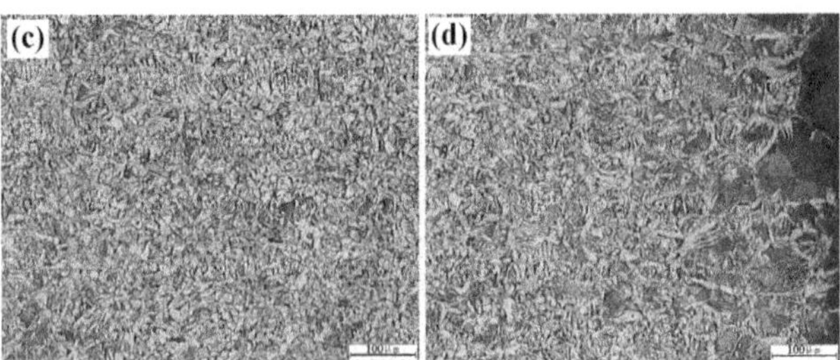

Figure 24. Microstructures of steel droplet FM sample 2: (a) at the center, (b) near the surface.

The surface is kept a relative longer time in γ phase zone than that in the center region. The carbon atoms may diffuse to the γ phase zone as the solid solution of C in the γ phase is much higher than that in the α phase. This may lead to the segregation of carbon near the surface region. For the TM sample 2, the carbon content of most local regions is far away from the eutectoid steel, and there are more ferrite at the center and more pearlite near the surface, so the microstructure is not so fine and uniform.

3.3. P distribution

The distribution of alloying elements P and C were detected along the diameter of each droplet sample. For sample 1, When the P content was less than about 0.1 mass % in low carbon steels, P did not show obvious segregation in rapid solidified droplet samples; the distribution of C also showed nearly uniformity throughout the whole sample, including both the 0.2m and 50m droplet samples.

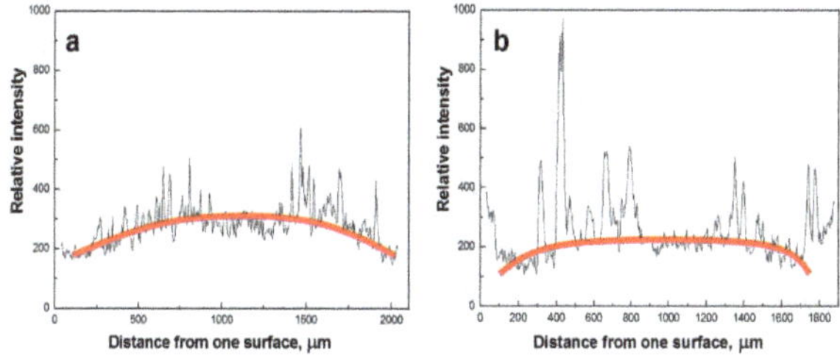

Figure 25. Distribution of P from one surface to another throughout the droplet samples 3: (a) for 0.2m droplet samples, and (b) for 50m droplet samples.

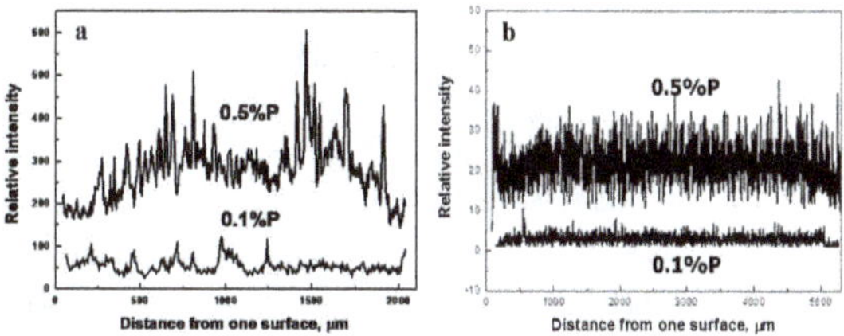

Figure 26. P distribution profiles along the diameter of droplet sample 1# (0.1%P) and 2# (0.5%P) (a, TM samples and b, FM samples).

Fig. 25 shows the distribution of P in sample 3 with higher P content. When the P content rose up to about 0.5 mass% in the samples, both 0.2m droplet samples (Fig. 25a) and 50m droplet samples (Fig. 25b) showed P center segregation and nearly uniformity C distribution. The difference in falling height did not affect the distribution of alloying elements obviously.

The segregation of P was observed in TM samples while not in FM samples, as shown in Fig. 26. This is regarded as relating to recalescence and the diffusion of C and P atoms during the solidification process [24,25].

3.4. Distribution of C with P addition

The distribution of C was detected along the diameter of each droplet sample. For sample 1, the distribution of C is nearly uniform throughout the whole sample, including both TM and FM droplet samples, as shown in Fig. 27.

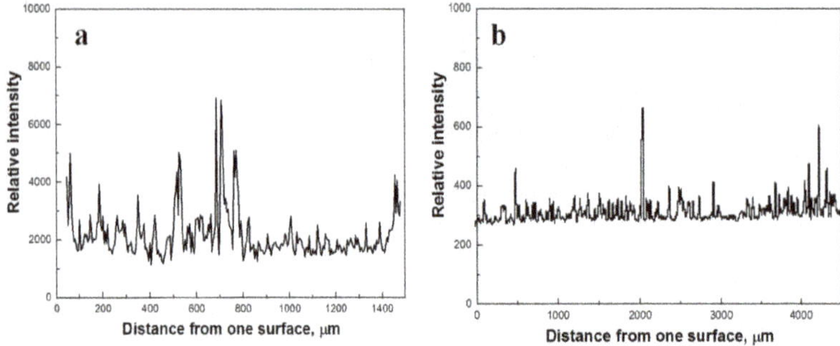

Figure 27. Distribution of C throughout the diameter of droplet TM 1 (a) and FM1 (b) samples.

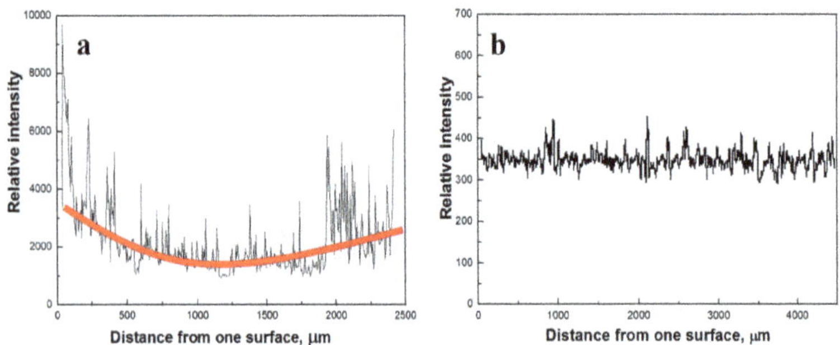

Figure 28. Distribution of C f throughout the diameter of droplet TM 2 (a) and FM 2(b) samples.

For the TM sample 1, the relatively higher C content was observed both near the surface and at the center as shown in Fig. 27 (a). The fluctuation of C distribution indicates that the small volume samples are affected greatly by the cooling and surrounding conditions, including the release of latent heat of crystallization. For the FM sample 1, the distribution of C is more uniform, as shown in Fig. 27 (b), this may relate to the greater amount of latent heat of crystallization.

With higher C content, the segregation of C was observed in TM sample 2, as shown in Fig. 28(a). The C content near the surface is higher than that at the center. The distribution of C is corresponding to the microstructure of the TM sample 2, as shown in Fig. 23, where there are more pearlites near the surface and there are more ferrite near the center.

In the FM sample 2, the uniform C distribution was observed as well, as shown in Fig. 28 (b), which also corresponds to the microstructures, as shown in Fig. 24. This may suggest that the intensity of cooling is equivalent to the latent heat of crystallization.

3.5. Effect of B on the distribution of P

The distribution of C and P were detected along the diameter of droplet sample 1 and sample 3, as shown from Fig. 29 to Fig. 32.

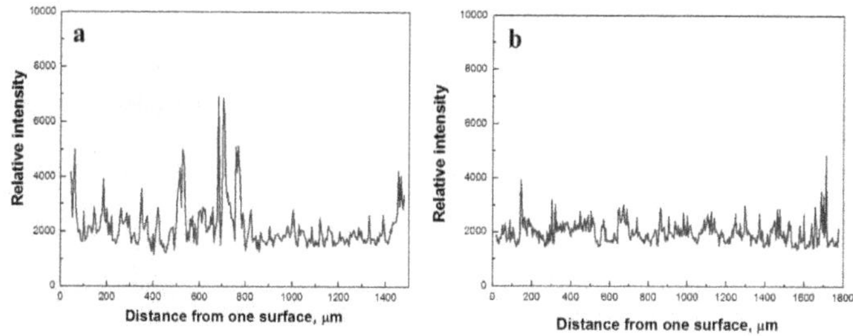

Figure 29. Distribution of C from one surface to another throughout the droplet samples TM 1 (a) and TM3 (b).

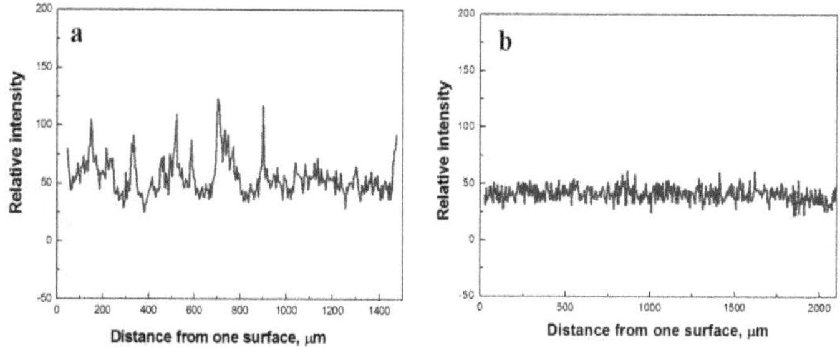

Figure 30. Distribution of P from one surface to another throughout the droplet samples TM 1 (a) and TM 3 (b).

For all the samples, the distributions of both C and P were much uniform in B-bearing samples than those in B free samples. Fig. 29 shows the C distribution profile along the diameter of the droplets TM1 (Fig. 29a) without B addition and TM 3 (Fig. 29b) with B addition, with the same relative intensity as vertical coordinate, it is obviously shows that B atoms can promote the uniformity of C in the rapid solidified steel.

B also can promote the uniformity distribution of other element such as P in rapid solidified steel sample, as shown in Fig. 30. The effect of B on the distribution of C and P in rapid solidified steel samples is also notable as shown in Fig. 31 and Fig. 32 when the volume of the samples increased.

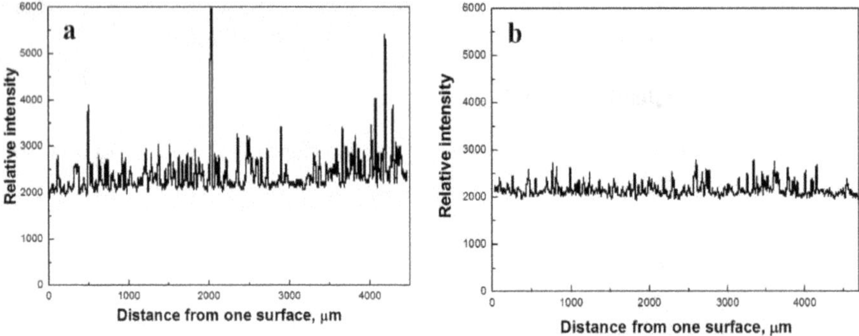

Figure 31. Distribution of C from one surface to another throughout the droplet samples FM 1 (a) and FM 3(b).

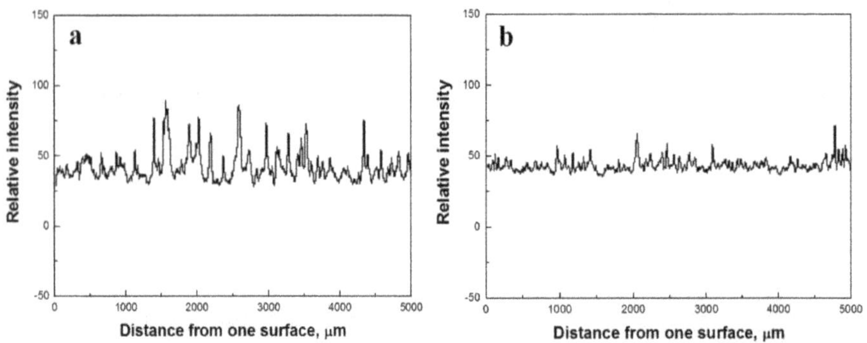

Figure 32. Distribution of P from one surface to another throughout the droplet samples FM 1 (a) and FM3 (b).

It also should be noticed that, the distribution of C and P becomes more uniformly when the volume of the samples increased by comparing that in sample TM 1 (Fig. 29a, Fig. 30a) and FM 1 (Fig. 31a, Fig. 32a) respectively. For TM samples, the volume of the samples is smaller and it is easily affected by the solidification conditions, and the latent heat of solidification increased with the larger volume of the FM samples, which present a more stable system during rapid solidification, and this may induce the more uniform distribution of alloying elements C and P. At the same time, the distribution of C and P in B-bearing FM samples (Fig. 31b, Fig. 32b) shows also more uniform than that in B free TM samples (Fig. 29b, Fig. 30b).

3.6. Micro-hardness

The micro-hardness (HV) of the original cast samples and all the TM and FM samples were measured in this experiment. For each samples, the differences between the maximum and minimum test value was within 15HV, which may be caused by the segregation of elements or the precipitation of compound phases, and the mean value was shown in Table 5. It can be seen from Table 5 that the micro-hardness of rapid solidified samples are much higher than those of original bulk samples. The micro-hardness of rapid solidified FM samples with larger volume show lower values than those of TM samples.

Sample No.	Bulk	TM	FM
0	97.93	133.85	125.96
1	118.22	157.33	150.49
2	203.13	239.01	227.36
3	135.31	184.15	159.86
4	130.72	171.95	167.41

Table 5. The micro-hardness (HV) of each sample.

Under each condition, the samples with higher P content show obviously higher micro-hardness than those with less P content, and when the C content increase, the samples show higher micro-hardness correspondingly.

The micro-hardnesses of B-bearing samples are higher than those of B-free samples, while this tendency weakened when the volume of droplet samples increased. So adding certain account of B in the steel can strengthen the material, even under the rapid solidified conditions.

4. Conclusions

(1) Finer as-cast microstructures have been obtained in twin-roll casting strips and droplet samples than those of normally cast bulk samples. The microstructures of 50m droplet samples are a little finer than those of 0.2m droplet samples. The solidification speed is approximately the same from the surface to the center of each sample.

(2) When the C content is rather low, the microstructures of the TM ($2\times2\times2mm^3$) droplet samples are a little finer than those of the FM ($5\times5\times5mm^3$) samples. When the C content increases to approach that of eutectoid steel, the microstructures of the FM samples are much finer than those of the TM samples. This may be related to the lower latent heat of crystallization when the carbon content increases. Meanwhile, the eutectoid steel and the microstructures may be refined during the cooling process after solidification.

(3) Much finer grains are observed with increasing P content, while the grain size decrease slightly when the P content is higher than 0.3% in mass. There are more α-ferrites precipitated with the increase of P content and the α-ferrites become globular. More α-ferrites present at the centre.

(4) In samples with lower C content (lower than 0.06% in these experiments), the scanning map and EPMA analysis of P show that P does not segregate significantly when the P content is less than 0.1wt%; and when the P content is high enough, P shows center segregation in both twin-roll cast strips and droplet samples. The negative segregation of P at the centre of the cast strips in thickness for high-P and high-C steels is observed. The P segregation mode is affected by the length of the mushy zone and the solidification outgrowth.

(5) The distributions of alloying elements in 0.2m drop tube samples are similar to those in 50m drop tube samples. The distributions of P and C are more uniform in FM samples than those in TM samples.

(6) In TM samples with high C, negative segregation of carbon is observed in high-C droplet samples, more C is distributed near the surface than that at the center, and more pearlites appear near the surface. The segregation of C is thought to relate to the solid solubility of alloying elements in different phases and the diffusibility of C during the solidification and recalescence process.

(7) The distribution of C and P is more uniform in B-bearing droplet samples than that of B-free ones for both TM and FM samples, which indicate that B atoms promote the uniformity of other alloying elements such as C and P.

(8) The micro-hardness of both twin-roll cast strips and droplet samples are significantly higher than those of the bulk solidified samples. Both C and P show strong solution strengthening, especially under rapid solidification conditions. With an increase in P content, the micro-hardness, strength and corrosion-resistance are improved with a sacrifice of plasticity. The micro-hardness of B-bearing samples is higher than those of B-free samples, while this tendency weakens when the volume of droplet samples increased.

(9) During the strip casting process, P is redistributed in the melting pool under the stress of liquid flow and squeezing of the rolls, and more P-rich melt prefer to distributed near the meniscus, which can enter the solidified skull under the rapid solidification condition and lead to the higher solute content at solidification front, and finally form the higher content of P near the surfaces than at the centre of the strip.

Acknowledgements

This work is financially supported by the Major State Basic Research Development Program of China (No.2004CB619108), the National Natural Science Foundation of China (Project No. 51074210), and the open fund of Key Laboratory of Materials Forming and Microstructure Properties Control, Liaoning Province (No. USTLKL2012-01).

Author details

Na Li[1*], Shuang Zhang[1], Jun Qiao[1], Lulu Zhai[1], Qian Xu[1], Junwei Zhang[1], Shengli Li[1], Zhenyu Liu[2], Xianghua Liu[2] and Guodong Wang[2]

*Address all correspondence to: huatsing2006@yahoo.com.cn

1 School of materials and metallurgy, University of Science and Technology Liaoning, Anshan, China

2 The State Key Laboratory of Rolling and Automation, Northeast University, Shenyang, China

References

[1] Hirata, K., Umezawa, O., & Nagai, K. (2002). Microstructure of cast strip in 0.1mass% C steels containing phosphorus [J]. *Materials Transaction*, 43(3), 305-310.

[2] Misawa, T., Kyuno, T., Suetaka, W., & Shimodaria, S. (1971). The mechanism of atmospheric rusting and the effect of Cu and P on the rust formation of low alloy steels [J]. *Corrosion Science*, 11(1), 35-48.

[3] Chen, Y. Y., Tzeng, H. J., Wei, L. I., & Shih, H. C. (2005). Mechanical properties and corrosion resistance of low-alloy steels in atmospheric conditions containing chloride [J]. *Materials Science and Engineering A*, 398(1-2), 47-59.

[4] Yoshida, N., Umezawa, O., & Nagai, K. (2004). Analysis on refinement of columnar γ grain by phosphorus in continuously cast 0.1 mass% carbon steel [J]. *ISIJ Int.*, 44(3), 547-555.

[5] Yoshida, N., Umezawa, O., & Nagai, K. (2003). Influence of phosphorus on solidification structure in continuously cast 0.1% carbon steel [J]. *ISIJ Int.*, 43(3), 348-357.

[6] Li, N., Liu, Z. Y., Lin, Z. S., Qiu, Y. Q., Liu, X. H., & Wang, G. D. (2006). Solidification structure of low carbon steels strips with different phosphorus contents produced by strip casting [J]. *Journal of Materials Science and Technology*, 22(6), 755-758.

[7] Emoto, K., Nozaki, T., & Yanazawa, T. (1986). Restructuring Steel Plant for Nineties [M]. *The Institute of Metals*, London, 151-160.

[8] Maruyama, T., Matsuura, K., Kudoh, M., & Itoh, Y. (1999). Peritectic Transformation and Austenite Grain Formation for Hyper-peritectic Carbon Steel [J]. *Tetsu-to-Hagané*, 85(8), 585-591.

[9] Maruyama, T., Kudoh, M., & Itoh, Y. (2000). Effects of Carbon and Ferrite-stabilizing Elements on Austenite Grain Formation for Hypo-peritectic Carbon Steel [J]. *Tetsu-to-Hagané*, 86(2), 86-91.

[10] Tavares, R. P. (1997). Vertical twin-roll caster:metal-mould heat transfer, solidification and product characterization[D], Ph.D. Thesis, McGill University.

[11] Kubaschewski, O. (1982). IRON-Binary Phase Diagrams [M], Springer-Verlag, Berlin/Heidelberg, and Verlag Staheisen mbH, Düsseldorf.

[12] Tekko-Vinran, (1981). Handbook of Iron and Steel [M], 3rd ED., (ISIJ, Maruzen, Tokyo), 93.

[13] Leslie, W. C. (1981). The Physical Metallurgy of Steels [M], McGraw-Hill, London.

[14] Shin, Y. K., Kang, T., Reynolds, T., & Wright, L. (1995). Development of twin-roll strip caster for sheet steels [J]. *Ironmaking and Steelmaking*, 22(1), 35-44.

[15] Zhou, G. P., Liu, Z. Y., Yu, S. C., Chen, J., Qiu, Y. Q., & Wang, G. D. (2011). Formation of Phosphorous Surface Inverse Segregation in Twin-Roll Cast Strips of Low-Carbon Steels [J]. *Journal of Iron and Steel Research International*, 18(2), 18-23.

[16] Zhu, H. Q., Tang, Y. J., Guo, S. R., Zhang, Z. Y., Zhu, Y. X., Hu, Z. Q., & Shi, C. X. (1994). Effects of P, Zr and B on microstructure and segregation of directionally solidified IN738 superalloy [J]. *Acta Metallurgica Sinica (in Chinese)*, 30(7), A312-320.

[17] Miao, Y. C. (2001). Numerical Simulation of Solidification of Twin-roll Strip Casting Process [D], Ph.D. Thesis, Northeastern University, Shenyang, China.

[18] Li, N., Liu, Z. Y., Zhou, G. P., Liu, X. H., & Wang, G. D. (2010). Effect of phosphorus on the microstructure and mechanical properties of strip cast carbon steel [J]. *International Journal of Minerals Metallurgy and Materials*, 17(4), 417-422.

[19] Ortrud, K. (1982). Iron-Binary Phase Diagrams, Springer-verlag, Berlin.

[20] Flemings, M. C. (1974). Solidification Processing [M], McGraw-Hill, New York.

[21] Wang, H. Y., Liu, R. P., et al. (2005). Morphologies of Fe-66.7at.%Si alloy solidified in a drop tube [J]. *Science in China Series G: Physics Mechanics & Astronomy*, 48(6), 658-666.

[22] Yiu, Wingchan. (1989). Finite element simulation of heat f low in continuous casting [J]. *Advances in Engineering Software*, 11(3), 128-135.

[23] Zeoli, N., Gu, S., & Kamnis, S. (2008). Numerical modelling of metal droplet cooling and solidification [J]. *International Journal of Heat and Mass Transfer*, 51, 4121-4131.

[24] Li, N., Zhang, J. W., & Zhai, L. L. (2011). Segregation of C and P in steel droplets solidified in drop tube [J]. *Journal of Iron and Steel Research International*, 18(1-1), 282-286.

[25] Li, N., Sha, M. H., Xu, Q., Zhang, S., & Li, S. L. (2012). Rapid Solidification of Steel Droplets with Different Carbon Content in Drop Tube [J]. *China Foundry*, 9(1), 20-23.

Chapter 11

Research on Simulation and Casting of Mechanical Parts Made of Wear-and-Tear-Resistant Steels

Ioan Ruja, Constantin Marta,
Doina Frunzăverde and Monica Roşu

Additional information is available at the end of the chapter

1. Introduction

The conversion of hydraulic energy into electrical energy is not polluting, supposes relatively small upkeep expenses, and there are no problems related to fuel; it constitutes thus a long-term solution. Hydroelectric power plants have the lowest exploitation costs and the longest life duration compared to other types of electric power plants. These hydroelectric power plants use several types of turbines, and the Kaplan turbine is the most frequently used. The Kaplan turbine is a hydraulic turbine with axial rotation, with an adjustable-blades rotor, used in small-fall hydroelectric power plants, i.e. H= 10 - 50 metres and Q= (700-800) m^3/s. At present, the equipment of hydroelectric power plants which meets the above parameter requirements is mainly endowed with Kaplan turbines. The researches of specialists in the construction of these turbines are focused on the turbines' configuration and physical dimensions, on reaching high hydro-energetic parameters, as well as on the increase of these turbines' life duration. The preoccupations of the research team who conducted the present study fall within this context, for the purpose of getting optimum technical results by improving the casting conditions of the main parts in the structure of the Kaplan turbine, more precisely the rotor block and the blades of the turbine rotor.

Some of the steel makes used for the casting of these parts are 1.4314 GX4CrNi13-4 or 1.4414GX3CrNiMo13-4 [1], [2]. The chemical compositions of the two types of steel are shown in tables 1 and 2 respectively:

C%	Si%	Mn%	P%	S%	Cr%	Mo%	Ni%
Max	Max	Max	Max	Max	12.00	Max	3.50
0.06	1.00	1.00	0.035	0.025	13.50	0.70	5.00

Table 1. Chemical composition of GX4CrNi13-4 (1.4313) steel.

	C%	Si%	Mn%	P%	S%	Cr%	Ni%	Mo
Min.	0	0	0	0	0	12.00	3.50	0.30
Max.	0.05	0.70	1.50	0.040	0.015	14.00	4.50	0.70

Table 2. Chemical composition of GX3CrNiMo 13-4 (1.4314) steel.

These two steel makes are used because they have a good resistance to corrosion, high tenacity and strength, as well as a high weldability. In order to optimise the casting technology we used the AnyCasting software allowing the researcher to elaborate several casting variants and to obtain a part closer to the performances expected in exploitation. These variants are obtained mainly by setting several casting parameters such as casting speed, filling rate, casting temperature, temperature of the casting mould, the meshing of the part and of the liquid pressure in the casting mould.

The objectives aimed at are: enhancing the efficiency of the mould and position of feeders, locating the casting defects, i.e. closed and open cavities, pores, fissures, air bubbles, and their directing to the areas decided by the designer by orienting the thermal flux; reducing the number of casting defects and decreasing the quantity of liquid metal. From the practical experience we may list over 70 casting defects specific to this type of part (different defects or defects of the same type, but of different sizes). These defects were found throughout the entire volume of the part. The distribution of defects is the following: in the upper area, in the central zone, in the zone of the block windows and in the lower section. The defects are open macro-cavities in the upper section of the block, technologic defects in the form of closed secondary cavities in the medium and inferior section of the part. The defects in the block's superior section are determined by the casting conditions and the inappropriate design of the mould and feeders. The closed secondary cavities in the middle and inferior section of the part are determined by the geometry of the part, as well as by the presence of thermal knots. The presence of the closed defects in the central and lower area does not allow a good equilibration of the block. Another type of defects are the micro- cavities, air bubbles and porosities found throughout the entire mass of the part, determined by the human factor. These defects occur because of the inadequate elaboration of steel, faulty deoxidisation, insufficiently dried casting mould and paint at the beginning of casting. These aspects cannot be highlighted yet by the simulation software.

2. Presentation of the AnyCasting software

The experimental research was generally utilised in order to find the effect that the patterns, i.e. the filling and solidification types, have on the qualities of the product. Nevertheless, due to the specific features of the casting process, there are limits as regards its visualisation. Nowadays, the research for enhancing the casting process is performed both through experimental casting projects and by numerical simulation. Simulation may be very useful for the improvement of the initial project and for the validation of the modifications in the project, and it may provide information on the filling and solidification patterns even in the areas that are not visible, as they are covered by melt or mould, grace to the options of the software allowing the sectioning of the model in certain areas [3].

Based on this information, we may identify the problems occurring in the design of the manufacture process, making it possible to reduce costs and to increase productiveness.

The software is made of the following modules:

2.1. AnyDBASE

AnyDBASE is a database programme used for managing the properties of ferrous and non-ferrous alloys, of the properties of the moulds, paints, refractory materials used etc. that have to be selected in AnyPRE. AnyDBASE is conceived so that it may allow the accessing of two types of databases: the general database and the user database, the latter being utilised for modifying or adding materials. The general database contains the properties of ferrous, non-ferrous, non-metallic and functional alloys according to different international standards (KS, JIS, ASTM, etc.). In the user database we may create and manage a personally customised database.

2.2. AnyPRE

AnyPRE performs processing, by reading the CAD data, allows us to import the.stl extension files containing the part geometry and then turns them into.gsc files. The.gsc files store information related to the initial simulation data and allow the setting of simulation information in order to run it in anySOLVER. AnyPRE offers functions of network generation, generating networks with tens of millions of elements. We may also choose a variable meshing in certain areas of the part by using the variable mesh function of anyPRE. Moreover, it allows the selection of the processes and groups of working materials and the setting of heat, gating conditions etc. In conclusion, AnyPre creates three files, which determine the results of the simulation:

- project with the *gsc extension – it contains the data of the simulation used in anyPRE;

- project with the *msh extension – it contains information on the mesh networks;

- projects with the *prp extension – it comprises the simulation conditions.

2.3. AnySOLVER

AnySOLVER reads the input files from anyPRE and simulates the flowing and the temperature fields of the designed process. All aforementioned processes are calculated by a single solver.

2.4. AnyPOST

AnyPOST reads the results created in anySOLVER and shows the results graphically. anyPOST offers the opportunity to visualise the basic results such as: filling time, solidification time, outline (temperature, pressure, speed), the speed vectors in two or three dimensions, and creates plots based on the sensor results. We may also verify the various solidification defects in two or three dimensions using the combined results function.

3. Applications in AnyCasting

We shall proceed by presenting applications of simulations on the assembly of the cast rotor block part and the assembly of the turbine blade.

3.1. Applications in AnyCasting [4] on the set of the rotor block part

After the realisation of the solid in Solid Works, it is imported into AnyPre, where we introduce all the data necessary for simulation, the material being selected from AnyDBase [5] together with the physical-chemical properties. Four simulations were performed:

1. Simulation of the casting of the assembly of the rotor block part on the basis of a classic technology with single inferior feeding, without coolers and with feeders only in the upper section;

2. Simulation of the casting of the set of rotor block part with a technology using external coolers in the lower section of the block and feeding from a three-level casting network;

3. Simulation of the casting of the assembly of the rotor block part based on a technology with external coolers located in the lower section and under the area of feeders, and with exothermal powders;

4. Simulation of the assembly of the rotor block part with a technology using external coolers in the inferior section and under the feeders' zone, exothermal powders and feeder-covering powders, as well as blind feeders.

We chose the four variants in order to observe the influence of coolers and of blind feeders, as well as the evolution of casting defects in each variant, the influences that will define the type and position of defects.

3.1.1. Simulating the casting of the set of rotor block part with a classic technology with single feeding at the lower part, without coolers and with blind feeders situated only in the upper section

Figure 1 presents the geometry of the block part corresponding to variant 1. The mould has a single feeding network by indirect casting from the lower part, technologic feeders being placed normally at the top. The mould is made of sand without coolers and no exothermal or insulating powders are used. The casting temperature is 1580°C, and the casting time is 240 seconds.

Figure 1. Geometry of the block assembly.

From the *set probabilistic defect parameter* menu we select the *defect parameter* criterion which presents the defect in the most accurate manner, and its position in the part volume. In the postprocessor we will see the results of simulation by types of defects, i.e. the macro-cavities located in the upper part of the casting assembly, that is in the area of feeders and in the upper section of the part, followed by the defects from the central area and those situated in the block inferior section. Figure 2 shows the position of the defects through the analysis of the *temperature variation* criterion during the filling and solidification of the cast set. The macro-defect under the form of open cavities is located in the superior part of the set and exceeds the joining area of the feeders with the body of the part [6]. Defects of this type are located on the surface of the part, may be seen with the naked eye and can be repaired by welding.

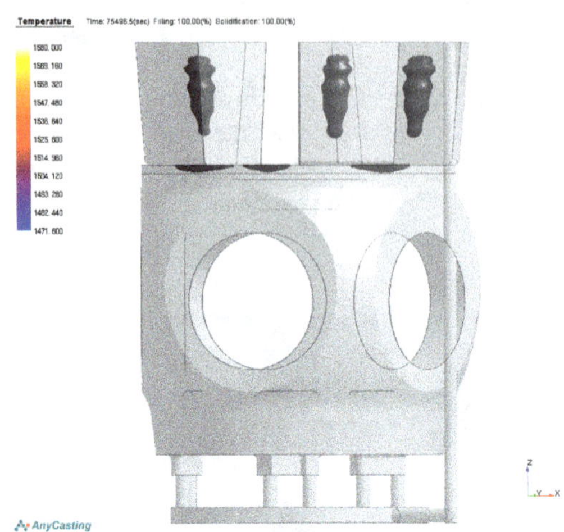

Figure 2. Position of macro-cavities in the block.

The same defect occurs also in Figure 3, the transparent visualising mode allows visualising the defects both from the viewpoint of position and from the perspective of mould and volume. We may also remark in Figure 3 that both defects are identical as regards position and shape with those in Figure 2. The analysis of this type of defects and the measures to be taken for their elimination were presented in detail in [6].

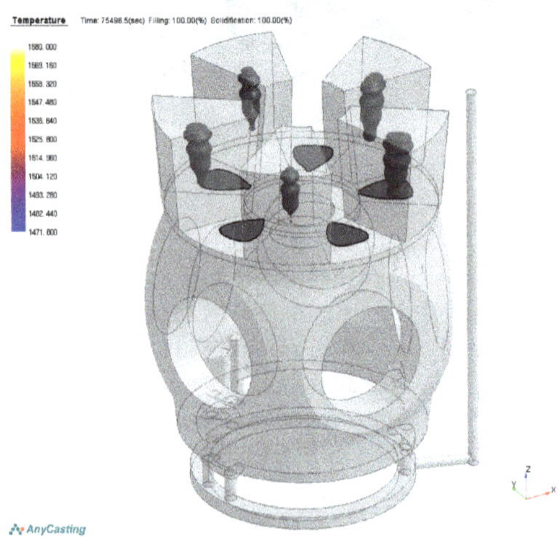

Figure 3. Position of macro-cavities in the block - 3D.

The defects taking the shape of closed secondary cavities in the middle area and in the lower section of the part are defects that cannot be observed by free visualisation and require non-destructive investigations, sometimes even by cutting the part for highlighting them. The presence of this type of defects makes it difficult to put the rotor block set into operation. These kinds of defects cannot be remedied and they generally trigger the discarding of the entire cast part. In Figure 4, section through the block, based on the *local solidification time* criterion, we see the variation of the solidification time in each zone, and the way in which each area of the casting assembly solidifies. We remark that the last zone to solidify is the area under the feeders and the lower zone of the part; between the block window and the feeding network. The alloy solidifies the fastest in the peripheral areas, from bottom to top, in the sense opposed to the casting direction, then the feeders' area. In the upper section of the part, under the feeder, thermal knots appear during solidification, determined by the modification of the part geometry, and the presence of these thermal knots indicates the occurrence of casting defects. From the analysis of solidification in the same Figure 4 we remark in section that in the central area (see the right zone) the prediction of the occurrence of other defects, presented in Figures 6 and 7.

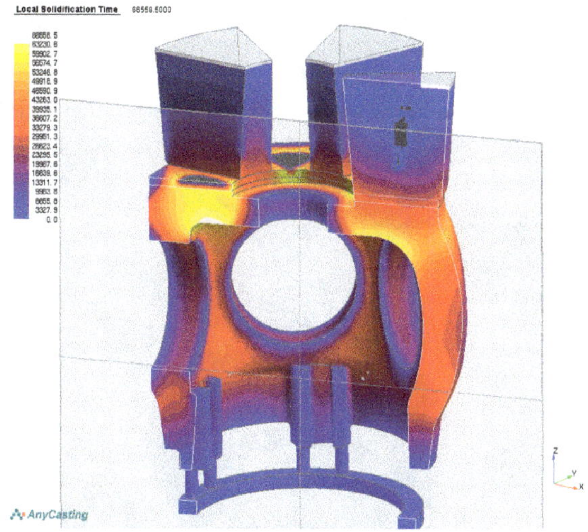

Figure 4. Local solidification time - section.

In Figure 5, by analysing the *defect parameter* probabilistic criterion, we observe the defects in the form of micro porosities and secondary or closed cavities in the part. The position of defects under the window in Figure 5 confirms the prediction of defects anticipated in Figures 3 and 4. From the analysis of the colour code bar, situated on the left side, in Figure 5, these defects' probability of occurrence in the area under the feeders has the maximum value 1,000, which means that the presence of these defects is inevitable.

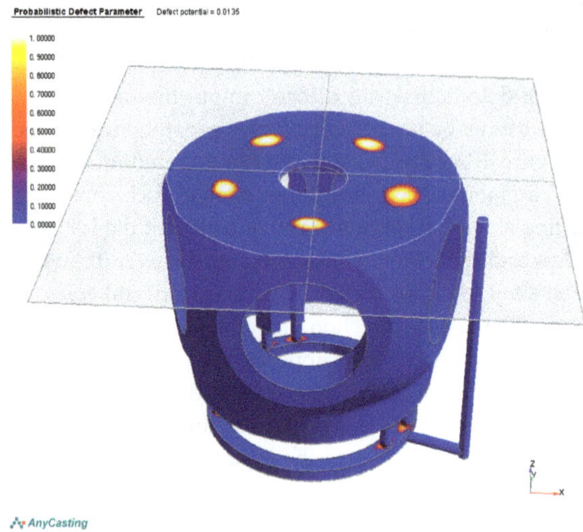

Figure 5. Micro porosities and secondary cavities.

These defects are located under the feeders and reach the medium section of the part, Figure 6, in the windows area and in the lower section, Figure 7. The position and shape of defects is a confirmation of the prediction provided by the criterion presented in Figure 4.

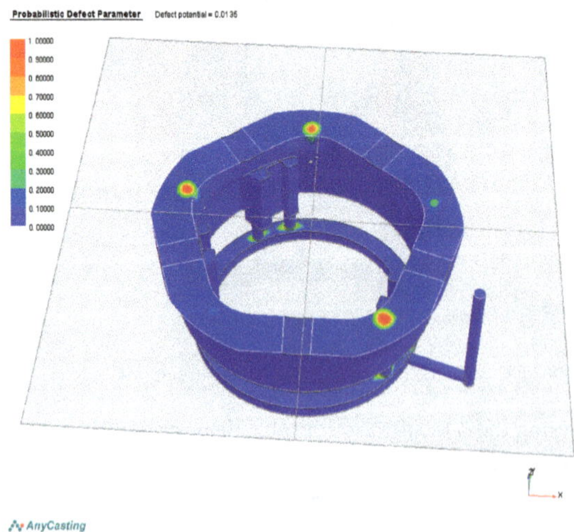

Figure 6. Micro porosities and secondary cavities.

Figure 6 confirms the presence of casting defects also in the lower section of the part, especially in the zone of the block windows. In this case too the probability of the apparition of defects is approximated based on the colour code displayed on the left side of the figure and it is estimated at 1,000 again. Figure 7 shows, by means of the criterion *variation of the temperature gradient during solidification*, the possibility of casting defect occurrence in the body of the part, i.e. in the superior and middle section, in the windows area and in the inferior side of the assembly.

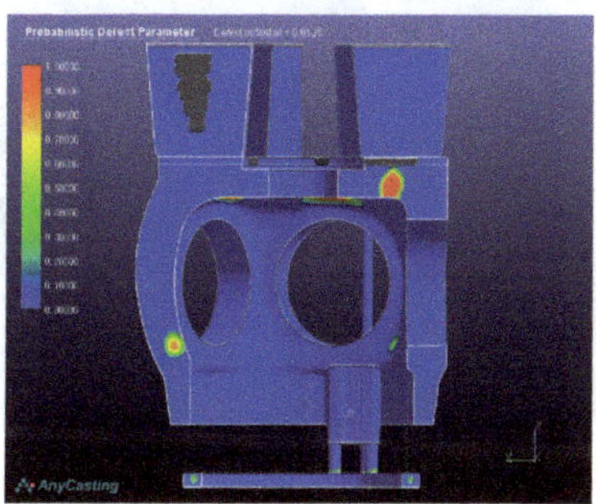

Figure 7. Temperature gradient.

In Figure 8, the *iso solid fraction curvature* criterion defines the position of defects in the middle zone, the windows zone and the lower section. In order to understand the criterion, we consider that the blue surfaces solidify the fastest; whereas the red surfaces are the slowest to solidify.

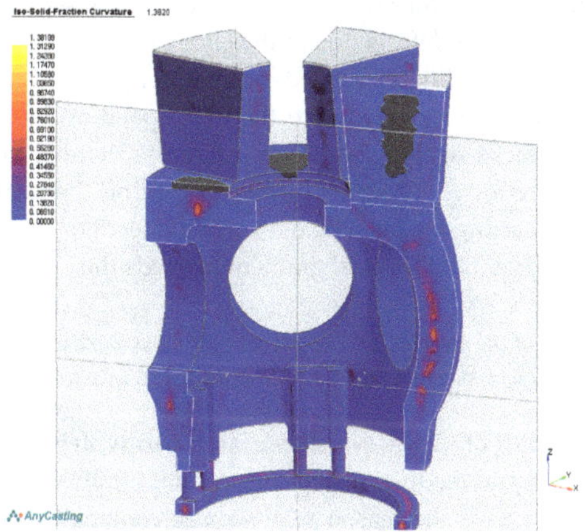

Figure 8. The *iso solid fraction curvature* criterion.

In Figure 9, highlighting micro porosities and secondary cavities, the defects in the window area are evident.

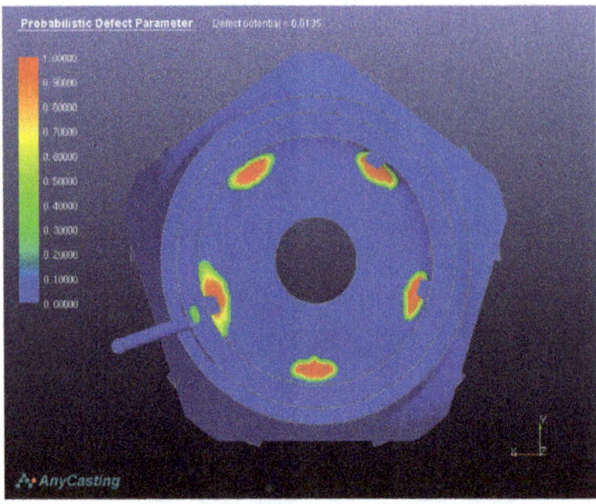

Figure 9. Micro porosities and secondary cavities.

3.1.1.1. Partial conclusions

By simulation we highlighted the defects formed in the cast assembly, i.e. the open macro-cavities located in the upper section of the part, as well as other defects which may occur in the central area, the windows zone and the lower side of the part. The apparition of defects in the central and lower area is triggered by the part geometry.

The thermally narrowed areas produce cavities and porosities caused by the presence of the thermal knots. The elimination of defects due to the change of the part geometry is done by adding external coolers or by adding extra feeders. In order to avoid the formation of cool drops in the walls of the casting mould due to the high speed of the feeding jet, we shall choose a positioning of attacks so that they concentrate the thermal gradient towards the feeders. A possible measure for avoiding cool drops is feeding the casting mould in steps. The last attack level of the casting leg must be placed at the feeders' basis, in order to ensure the heat flow necessary to orient the thermal gradient towards the part's upper side.

3.1.2. Simulating the casting of the assembly made of the rotor block part with a technology with external coolers on the block lower side and feeding with a three-level casting network

In the case of casting the rotor block set, which is a large part with the mass of approximately 12,000 kg, using only a liquid feeding in steps, we cannot totally control the orientation of the thermal gradient, and this is the reason why we use coolers. Coolers may be external coolers (their thickness should not exceed half of the thickness of the part wall in the application area) or internal soluble coolers.

The soluble internal coolers are very efficient, but much more difficult to control. For an efficient use the coolers must be heated before use to a temperature of around 80°C, in order to avoid the boiling of the jet of liquid metal in contact with them. The use of cold coolers risks triggering violent reactions, leading to the erosion of the mould in the respective area and the occurrence of vortexes which trigger the oxidation of the liquid, with negative influences on the part quality. Figure 10 presents the geometry of the rotor block casting set corresponding to the casting variant 2.

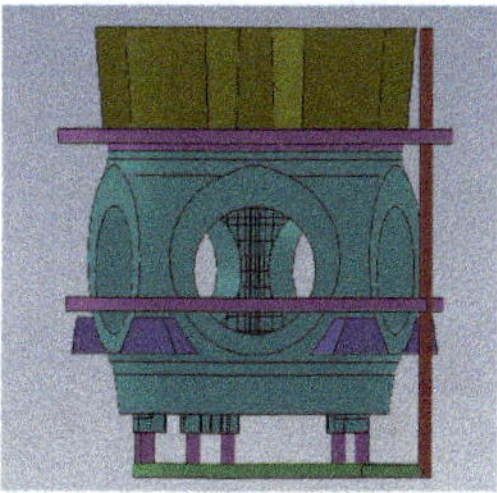

Figure 10. Block shape.

In this variant, the part is fed by a three-step casting network for a feeding meant to avoid the formation of cold drops or oxides in the part, and external coolers for directing the thermal gradient towards the areas intended by the designer. Figure 11 shows the cavities occurring in the upper section of the casting set.

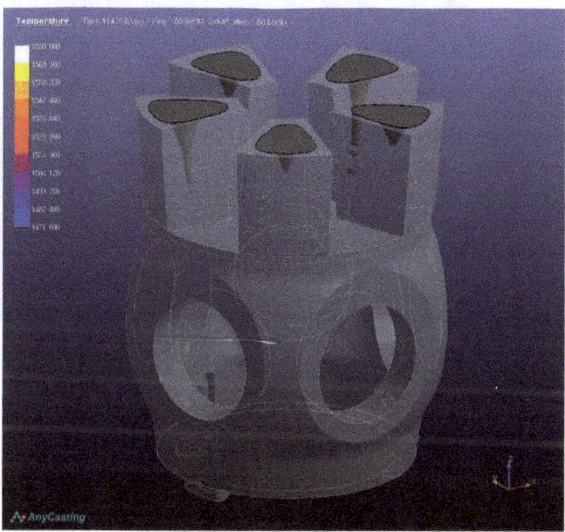

Figure 11. Cavity porosity - 3D.

The transparent display mode allows the visualisation of the defects throughout the entire volume of the cast assembly after solidification ends and the temperature in the mould is much below the steel solidus temperature. Moreover, in the ligth grey areas located in the lower side of the block windows we remark the presence of local cavities defects. Defects of this type were present also in the casting variant 1. The absence of defects in the lower zone, under the block windows, indicates that the external cooler manifests its effect by the fact that solidification in the respective zone takes place during the mould filling.

In Figure 12, grace to the *solidification temperature* criterion, we can follow the temperature variation during solidification at 100% filling and the beginning of solidification. In the lower left side we remark the effect of the cooler by the fact that the crust temperature is lower than the temperature in other zones of the assembly.

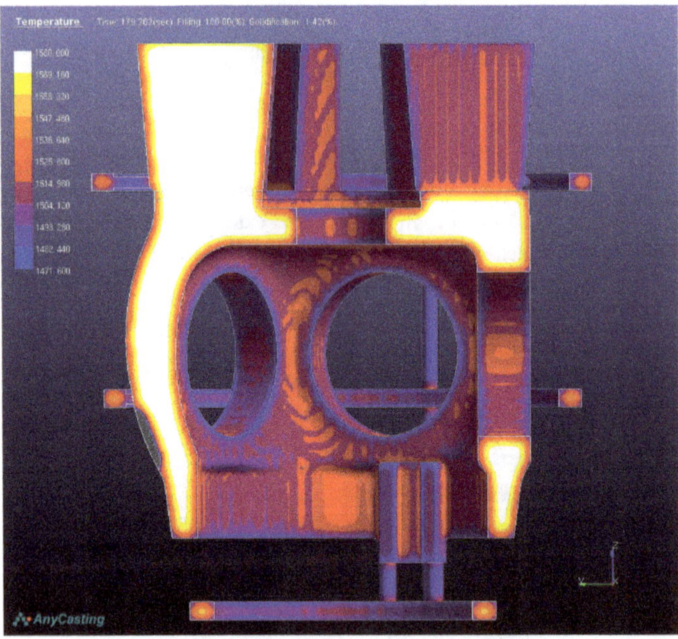

Figure 12. Solidification temperature - at 14% solidification.

Figure 13 presents, also through the *solidification temperature* criterion, the temperature variation during solidification at 100% filling and solidification completed in proportion of 90.7%. We can see that the casting defects are located in the feeders and in the zone under the feeders towards the lower section of the part. The temperature in the coloured zones is still higher than the solidus temperature, which proves that these areas are the last to solidify.

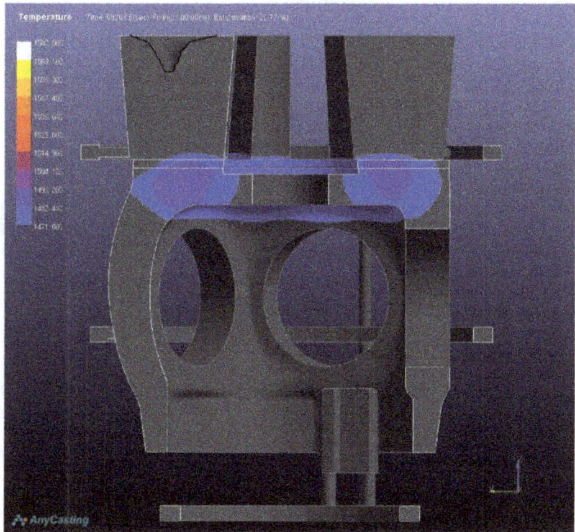

Figure 13. Solidification temperature - at 90% solidification.

Figure 14, through the *solidification temperature* criterion - 14% solidification, shows the position of the defect at the beginning of solidification. In the dark red zone a casting defect will occur under the form of a secondary cavity. In this phase the occurrence probability ranges between 0.300 and 0.400. The colour analysis is done based on the colour bar displayed on the left side of the figure. Like in the previous case, this zone will solidify the last.

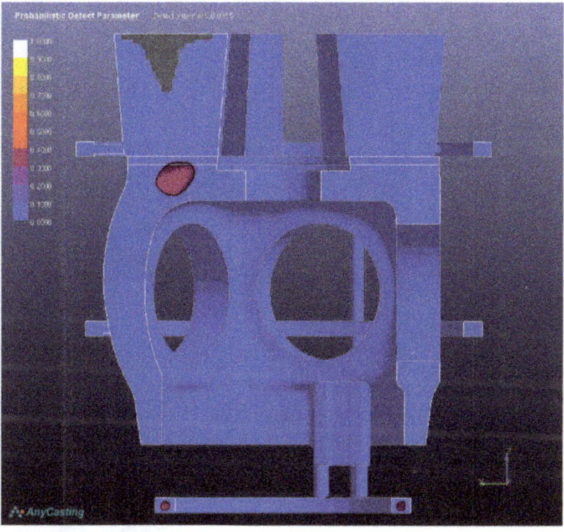

Figure 14. Retained melt volume - 14% solidification.

Figure 15, still through the *solidification temperature* criterion at the end of solidification confirms the fact that the casting defect will be located under the feeders, in the upper zone of the part, and the probability of its occurrence in this case is of 0.400 on a scale with the maximal value 1.000. The analysis is performed also with the help of the colour bars displayed on the left side of the figure, the colour of the area being whitish.

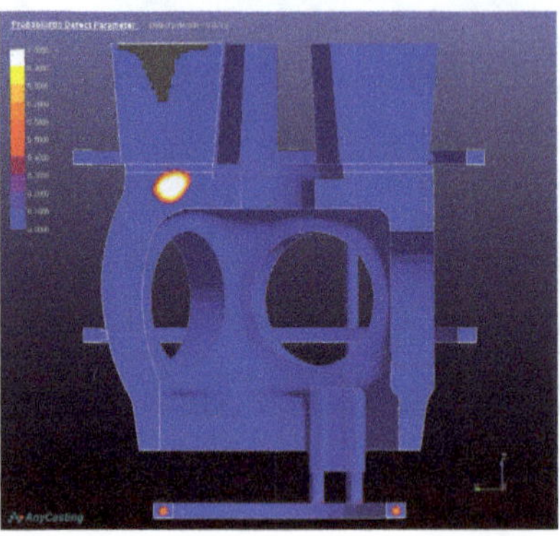

Figure 15. Retained melt volume - 90% solidification.

Figure 16, by the analysis of the *retained melt modulus* criterion at the start of solidification, shows an overall image of the cast part solidification along time. This figure presents the heat-influenced areas in different colours. The defects are already occurring in the superior part, in the feeders, under the feeders towards the central zone of the part. At the part lower end, the coloured area under the windows starts to solidify more rapidly, and the position of the defect is already anticipated in the upper side of the part.

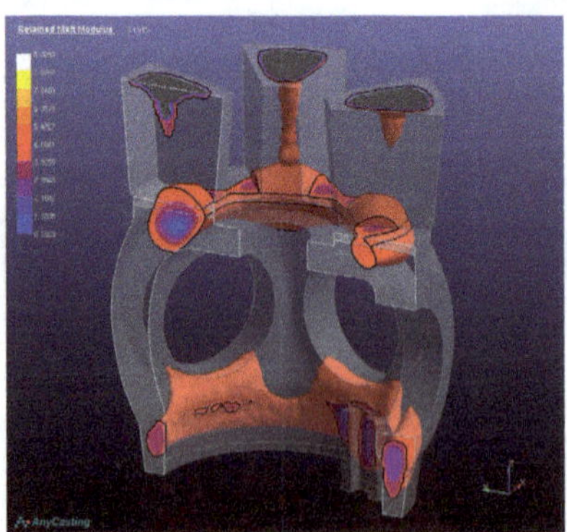

Figure 16. Retained melt modulus - 14% solidification.

In Figure 17 we can see the efficiency of the cooler in the lower part in the yellow areas; the defect appears under the windows and in the windows zone. Figure 17 clearly shows the part of liquid separated in the lower section and not compensated by re-feeding with liquid metal. In this figure also the defects in the block window are highlighted in white.

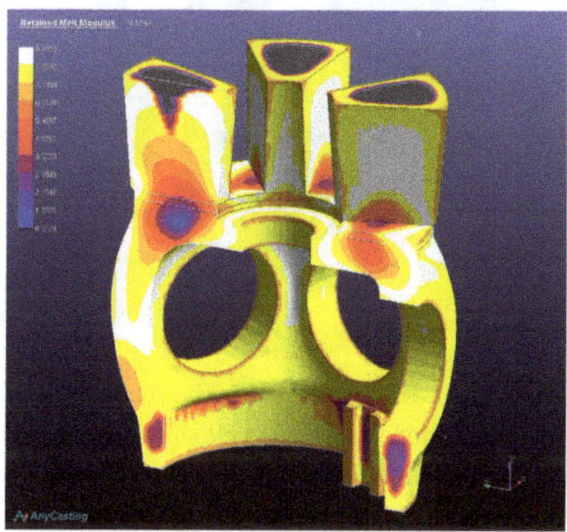

Figure 17. Retained melt modulus - 90% solidification.

3.1.2.1. Partial conclusions

In this simulation variant we remark the influence of the cooler by the fact that defects in the lower side of the part disappear. The part is fed by means of a three-step casting network destined to avoid the formation of cool drops or of oxides in the part. The external coolers determine the directing of the thermal gradient towards the zones chosen by the designer. However, they do not solve the problems enitrely.

3.1.3. Simulating the casting of the rotor block part assembly using a technology with external coolers in the lower section and under the feeders area, as well as exothermal powders

It is less economical, but nevertheless very efficient, to use exothermal mixtures such as plates, bushes, exothermal powders, izolex or pearlite insulating powders. Besides these we may also utilise covering unguent powders in order to facilitate the liquid flow along the mould walls, and at the same time in order to compensate the heat losses in the upper side of the cast part, which leads to a longer heat preservation in feeders, gradually reaching an accented temperature gap between the upper and lower sections of the cast part (with influence on the thermal gradient). In this example, we present the casting simulation applied to the rotor block part using three-step feeding and external coolers in the lower side of the part, under the feeders' area, and exothermal and insulating materials in the upper side of the part, as shown in Figure 18. The simulation respects the same conditions regarding the physical-chemical properties of the liquid, but makes the aforementioned technological changes.

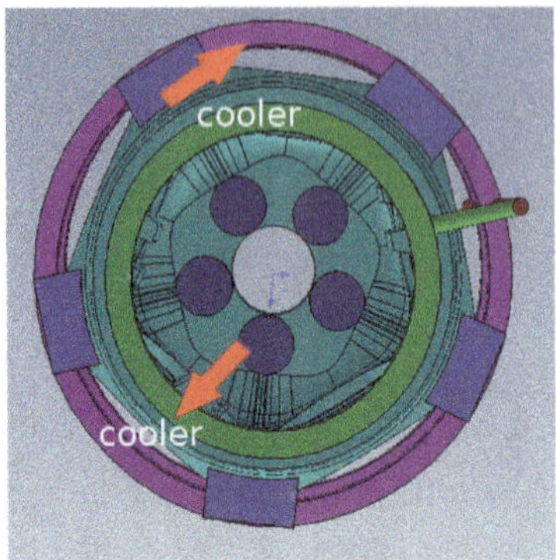

Figure 18. Casting mould equipped with coolers.

This approach solves the problem of cavities under the feeders, but the defects analysed in the casting variants 1 were not eliminated. Figure 19 shows in section the position of cavity and porosities; the open cavity macro-defect is present in the upper side of the feeder, whereas the porosity defect occurs in the lower side of the rotor block window. Figure 19 also presents the depth reached by the defect in the part volume and at the same time we can see the effect of insulating powders, as the cavity volume is smaller than in the previous cases. We can see as well the defects in the block windows.

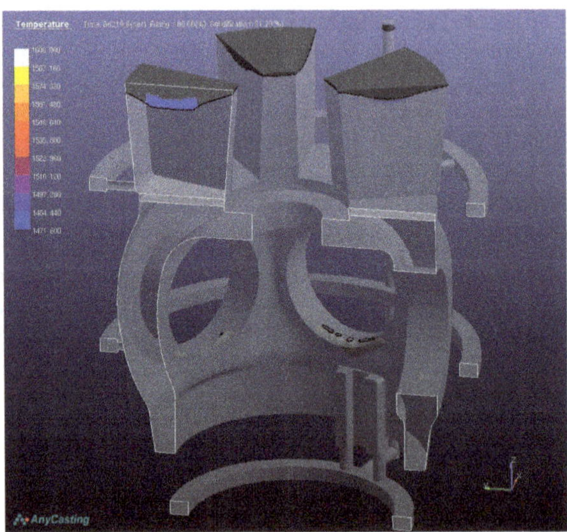

Figure 19. Position of cavity and porosities - section.

The same defect types are highlighted in Figure 20 too, along with all the defects appeared in the block window. The blue zones indicate the volume of these defects too.

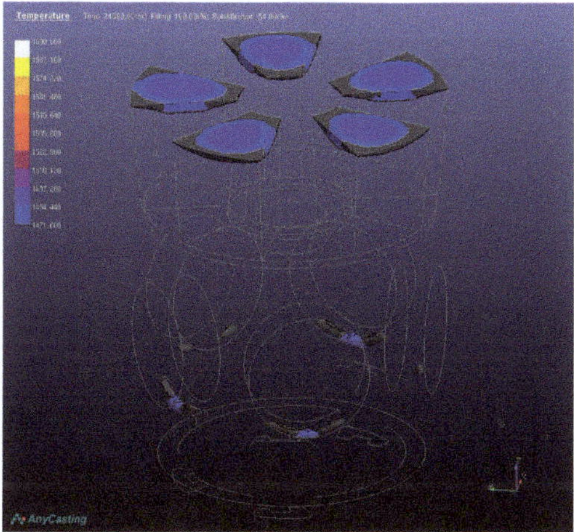

Figure 20. Position of cavity and porosities - transparent.

Figure 21 shows the evolution of casting defects depending on the local solidification time, in fact the solidification duration of each part zone is shown in section. We remark that defects are concentrated in the upper area of the feeder and in the window area.

Figure 21. Local solidification time.

In Figure 22, by the temperature variation during solidification and the evolution of defects during the solidification of the cast set, we remark that the defects have the same positions as in the previous figure.

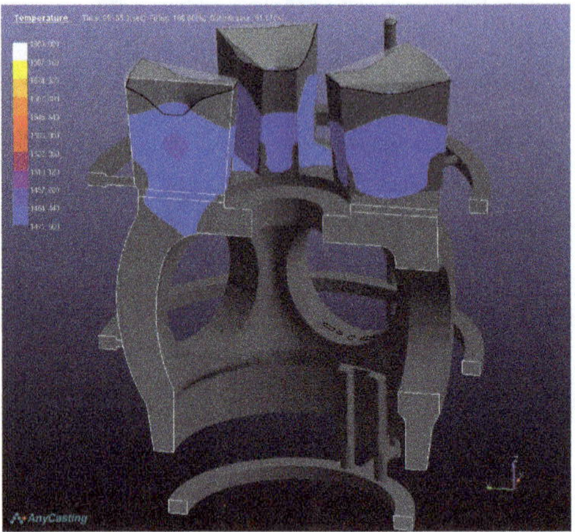

Figure 22. Temperature solidification.

In Figures 23 and 24, we point out the influence of the *defect parameter* probabilistic criterion in accordance with the cooling rate and the *retained melt modulus* criterion, indicating at the same time that the defect occurrence probability is maximal in the light-coloured areas.

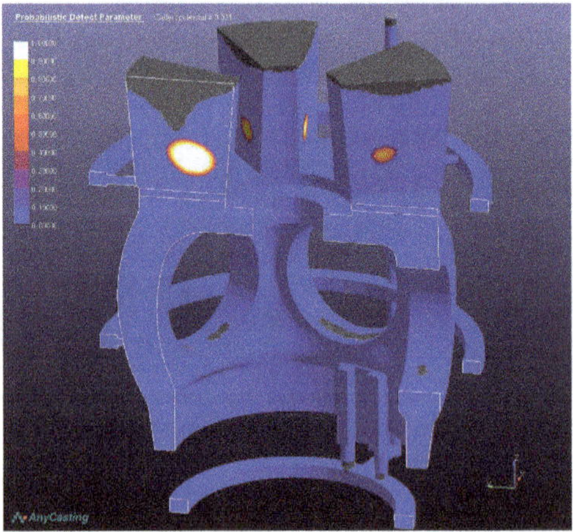

Figure 23. Cooling rate.

Figure 24, by means of the *retained melt modulus* criterion, identifies the defect formation probability based on the colour bar displayed on the left as being maximal in the technological feeders. We can see in light grey the cavities in the windows zone.

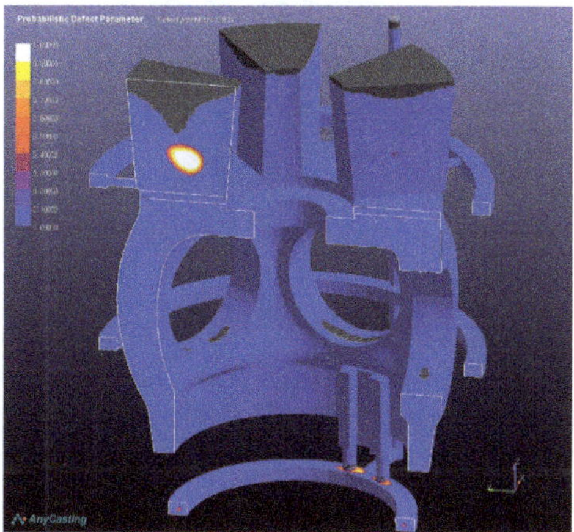

Figure 24. Retained melt modulus.

As a final measure for obtaining a defect-free cast part with all the sufficiency conditions fulfilled, we use in addition blind feeders and exothermal and insulating powders.

3.1.3.1. Partial conclusions

The presence of coolers in the lower and upper sides determines the gradient shift towards the superior side, which results in the elimination of defects in the lower side.

The utilisation of exothermal and covering powders leads to the reduction of the cavity in the area under the feeders towards the upper side of the part. The defects in the zone of the block windows remain still apparent.

3.1.4. Simulating the casting of the rotor block assembly part with a technology using external coolers in the lower section and under the feeders' area, exothermal and feeder-covering powders, as well as blind feeders

Figure 25 presents the casting assembly equipped with coolers and blind feeders.

In this variant we use external coolers in the upper side of the set, external coolers in the contact area between the feeders and the body of the part, and additional blind feeders in the windows zones as well as covering and exothermal powders. This variant, although more expensive, determines the reduction of casting defects, eliminating especially the interior defects, either in the middle or lower section.

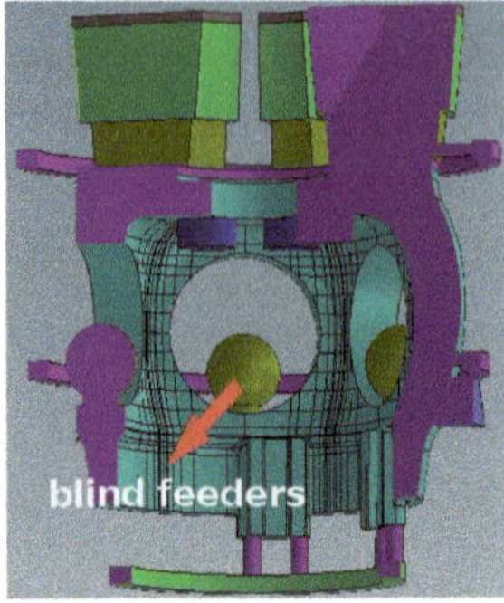

Figure 25. Casting mould equipped with coolers and blind feeders.

Figure 26 presents the position of cavities and porosities, the macro-defects in the upper side of the part and the superior section porosities located in the blind feeders, eliminating the defects in the windows zone.

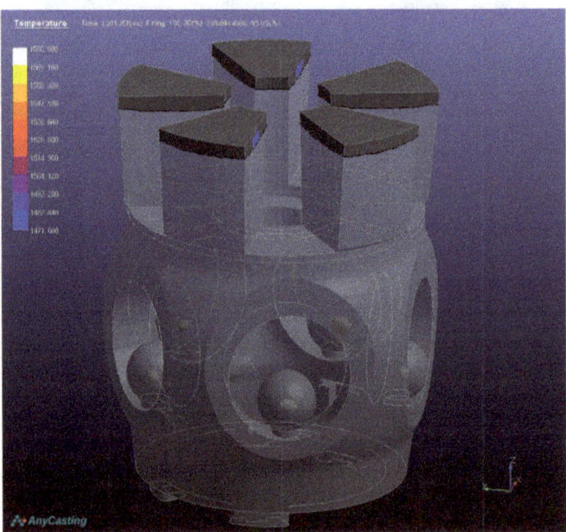

Figure 26. Position of cavities and porosities.

In Figure 27, grace to the *time solidification variation* criterion, we can anticipate the position of defects at the end of solidification, these defects being located in the upper side of the part and in the blind feeders. The efficiency of the exothermal and insulating powders is obvious by the fact that the macro-cavity in the superior side is smaller in volume than in casting variants 1 and 2. The block windows remain free from cavity defects.

Figure 27. Solidification time.

Figure 28, through the analysis of the *cooling rate* criterion, presents the way in which the part solidifies, and accordingly the blue areas solidify the slowliest whereas the light-coloured zones solidify the quickest due to the effect of the added external coolers.

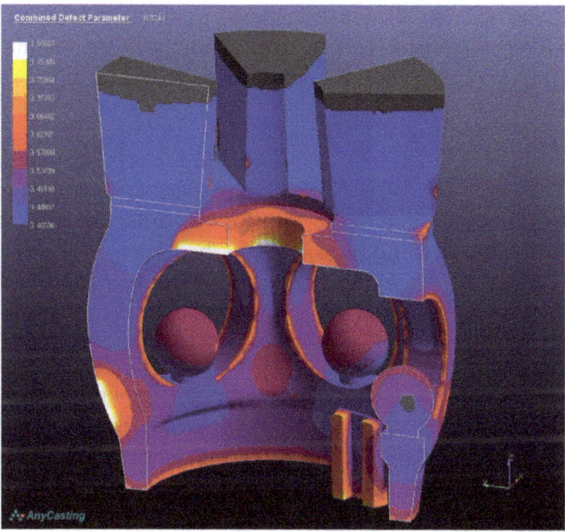

Figure 28. Cooling rate.

Still on the basis of this criterion we remark the effect of exothermal and insulating powders in the upper part of feeders. In both cases there are no defects occurring in the windows zone, the cavities being concentrated in the blind feeders.

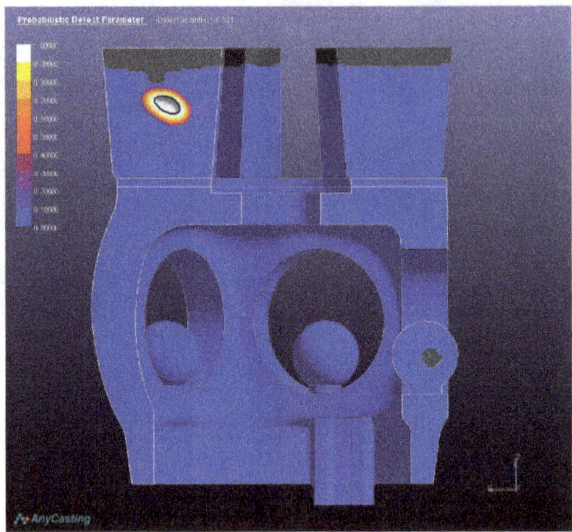

Figure 29. Retained melt volume.

In Figure 29, based on the analysis of the *retained melt volume* criterion, we see very clearly the positioning of casting defects in the upper part of the external feeder, and in grey we identify the positioning of cavity defects in the blind feeder. The defect in the window area was shifted by the thermal gradient, i.e. by the modification of the solidification conditions, towards the centre of the blind feeder. All criteria used at this point indicate a healthy part without major defects.

3.1.4.1. *Partial conclusions*

In this variant, the casting defects, grace to the directing of the thermal gradient, are eliminated from the body of the part and transferred to the feeders in the upper section of the part, whereas the defects from the windows zone are directed into the blind feeders.

3.1.5. *Practical results of the casting of the rotor block assembly part*

We continue by presenting the practical results of the casting of the rotor block set part. The casting was performed using the results of the simulation in the first variant. After the verification of the cast part, the following resulted: the internal defects under the form of closed secondary cavities in the middle areas and the lower section were highlighted only by non-destructive investigation methods. Figure 30 shows the open defects (porosities) occurred in the lower area of the rotor block window. These defects are small in size and are located on the surface of the part, being apparent by direct visualisation.

Figure 30. Casting defects in the block window.

In Figure 31, porosity-type casting defects occur again in the zone of the rotor block window. The rotor block window open porosities do not occur in all windows. On the exterior surface of the part porosities are visible. The porosities on the exterior surface are due to the mould and paint insufficiently dried before casting. In Figure 31 we can also remark that the superior side of the block after mechanical processing is irregular due to the presence of macro-cavities casting defects. All defects presented in Figures 30 and 31 can be repaired by welding.

Figure 31. Porosity-type casting defects in the block window and on the exterior surface.

Figures 32 and 33 present the part cast according to simulation variant 4. We can remark in the two figures, by direct visualisation, that the upper side of the part; i.e. the feeders' zone, is very healthy and does not exhibit any traces of cavities inside the part, nor does it show other types of defects or porosities.

Figure 32. View of the rotor block upper side.

Similarly to Figure 33, the part is clean and free of casting defects visible on the surface of the part. On the part's exterior surface we can identify the traces of the mould casting network.

Figure 33. Overall view of the rotor block.

3.2. Conclusions

After having compared the results obtained by simulation with the results obtained by actually casting the part, we came up with the following conclusions:

- the most efficient casting variant proved to be the casting according to the simulation for the rotor block assembly casting using the technology with external coolers in the lower zone and under the feeders, exothermal and feeder-covering powders, as well as blind feeders, in the sense that this variant provokes the less defects;

- the information about the type of casting defects occurred in the rotor block assembly and the blades, and about the casting technologies for these complex parts are very scarce and not thoroughly treated in the literature.

- the contribution of simulation is obvious, due to the fact that we can highlight defects classified in defect catalogues that are updated rather rarely, at long intervals of time that can even exceed 50 years.

- the selection of simulation parameters may lead to less expensive solutions and to manufacturing higher quality parts.

4. Applications in AnyCasting on the casting of the Kaplan blade part assembly

The chemical composition of the steels used in casting the blades are shown in Tables 1 and 2. The objectives to reach by simulation refer to obtaining a blade reliable in charge, with a low number of defects and water-resistant. After elaborating the geometry of the part in Solid Works a file is generated with the extension *stl. This file is imported into AnyPre, where we introduce all the data necessary for simulation, the material being selected from AnyDBase together with the physical-chemical properties. Figure 34 presents the geometry of the ensemble of blade part and the casting position. We designed a system of indirect casting with inferior liquid feeding and distribution by several feeders. In the upper side the casting mould is equipped with 2 feeders for the retention of gases and oxides formed during the mould filling. The mould is made of sand, the casting temperature being 1580°C.

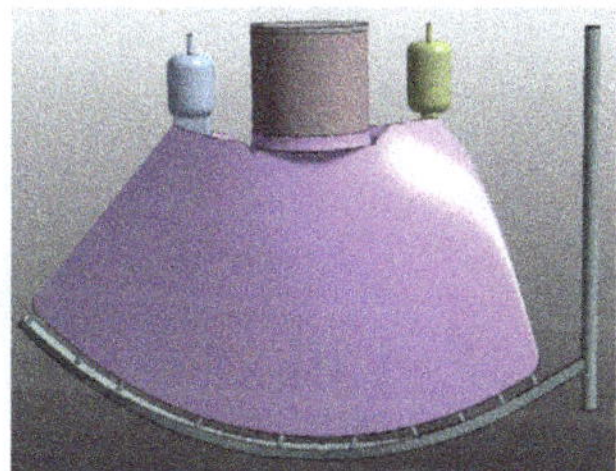

Figure 34. Geometry of the blade part assembly.

The most frequently encountered defects in the case of the blade casting are porosities, air bubbles and oxides, which appear on the surface. Specialists generally grant more attention to the solidification process to the detriment of the mould filling phases. In this case we shall especially analyse the mould-filling manner based on the *oxide inclusion* criterion. In Figure 35 we can see the penetration of the liquid alloy jet into the mould at 5% filling from the volume of the mould.

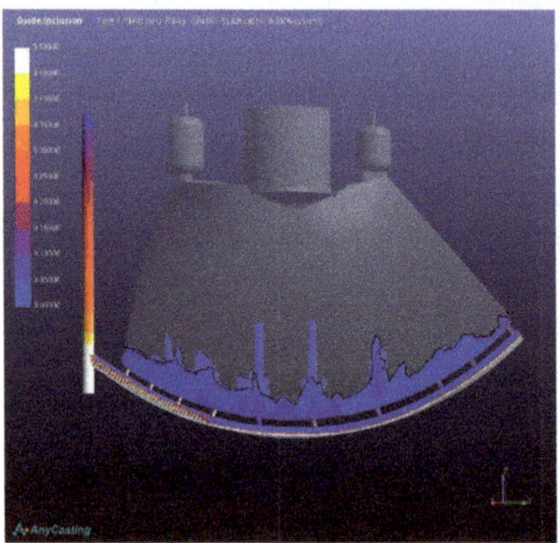

Figure 35. Oxide inclusion - 5.1% filling.

At the entry into the mould through the ingate the jet has a high speed due to the ferro-static pressure in the casting system. In figure 36 the jet of liquid metal is much calmer due to the counter pressure exercised by the quantity of liquid already existing in the mould.

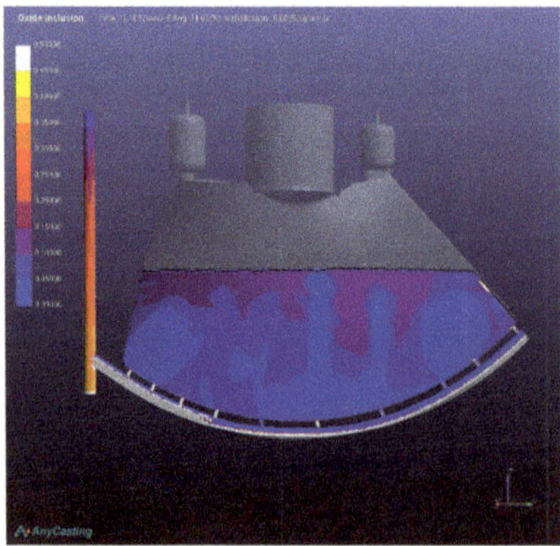

Figure 36. Oxide inclusion - 34% filling.

Figure 36 also shows that when we have filled 34% of the mould volume the oxides are raised together with the movement of the liquid in the mould.

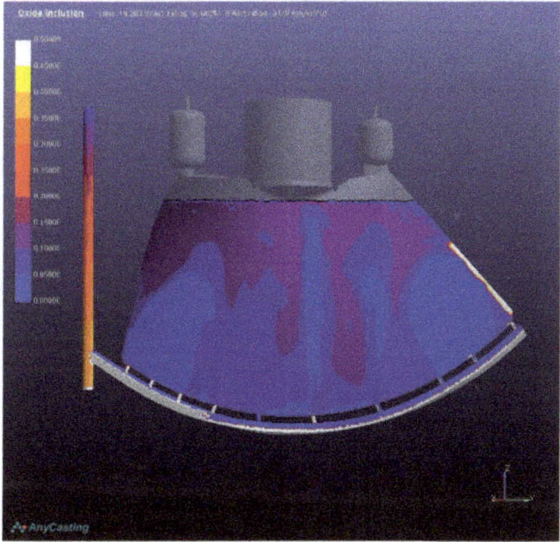

Figure 37. Oxide inclusion - 55% filling.

In Figure 37, with the help of oxide inclusion at a filling of 55% from the mould volume and in Figure 38, grace to the *oxide inclusion 100%* criterion, we have already defined the position of oxides and inclusions on the blade surface.

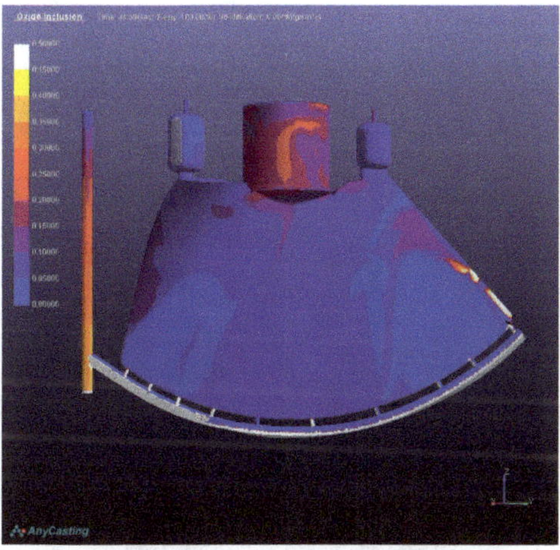

Figure 38. Oxide inclusion -100% filling.

The highest concentration of porosities is found on the active surface of the blade, towards its extremities in the lower, middle and upper sections. Figures 39 and 40 present the same phenomenon, this time in the section of the blade. We proceeded to the blade sectioning in order to observe if porosities are present also inside the blade.

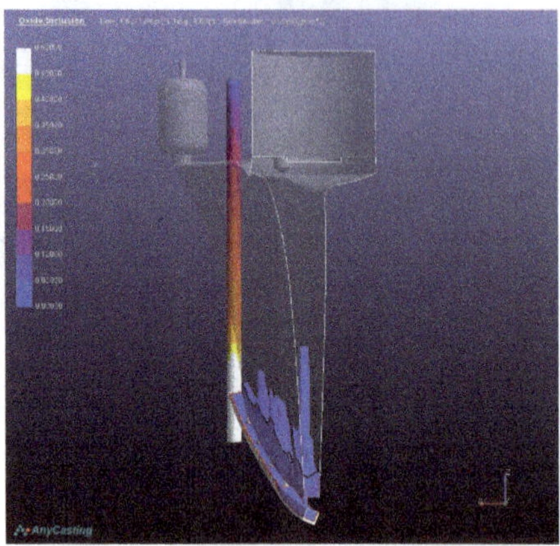

Figure 39. Oxide inclusion - 5.1% filling - vertical section.

Still in Figure 39 we observe the jet entering the sectioned mould, and we see that the central area has the highest speed, phenomenon apparent also in Figure 35. Figure 40 shows that the liquid jet feeding the mould exhibits oxides on the contact surface with the atmosphere from the mould.

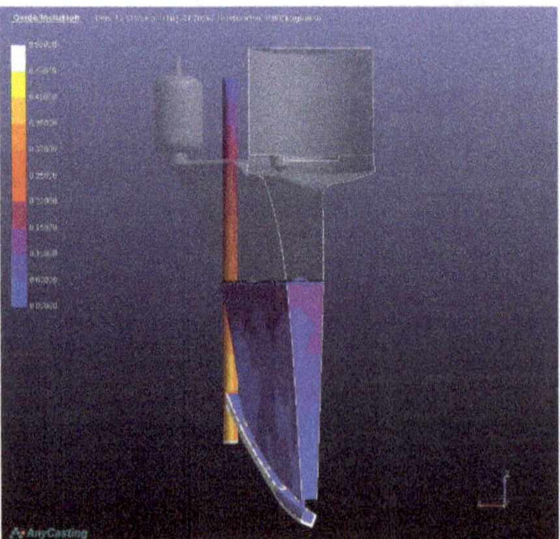

Figure 40. Oxide inclusion - 34% filling - vertical section.

Figures 39-42 illustrate the same criterion, this time in section. This mode of presentation enables the visualisation of the metal flow in the mould, a very important aspect for manufacturing defect-free parts and assemblies.

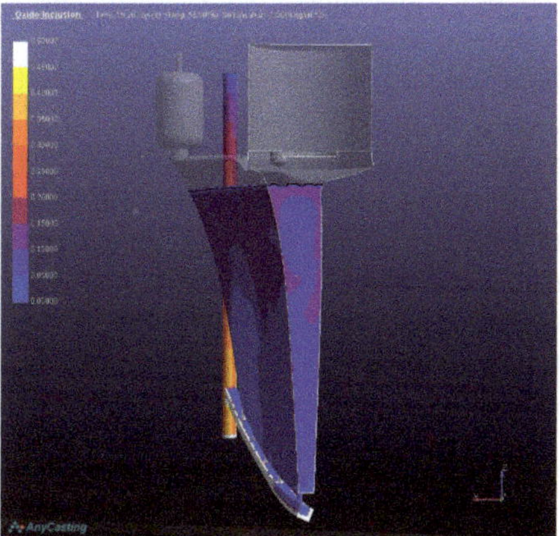

Figure 41. Oxide inclusion - 55% filling - vertical section.

The oxidation reactions are powerful due to the high temperature of the jet and favoured if the mould is insufficiently dry so that it is free from interior moist. The drying is performed before casting in order to avoid the apparition of bubbles and porosities caused by the gases in the mould. A parameter that is not taken into account in simulation or in the real conditions is the air quantity driven by the metal jet into the casting network. Figure 41 showing the situation at 55% filling, and Figure 42 exhibiting the oxide inclusion at 100% filling define the final position of oxides, bubbles and porosities in the blade.

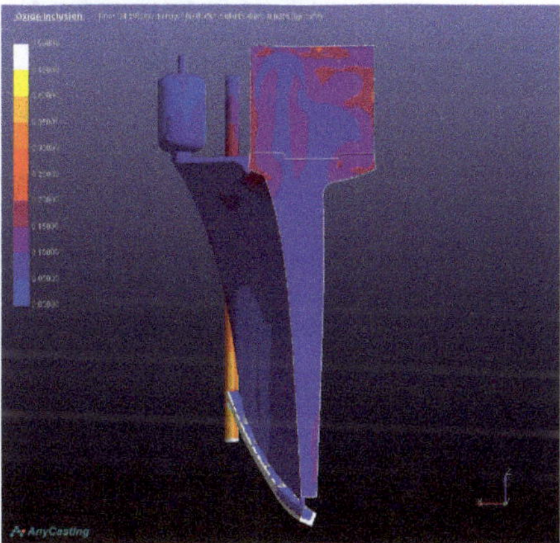

Figure 42. Oxide inclusion - 100% filling - vertical section.

An important cause determining the occurrence of porosities and air bubbles is the failure to observe the technology of steel elaboration and deoxidisation. Another criterion affecting the quality of the cast parts is determined by the chromium segregation. Chromium has a high affinity to carbon and forms chromium carbides located in the volume of the part.

In Figure 43, showing the chromium segregation at 50% solidification, we can observe the areas where chromium segregates under the form of carbides. The zone where the segregation is the most apparent according the colour code bar is situated on the exterior sides of the blade and in its upper section, the maximum concentration being 29%. Segregation evolves in time so that at the end of solidification, according to Figure 43 showing the chromium segregation at la 100% solidification, chromium is found throughout the volume of the blade at an average concentration of 15%. The chromium concentration in the change zone of the upper section, i.e. between the blade flange and the blade block, has the value 8.9% Cr (see the blue zone, Figures 44 and 46).

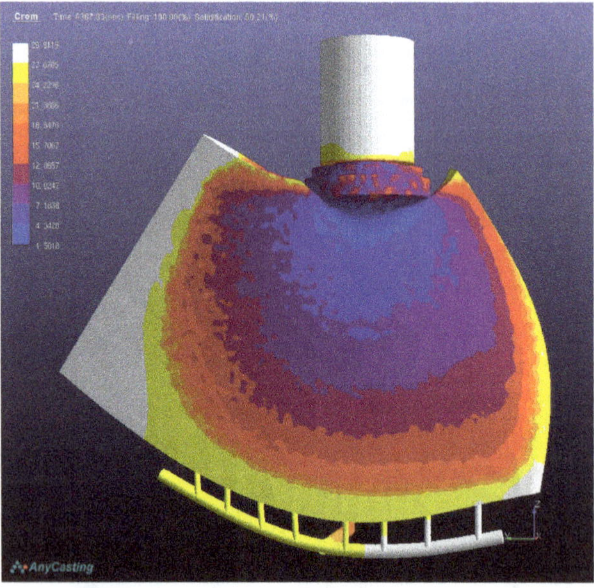

Figure 43. Chromium segregation at 50% solidification.

Figure 44 exhibits in frontal view the distribution of the chromium content on the blade surface, the area with the lowest chromium being the passage zone between the blade flange and the blade block.

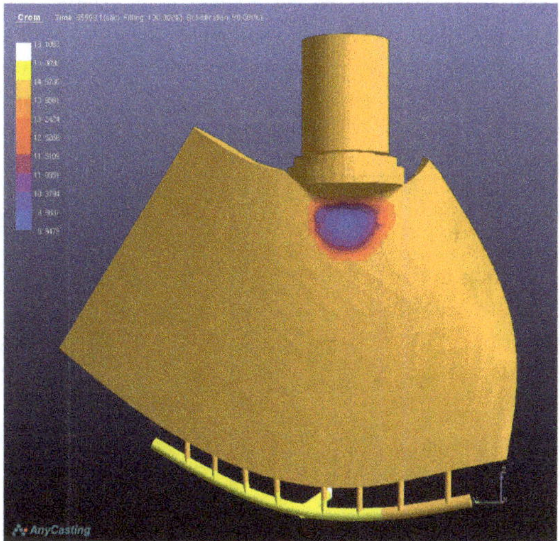

Figure 44. Chromium segregation at 100% solidification.

The same chromium distribution is found also in Figure 45 highlighting the chromium segregation at 50% solidification in section, and in Figure 46, presenting the chromium segregation at 100% solidification in section.

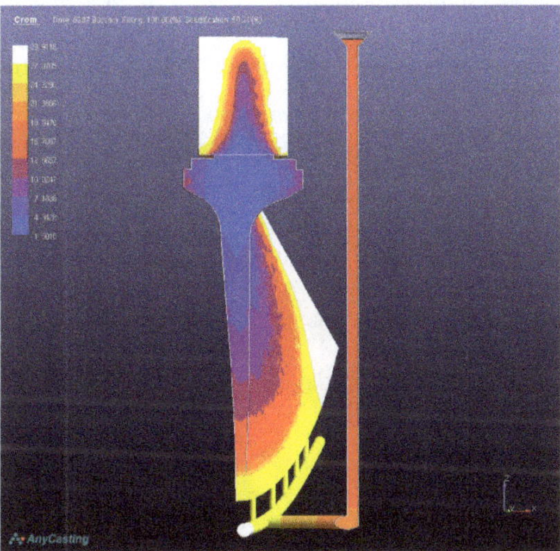

Figure 45. Chromium segregation at 50% solidification - vertical section.

In Figure 45, in section, we remark that the zones with the maximum chromium content are highlighted in white, while the areas with minimum chromium content appear in blue. According to the colour bar, we can determine the chromium content in each section of the blade. Similarly, Figure 46 in section shows the distribution of the chromium concentration throughout the entire volume of the blade at the end of solidification.

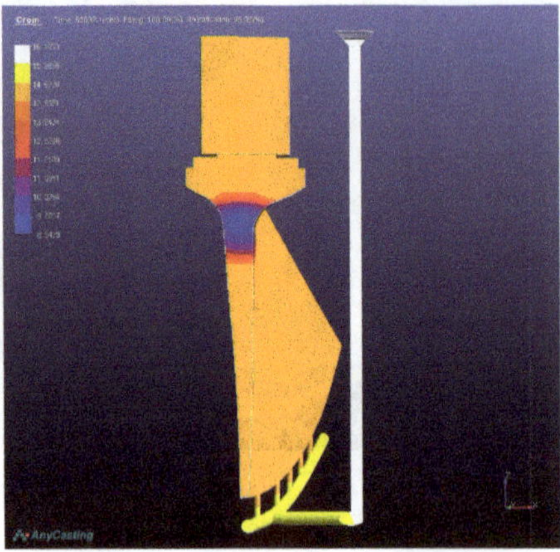

Figure 46. Chromium segregation at 100% solidification - vertical section.

Carbon has a high solubility in liquid state, the last part separating from the solution has a higher carbon content compared to the rest of the basic metal mass, which is iron. Figure 47 (carbon segregation at 50% solidification) and Figure 48 (carbon segregation at 100% solidification) reflect the carbon concentrations in the blade volume. In the case of carbon, we remark an opposite segregation compared to chromium in the analysed area.

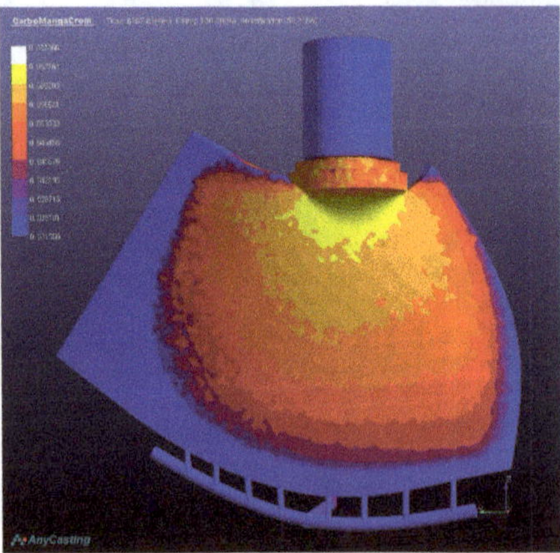

Figure 47. Carbon segregation at 50% solidification.

The increase of the carbon percentage and the decrease of chromium concentration by segregation in the passage area between blade and blade axis have a negative influence on mechanical properties. A passage takes place from an austenitic-martensitic structure towards a martensitic-ferritic structure, provoking the decrease of elongation, toughness and tensile strength.

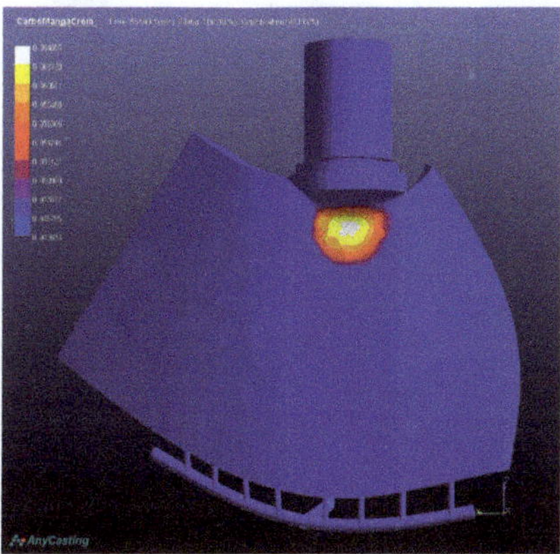

Figure 48. Carbon segregation at 100% solidification.

In Figure 48 we remark that the carbon distribution on the blade surface is minimal in the blue zones and maximal in the white zone. The analysis is made by comparing the colours against the colour bar displayed on the left. The inverse carbon segregation also occurs in Figure 49 (carbon segregation at 50% solidification in section) and in Figure 50 (carbon segregation at 100% solidification in section). The increase of the carbon content and the drop of the chromium content by local segregation lead to an accented possibility of manifestation of inter-crystalline corrosion.

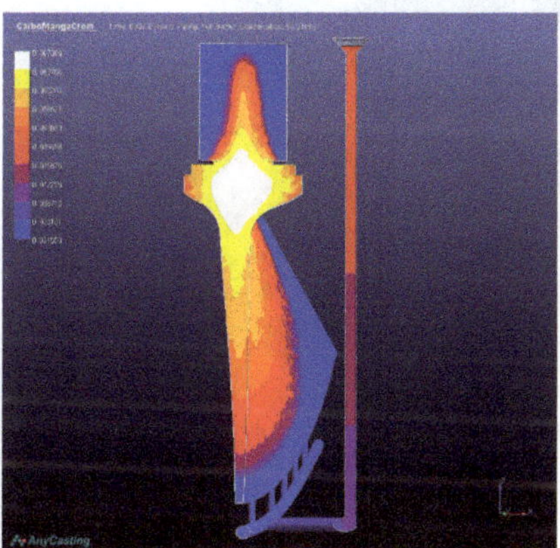

Figure 49. Carbon segregation at 50% solidification - vertical section.

Figure 50 shows the carbon distribution in the part volume as being maximum in the white area, at a value of 0.0645, and minimum in the blue zone, at a value of 0.043.

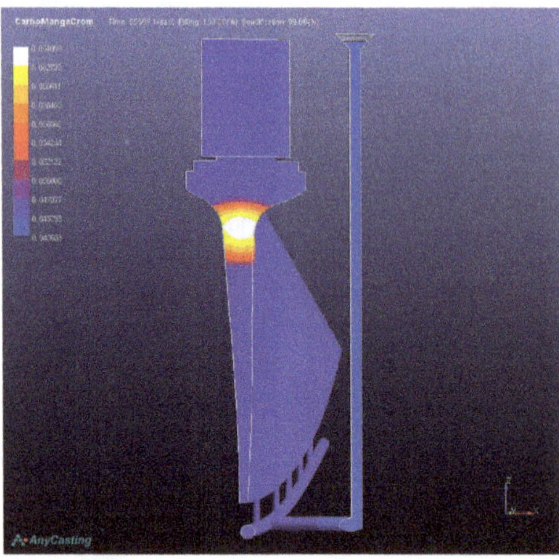

Figure 50. Carbon segregation at 100% solidification - vertical section.

4.1. Practical results of the casting of the Kaplan blade part set

We continue by performing analyses with penetrating liquids on the rotor blade in view of identifying casting defects. After the mechanical processing we proceeded to the application of the penetrating substance and of the revelatory. In Figure 51 we can see the effect of penetrating substances on the blade surface.

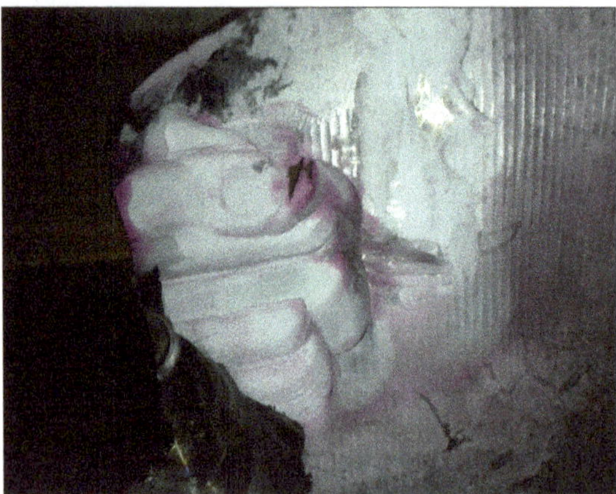

Figure 51. Detail in the analysis of the effect of penetrating substances on the blade surface.

Figure 52 exhibits the position of the open air-bubble casting defect in the active exterior side, located near the attack board at around 250 mm under the blade's upper side. The air bubble has smooth walls, does not present dendrite growths, being provoked by the gases from the liquid metal jet, which were not purged during the mould filling and solidification. Around the main bubble and to the lower left side of Figure 52 the reddish coloured zones are in fact several defects such as pores, oxides and rugositites.

Figure 52. Position of the bubble open defect and of pores and oxides.

The verification of the conclusions drawn from the blade simulation was done by measurements of inductance, using adequate sensors. The distribution resulted for the of inductance values on the blade surface can be studied in Figure 53, which shows precisely the distribution of inductance values on the blade surface [7].

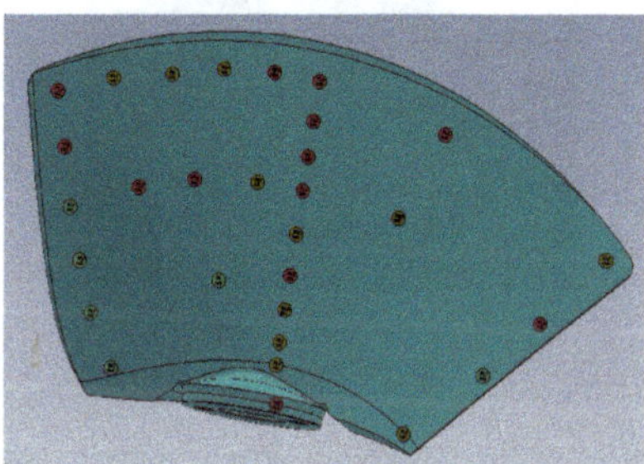

Figure 53. Distribution of inductance values on the blade surface.

The inductance values may be grouped according to colour:

- red – values between 90.05 mH and 93.9 mH; green – values ranging between 86.7 mH and 88.0 mH; yellow – values in the range 80.26 mH and 85.5 mH. In Figure 54 we can see in detail the values of inductance measured in the passage area between the blade flange and the blade block.

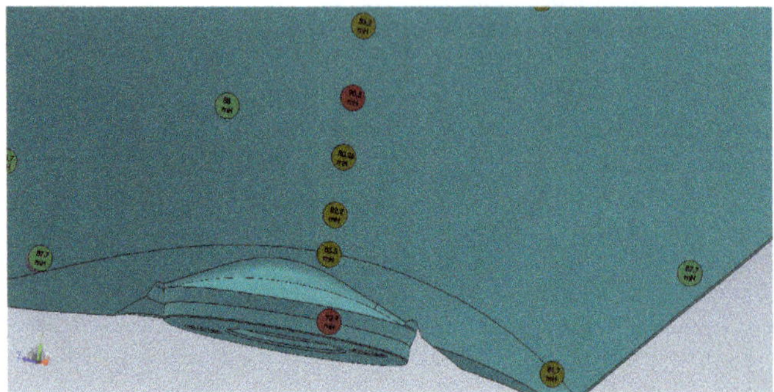

Figure 54. Values of inductance measured in the passage area between the blade flange and the blade block.

The inductance values measured in the passage area indicate modifications in the chemical composition and structure of the part volume, which confirms by the results obtained by simulation (see Figures 42-50).

In the figure 55 we can see the martensitic structure with delta ferrite transformations (δ), according to the Scheffler graphic for the chemical composition realized, with the remark that this structure was removed from under the area of the spindle flange at approximately 500 mm. The presence of δ ferrite was influenced by a slower cooling speed in that area.

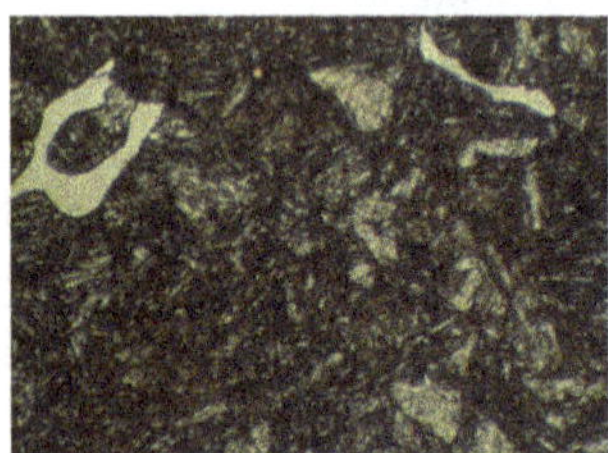

Figure 55. Needle martensitic structure with δ ferrite -100X.

In figure 56 we can observe a needle martensitic structure, with separations of carbides, typical for the area from the run board, with a smaller thickness of the blade and a greater cooling speed. This area presents enhanced chromium separations.

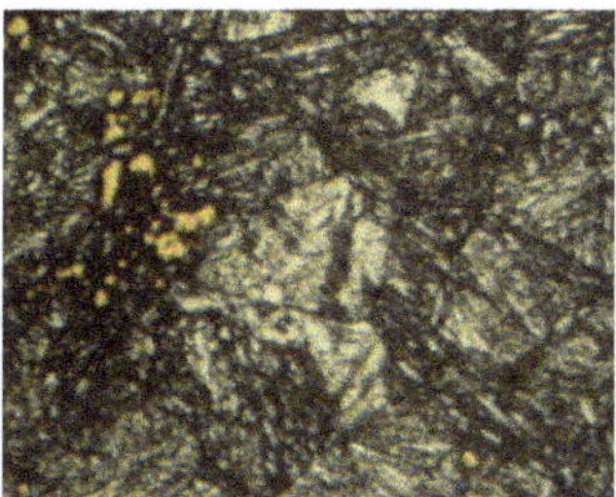

Figure 56. Needle martensitic structure with complex carbides -200X.

5. Conclusions

Beside the defect presented in Figure 52, other defects are located on the active surface of the blade towards its extremities in the lower left and right sides, middle section and upper side, represented by pores with sizes ranging between 2-3 mm. Other types of defects are oxides, rugositites, fissures on the blade, cracks of the joining between the blade flange and block, preferential segregation of the accompanying and alloying chemical elements (Cr and C). The analysis of the simulation results may lead to the obtaining of solutions meant to enhance the quality of the cast product. Some refer to the technology of steel elaboration and others refer to the improvement of the blade casting process. A measure already presented was feeding the mould by a single feeding network using the system of multi-ingate indirect casting and providing additional feeders in the upper part. The role of these feeders is to retain the driven inclusions and the oxides resulted from the process of steel elaboration and those resulted from the casting process following the secondary oxidation of the liquid metal jet. The module of these feeders can be increased if the elaboration and casting technology are not sufficiently mastered. The conclusions drawn from the research on the improvement of the casting technology for the above parts lead to the manufacture of higher-performance Kaplan turbines from the viewpoint of mechanical properties, with higher resistance to inter-crystalline corrosion and with favourable consequences on life duration. The future researches will be focused on the extension of simulation for a detailed observation of the filling processes, the liquid flow and the solidification of the alloy based on all the criteria specific to this software.

Author details

Ioan Ruja, Constantin Marta, Doina Frunzăverde and Monica Roşu*

*Address all correspondence to: m.rosu@uem.ro

University "Eftimie Murgu" of Reşiţa, Romania

References

[1] Standard: EN 10088 - 2/2005.

[2] Standard: EN 10088 - 3/2005.

[3] AnyCasting (2009). Advanced Casting Simulation Software, version 3.10.

[4] Marta, C. (2011). Aplicatii in AnyCasting / Applications in AnyCasting, Timisoara; Stampa.

[5] AnyCasting (2009). DataBase, Advanced Casting Simulation Software, version 3.10.

[6] Marta, C., Suciu, L., & Nedelcu, D. (2009). Simulation of the kaplan rotor hub casting and comparison with the experimental results. *Metalurgia International*, 14(10), 33-38.

[7] Ignea, A. (1996). Masurarea electrica a marimilor neelectrice /Electrical measurement of non-electrical dimensions. Timisoara: Editura de Vest.

Permissions

The contributors of this book come from diverse backgrounds, making this book a truly international effort. This book will bring forth new frontiers with its revolutionizing research information and detailed analysis of the nascent developments around the world.

We would like to thank Prof. Malur Srinivasan, for lending his expertise to make the book truly unique. He has played a crucial role in the development of this book. Without his invaluable contribution this book wouldn't have been possible. He has made vital efforts to compile up to date information on the varied aspects of this subject to make this book a valuable addition to the collection of many professionals and students.

This book was conceptualized with the vision of imparting up-to-date information and advanced data in this field. To ensure the same, a matchless editorial board was set up. Every individual on the board went through rigorous rounds of assessment to prove their worth. After which they invested a large part of their time researching and compiling the most relevant data for our readers. Conferences and sessions were held from time to time between the editorial board and the contributing authors to present the data in the most comprehensible form. The editorial team has worked tirelessly to provide valuable and valid information to help people across the globe.

Every chapter published in this book has been scrutinized by our experts. Their significance has been extensively debated. The topics covered herein carry significant findings which will fuel the growth of the discipline. They may even be implemented as practical applications or may be referred to as a beginning point for another development. Chapters in this book were first published by InTech; hereby published with permission under the Creative Commons Attribution License or equivalent.

The editorial board has been involved in producing this book since its inception. They have spent rigorous hours researching and exploring the diverse topics which have resulted in the successful publishing of this book. They have passed on their knowledge of decades through this book. To expedite this challenging task, the publisher supported the team at every step. A small team of assistant editors was also appointed to further simplify the editing procedure and attain best results for the readers.

Our editorial team has been hand-picked from every corner of the world. Their multi-ethnicity adds dynamic inputs to the discussions which result in innovative outcomes. These outcomes are then further discussed with the researchers and contributors who give their valuable feedback and opinion regarding the same. The feedback is then collaborated with the researches and they are edited in a comprehensive manner to aid the understanding of the subject.

Apart from the editorial board, the designing team has also invested a significant amount of their time in understanding the subject and creating the most relevant covers. They scrutinized every image to scout for the most suitable representation of the subject and create an appropriate cover for the book.

The publishing team has been involved in this book since its early stages. They were actively engaged in every process, be it collecting the data, connecting with the contributors or procuring relevant information. The team has been an ardent support to the editorial, designing and production team. Their endless efforts to recruit the best for this project, has resulted in the accomplishment of this book. They are a veteran in the field of academics and their pool of knowledge is as vast as their experience in printing. Their expertise and guidance has proved useful at every step. Their uncompromising quality standards have made this book an exceptional effort. Their encouragement from time to time has been an inspiration for everyone.

The publisher and the editorial board hope that this book will prove to be a valuable piece of knowledge for researchers, students, practitioners and scholars across the globe.

List of Contributors

Tian Limei and Gao Zhihua
Key Laboratory of Bionic Engineering of China Ministry of Education, Jilin University, Changchun, China

Bu Zhaoguo
Key Laboratory of Bionic Engineering of China Ministry of Education, Jilin University, Changchun, China
FAW Wuxi Fuel Injection Equipment Research Institute, Wuxi, China

Ryosuke Tasaki and Kazuhiko Terashima
Department of mechanical engineering, Toyohashi University of Technology, Japan

Yoshiyuki Noda
Department of mechanical system engineering, Yamanashi University, Kohu-city, Japan

Kunihiro Hashimoto
Sintokogio, Ltd., Japan

Sebastian F. Fischer and Andreas Bührig-Polaczek
Foundry-Institute, RWTH Aachen University, Aachen, Germany

Ram Prasad
Aero Metals Inc., USA

Elena Brandaleze, Leandro Santini, Alejandro Martín and Edgardo Benavidez
Department of Metallurgy & DEYTEMA, Facultad Regional San Nicolás - Universidad Tecnológica Nacional, Argentina

Gustavo Di Gresia
Ternium Siderar SAIC, Argentina

M. S. Ramaprasad
Foundry Consultant, Bangalore, India

Malur N. Srinivasan
Department of Mechanical Engineering, Lamar University, Beaumont, USA

Qing Liu, Xiaofeng Zhang, Bin Wang and Bao Wang
State Key Laboratory of Advanced Metallurgy, University of Science and Technology Beijing, School of Metallurgical and Ecological Engineering, University of Science and Technology, Beijing, China

Minoru Hatate, Tohru Nobuki and Shinichiro Komatsu
Kinki University, School of Engineering, Higashihiroshima, Japan

M. Srinivasan
Department of Mechanical Engineering, Lamar University, Beaumont, Texas, USA

S. Seetharamu
Materials Technology Division, Central Power Research Institute, Bangalore, India

Na Li, Shuang Zhang, Jun Qiao, Lulu Zhai, Qian Xu, Junwei Zhang and Shengli Li
School of materials and metallurgy, University of Science and Technology Liaoning, Anshan, China

Zhenyu Liu, Xianghua Liu and Guodong Wang
The State Key Laboratory of Rolling and Automation, Northeast University, Shenyang, China

Ioan Ruja, Constantin Marta, Doina Frunzăverde and Monica Roşu
University "Eftimie Murgu" of Reşiţa, Romania